Coordination in Human and Primate Groups

Margarete Boos · Michaela Kolbe ·
Peter M. Kappeler · Thomas Ellwart
Editors

Coordination in Human and Primate Groups

Editors

Prof. Dr. Margarete Boos
Georg-Elias-Müller-Institute of
Psychology
Georg-August-University Göttingen
Goßlerstrasse 14
37075 Göttingen
Germany
mboos@uni-goettingen.de

Prof. Peter M. Kappeler
Department of Behavioral Ecology and
Sociobiology
German Primate Center
Kellnerweg 6
37077 Göttingen
Germany
pkappel@gwdg.de

Dr. Michaela Kolbe
Department of Management
Technology, and Economics
Organisation, Work, Technology Group
ETH Zürich, Kreuzplatz 5, KPL G 14
8032 Zürich, Switzerland
mkolbe@ethz.ch

Prof. Dr. Thomas Ellwart
University of Trier
Department of Economic Psychology
D-54286 Trier
Germany
ellwart@uni-trier.de

ISBN 978-3-642-15354-9   e-ISBN 978-3-642-15355-6
DOI 10.1007/978-3-642-15355-6
Springer Heidelberg Dordrecht London New York

© Springer-Verlag Berlin Heidelberg 2011

This work is subject to copyright. All rights are reserved, whether the whole or part of the material is concerned, specifically the rights of translation, reprinting, reuse of illustrations, recitation, broadcasting, reproduction on microfilm or in any other way, and storage in data banks. Duplication of this publication or parts thereof is permitted only under the provisions of the German Copyright Law of September 9, 1965, in its current version, and permission for use must always be obtained from Springer. Violations are liable to prosecution under the German Copyright Law.

The use of general descriptive names, registered names, trademarks, etc. in this publication does not imply, even in the absence of a specific statement, that such names are exempt from the relevant protective laws and regulations and therefore free for general use.

Cover photo: Composition of Primates (upper photo) ©Peter M. Kappeler
           and humans (lower photo) ©Rainer Sturm / Pixelio (www.pixelio.de)

Cover design: deblik, Berlin

Printed on acid-free paper

Springer is part of Springer Science+Business Media (www.springer.com)

Preface

All members of our species are faced with cooperative decision making and group coordination on a daily basis. By definition, group coordination involves the coordination and reconciliation of potentially conflicting interests of individuals within a group to produce a joint solution. It is therefore cumbersome, time-consuming, and politically problematic. As psychologists, we are learning from cooperative projects with our primatologist colleagues (such as this book) that this weighing of the costs and benefits of group coordination defines the very causal roots of primate group living. Primatological studies reveal that cooperation and coordination are also involved in daily decisions of non-human primate groups, providing an important comparative perspective that is leading to a better understanding of general patterns and mechanisms of group coordination as well as aspects that are unique to humans.

We therefore invite everyone faced with decision making and the challenges that group coordination poses – from family to lecture hall – to explore the essays in this book. Even sole proprietors of entrepreneurial start-ups who regularly make decisions on their own could learn a thing or two from this book about the survival benefits of making those decisions in a cooperative setting instead. Together, these chapters provide a refreshingly comparative perspective on group coordination within both human and non-human primate groups and reveal a stunning diversity of behavioural mechanisms with surprising outcomes. Our goal is to contrast concepts and methods of coordination, which, of course, reveal many differences but also show some interesting similarities. For example, where humans would expect the most dominant, physically powerful male of a non-human primate group to make all decisions, we find that in many cases the needs of the younger and physically vulnerable group members influence pivotal decisions affecting the entire group as well. The survival imperatives underlying successful primate group coordination at the group level make the metaphorical applications to human group coordination boundless and eye-opening. One constant among humans and non-human primate groups appears axiomatic: No one member – no matter how intelligent or talented or multi-faceted – can approach successful group interactions from all perspectives and dispose of all data required for the coordination of the entire group.

The book is organized much like any approach to group coordination would be. Contributions to Part I deal with theoretical approaches, defining the task of group coordination. Chapters in Part II explore scientific concepts and methods of group coordination, offering state-of-the-art data on the subject from different psychological perspectives. Part III presents four aspects on coordination in non-human primate groups that are of great interest for understanding human coordination. The authors provide insights into mechanisms of primate group movement, introduce a variety of communicative signals in different modalities, impress psychologists with rudimentary forms of shared intentionality in great apes groups, and discuss the effects of heterogeneity in primate group composition. At first glance, the reader might think that coordination in non-human primate groups is lacking the essential and most salient aspects of human coordination such as verbal communication and written plans. However, these contributions reveal that there are indeed some important similarities that make this comparison valuable for research and theory.

As is always the case with studies on group coordination, each section approaches its particular focus with the assumption that no research project is ever complete and therefore outlines questions and ideas ripe for future research. Because this is one of the most dynamic areas of inter-disciplinary research, we do not claim that this volume provides an exhaustive summary. However, most readers open to an inter-disciplinary approach will in all likelihood encounter perspectives that they have never contemplated before.

Faced with compiling a book on as ambitious a subject as coordination and decision making by human and non-human primates, clearly the best way, and frankly the only way, to present the science on this topic was to do so as a group. This collaborative endeavour allowed us to experience some of the rather practical group coordination challenges firsthand (e.g. choosing contributors, working with and reconciling different ideas of how to edit a book together, coordinating the timing and input of the contributions themselves, etc.). But without a doubt, the richness of its final form benefits from these challenges – a testimony to group coordination itself.

This book is a direct outcome of interdisciplinary cooperation made possible by the Courant Research Centre "Evolution of Social Behavior" at the University of Göttingen in Germany. This centre was founded in 2008 with DFG (German Research Foundation) funding, and its constituent members study the social behaviour of human and non-human primates from an evolutionary perspective. The book's contributors were largely chosen among the participants of a workshop on implicit and explicit coordination in Göttingen in 2006 that proved pivotal to the establishment of this Courant Research Centre. We would therefore like to express our gratitude to the DFG and the University of Göttingen (which funded the workshop) for ultimately making the publication of this book possible. We would also like to thank the contributing authors, who carved time out of their already over-burdened schedule to compose works that reflect the diversity and creative thought that their fields of research demand. And we extend special thanks to Anette Lindqvist at Springer for her enduring patience as our editor, Margarita Neff-Heinrich for her outstanding English-for-the-sciences proofreading, Christine John and Dennis Erge-zinger for their diligence in dealing with matters of layout and graphics, and a warm "thank you" to the extensive support staff too numerous to mention; without their help, an endeavour such as this would have been impossible.

Göttingen, Germany	Margarete Boos and Peter M. Kappeler
Zurich, Switzerland	Michaela Kolbe
Trier, Germany	Thomas Ellwart
November 2010	

Contents

Part I Theoretical Approaches to Group Coordination

1 **Coordination in Human and Non-human Primate Groups: Why Compare and How?** 3
Margarete Boos, Michaela Kolbe, and Peter M. Kappeler

2 **An Inclusive Model of Group Coordination** 11
Margarete Boos, Michaela Kolbe, and Micha Strack

3 **Coordination of Group Movements in Non-human Primates** 37
Claudia Fichtel, Lennart Pyritz, and Peter M. Kappeler

4 **Dimensions of Group Coordination: Applicability Test of the Coordination Mechanism Circumplex Model** 57
Micha Strack, Michaela Kolbe, and Margarete Boos

5 **The Role of Coordination in Preventing Harm in Healthcare Groups: Research Examples from Anaesthesia and an Integrated Model of Coordination for Action Teams in Health Care** 75
Michaela Kolbe, Michael Burtscher, Tanja Manser, Barbara Künzle, and Gudela Grote

6 **Developing Observational Categories for Group Process Research Based on Task and Coordination Requirement Analysis: Examples from Research on Medical Emergency-Driven Teams** 93
Franziska Tschan, Norbert K. Semmer, Maria Vetterli, Andrea Gurtner, Sabina Hunziker, and Stephan U. Marsch

Part II Assessing Coordination in Human Groups – Concepts and Methods

7 Assessing Coordination in Human Groups: Concepts and Methods 119
Thomas Ellwart

8 Assessing Team Coordination Potential 137
Kristina Lauche

9 Measurement of Team Knowledge in the Field: Methodological Advantages and Limitations 155
Thomas Ellwart, Torsten Biemann, and Oliver Rack

10 An Observation-Based Method for Measuring the Sharedness of Mental Models in Teams 177
Petra Badke-Schaub, Andre Neumann, and Kristina Lauche

11 Effective Coordination in Human Group Decision Making: MICRO-CO: A Micro-analytical Taxonomy for Analysing Explicit Coordination Mechanisms in Decision-Making Groups ... 199
Michaela Kolbe, Micha Strack, Alexandra Stein, and Margarete Boos

Part III Primatological Approaches to the Conceptualisation and Measurement of Group Coordination

12 Primatological Approaches to the Study of Group Coordination 223
Peter M. Kappeler

13 Communicative and Cognitive Underpinnings of Animal Group Movement 229
Julia Fischer and Dietmar Zinner

14 Communicative Cues Among and Between Human and Non-human Primates: Attending to Specificity in Triadic Gestural Interactions 245
Juliane Kaminski

15 Coordination in Primate Mixed-Species Groups 263
Eckhard W. Heymann

Index 283

Contributors

Petra Badke-Schaub Faculty of Industrial Design Engineering, Delft University of Technology, Landbergstraat 15, 2628 CE Delft, The Netherlands, p.g.badke-schaub@tudelft.nl

Torsten Biemann Economics and Social Sciences, University of Cologne, 50923 Cologne, Germany, biemann@wiso.uni-koeln.de

Margarete Boos Georg-Elias-Müller-Institute of Psychology, Georg-August-University Göttingen, Goßlerstrasse 14, 37075 Göttingen, Germany, mboos @uni-goettingen.de

Michael Burtscher Department of Management, Technology, and Economics, Organisation, Work, Technology Group, ETH Zürich, Kreuzplatz 5, KPL G 14, 8032 Zürich, Switzerland, mburtscher@ethz.ch

Thomas Ellwart University of Trier, Department of Economic Psychology, D-54286 Trier, Germany, ellwart@uni-trier.de

Claudia Fichtel Behavioral Ecology and Sociobiology Unit, German Primate Center, Kellnerweg 6, 37077 Göttingen, Germany, claudia.fichtel@gwdg.de

Julia Fischer Cognitive Ethology, German Primate Center, Kellnerweg 4, 37077 Göttingen, Germany, fischer@cog-ethol.de

Gudela Grote Department of Management, Technology, and Economics, Organisation, Work, Technology Group, ETH Zürich, Kreuzplatz 5, KPL G 14, 8032 Zürich, Switzerland, ggrote@ethz.ch

Andrea Gurtner Applied University of Berne, Berner Fachhochschule, Fachbereich Wirtschaft und Verwaltung, Morgartenstrasse 2c, 3014 Bern, Switzerland, andrea.gurtner@bfh.ch

Eckhard W. Heymann Behavioral Ecology and Sociobiology Unit, German Primate Center, Kellnerweg 4, 37077 Göttingen, Germany, eheyman@gwdg.de

Sabina Hunziker Departement für Innere Medizin, University Hospital of Basel, Abteilung für Intensivmedizin, Kantonsspital, 4031 Basel, Switzerland

Juliane Kaminski Max Planck Institute for Evolutionary Anthropology, Deutscher Platz 6, 04103 Leipzig, Germany, kaminski@eva.mpg.de

Peter M. Kappeler Department of Behavioral Ecology and Sociobiology, German Primate Center, Kellnerweg 6, 37077 Göttingen, Germany, pkappel@gwdg.de

Michaela Kolbe Department of Management, Technology, and Economics, Organisation, Work, Technology Group, ETH Zürich, Kreuzplatz 5, KPL G 14, 8032 Zürich, Switzerland, mkolbe@ethz.ch

Barbara Künzle Department of Management, Technology, and Economics, Organisation, Work, Technology Group, ETH Zürich, Kreuzplatz 5, KPL G 14, 8032 Zürich, Switzerland, bkuenzle@ethz.ch

Kristina Lauche Nijmegen School of Management, Radboud University Nijmegen, Thomas van Aquinostraat 3, 6500 HK Nijmegen, The Netherlands, k.lauche@fm.ru.nl

Tanja Manser Industrial Psychology Research Centre, School of Psychology, King's College, University of Aberdeen, G32 William Guild Building, Aberdeen AB24 2UB UK, t.manser@abdn.ac.uk

Stephan U. Marsch Departement für Innere Medizin, University Hospital of Basel, Abteilung für Intensivmedizin, Kantonsspital, 4031 Basel, Switzerland, smarsch@uhbs.ch

Andre Neumann Faculty of Industrial Design Engineering, Delft University of Technology, Landbergstraat 15, 2628 CE Delft, The Netherlands, a.neumann@tudelft.nl

Lennart Pyritz Behavioral Ecology and Sociobiology Unit, German Primate Center, Kellnerweg 6, 37077 Göttingen, Germany, LennartPyritz@gmx.net

Oliver Rack School of Applied Psychology, University of Applied Sciences Northwestern Switzerland, Riggenbachstrasse 16, 4600 Olten, Switzerland, oliver.rack@fhnw.ch

Norbert K. Semmer University of Berne, Institute of Psychology, Muesmattstrasse 45, 3000 Bern 9, Switzerland, norbert.semmer@psy.unibe.ch

Alexandra Stein Grohgasse 5-7/35, 1050 Vienna, Austria, Alexa7@gmx.de

Micha Strack Georg-Elias-Müller-Institute of Psychology, Georg-August-University Göttingen, Goßlerstrasse 14, 37075 Göttingen, Germany, mstrack@uni-goettingen.de

Franziska Tschan University of Neuchâtel, Institut de Psychologie du Travail et des Organisations, Rue Emile Argand 11, 2000 Neuchâtel, Switzerland, franziska.tschan@unine.ch

Maria Vetterli University of Neuchâtel, Institut de Psychologie du Travail et des Organisations, Rue Emile Argand 11, 2000 Neuchâtel, Switzerland, maria.vetterli@unine.ch

Dietmar Zinner Cognitive Ethology, German Primate Center, Kellnerweg 4, 37077 Göttingen, Germany, dzinner@gwdg.de

Part I
Theoretical Approaches to Group Coordination

Chapter 1
Coordination in Human and Non-human Primate Groups: Why Compare and How?

Margarete Boos, Michaela Kolbe, and Peter M. Kappeler

Abstract This chapter integrates the six chapters in Part I of this book. They offer different treatments of the theoretical aspects of small group coordination, thereby providing a framework for how coordination behaviour can be studied from the perspectives of social psychology and primatology. Although we have a good working definition of group coordination and have scientifically established that groups of all primates, including humans, are adapted to improve survival, we are less informed about the behaviours that keep groups together and resolve conflicts. Chapter 2 helps to narrow this gap by integrating contemporary thought on coordination and offering an inclusive model for investigators to use in their analysis of both human and non-human primate groups. Chapter 3 informs us about how and why group movements of non-human primates offer a particularly rich arena with which to study primate group coordination. Chapter 4 presents a thorough analysis of a classic tool in group coordination theory (Wittenbaum and colleagues' Coordination Mechanism Circumplex) and how it can be used to understand behaviours of both an observable and tacit nature that occur before and during the actual coordination task. Chapter 5 takes another perspective – that of high-dynamic anaesthesia teams – to show how theories of coordination can be applied to prevent harm in the operating room. The final chapter offers an outline of how the analysis of the group

---

M. Boos (✉)
Georg-Elias-Müller-Institute of Psychology, Georg-August-University Göttingen, Goßlerstrasse 14, 37075 Göttingen, Germany
e-mail: mboos@uni-goettingen.de

M. Kolbe
Department of Management, Technology, and Economics, ETH Zürich, Organisation, Work, Technology Group, Kreuzplatz 5, KPL G 14, 8032 Zürich, Switzerland
e-mail: mkolbe@ethz.ch

P.M. Kappeler
Department of Behavioral Ecology and Sociobiology, German Primate Center, Kellnerweg 6, 37077 Göttingen, Germany
e-mail: pkappel@gwdg.de

M. Boos et al. (eds.), *Coordination in Human and Primate Groups*,
DOI 10.1007/978-3-642-15355-6_1, © Springer-Verlag Berlin Heidelberg 2011

task itself can be used to develop categories of group processes and performance, adapting hierarchical task analysis tool for in-depth structural analysis.

Animals as well as humans have inherent tendencies toward group behaviour, a trait considered to be one of the major evolutionary transitions. Group living provides advantages such as protection, efficient foraging, and synergy in task performance (Voland 2000; West 2004). However, living in any kind of group requires coordination of behaviour and/or meanings and/or goals (Arrow et al. 2000; Kappeler 2006; Steiner 1972; Stroebe and Frey 1982).

We define group coordination among human and non-human primates as the goal-dependent management of interdependencies by means of hierarchically and sequentially regulated action in order to achieve a common goal. Group coordination can be analysed regarding its *functions* (e.g. contribution to a group decision or to a joint movement), its *processes* (e.g. democratic or hierarchical), its *mechanisms* (e.g. explicit or implicit), and its *entities* (e.g. level of behaviour, meaning, or goal; Arrow et al. 2000; Chaps. 2 and 7). The core assumption of the social-evolutionary perspective on small groups is that group structure and interaction reflect evolutionary forces that have shaped social behaviours over thousands of years (Poole et al. 2004). Within this evolutionary approach, the contributions to this book and others in the literature of social psychology, primatology, and anthropology demonstrate how social coordination behaviour can be studied from the perspectives of social psychology and primatology. This in turn allows us to provide answers to the anthropological questions of how mechanisms of group coordination have evolved and whether there are unique characteristics of so-called human nature. This evolutionary approach includes a selectionist and adaptionist framework (Daly and Wilson 1999).

The adaptive reasons why most animals live in stable social groups are well studied (Conradt and Roper 2003; Kerth 2010), but the behavioural mechanisms used to maintain group cohesion and to solve conflicts of interest are only beginning to be explored. We will explain this research gap using the example of group cohesion. For most primate species, the maintenance of group cohesion is of primary importance for ecological reasons. Maintaining group cohesion is not a trivial problem because groups can be large and can also contain individuals with valid diverging individual interests. Perhaps more so than any other animal species, humans exhibit behavioural mechanisms that promote and facilitate cohesion at the group level. Social psychological research is concerned with how groups obtain this aforementioned cohesion (Baron and Kerr 2003; Festinger 1957; Forsyth 2006; Williams and Harkins 2003). With some exceptions, of course, in contrast to primatological research that attempts to identify behaviours that lead to cohesion in a group, the social psychological concept is far less behavioural oriented and is based instead on affective states, cognition, or common symbols that promote cohesion. For example, a widely accepted conceptualisation of group cohesion in social psychology holds that cohesiveness can be based on interpersonal liking, prestige of the group, and/or commitment to a common goal (Hogg and Abrams 1989). Thus, comparative studies of human and non-human primate groups could give way to the inclusion of more behavioural elements in psychological concepts of group cohesion, and at the same time test to what extent affective states, cognition, or common symbols giving rise to cohesion in human groups can also be identified among non-human primates.

As established above, evolution does not require groups only to maintain cohesion, but also to act collectively in order to achieve common goals. Therefore, mechanisms of making collective decisions have to be formulated. Studying the behavioural processes that underlie decisions on the group level such as where and when to forage or rest is therefore a prime example for studies of functional communication and decision processes (Conradt and List 2009; see also Chaps. 12, 13, and 15). Primatology is becoming increasingly interested in how primate groups coordinate their activities by making collective behavioural decisions (Kappeler 2006). As in humans, vocal communication in non-human primates appears to play an important role in mediating decisions at the group level (Trillmich et al. 2004; see also Chaps. 3 and 13). For instance, when separated from conspecifics, many primates give loud calls that can be heard over large distances (Fischer et al. 2001). These vocalisations seem to function as 'contact calls' that are exchanged between widely separated individuals or subgroups (Rendall et al. 1999; see also Chap. 15). Despite their occurrence in specific contexts, there is some doubt about whether contact calls have evolved specifically to maintain contact between separated individuals. Although listeners can use the calls to maintain contact with signallers, signallers may not call with the intent to inform others. In the case of baboons, however, it seems clear that individuals give contact barks because they have lost the sight of others and are feeling anxious (Fischer et al. 2001).

Although there exist such studies of decision making in non-human primate groups, and many coordination mechanisms such as vocalisation and gesture have been identified (see, e.g. Chap. 13), the explicit and implicit signals and rules of communal decision making remain rather poorly understood.

We do know, however, that human group decision making is a widespread phenomenon within families as well as within colleague groups, committees, juries, etc. (Boos 1996). Group decision making has been extensively studied in social psychology (see, e.g. Chaps. 7 and 11). Large numbers of experimental and field studies have been conducted to identify, for example, regularities of information exchange in groups, in order to learn about how initial member preferences are integrated into a final group decision as well as how conflicts of interest are resolved in a group (Gouran et al. 1993; Orlitzky and Hirokawa 2001; Stasser and Titus 1985). Whereas any overview of the vast literature on group decision making clearly lies outside the scope of this contribution, we would like to highlight an interesting pattern evident in human decision-making research: Human decision-making groups are often considered to be a tool for exchanging and integrating their members' diverse expertise and knowledge to gain a more complete understanding of a decision problem from different perspectives and for rationally choosing the best of the available options. In other words, we often conceptualise groups as functioning something like a 'think tank'. However, experimental and field studies of how human group decision making actually takes place often yield a different picture, namely that of maintaining options of least resistance rather than that of rationally elaborating the pros and cons of different alternatives. For example, it has been shown that once a significant majority has emerged in the group, the group selectively searches for information only supporting the majority-supported alternative instead of conducting an unbiased search for the advantages and disadvantages of extant alternatives (Schulz-Hardt et al. 2000). As further research has shown, it is not only the information search that happens in a biased manner, but also the use of information during decision making which is not only biased but strategic (Schauenburg 2004; Wittenbaum et al. 2004). Even more disappointing but not that surprising, dominant members of a group as in those with high formal status often have the strongest impact on the group decision, irrespective of the quality of their arguments (Boos and Strack 2008). Armed with the knowledge of these tendencies in human group decision making, tools developed by social psychologists are emerging to encourage a more thorough perusal of decision options (e.g. Hackman and Wageman 2005; Schweiger and Sandberg 1989).

This tendency of human groups to bolster an emerging dominant tendency in the group or to overestimate the performance of a member in a high position offers striking parallels to group decision making among some non-human primates as dominance hierarchies occur in most primate species. For example, when deciding which water hole to visit, hamadryas baboons appear to use similar 'majority rules' paradigms to reach a decision about the group's behaviour. Also, individuals with higher hierarchical status tend to overrule those of lower rank from food and mating opportunities. These hierarchical rankings are not always fixed, however, especially among males, and depend on intrinsic factors such as age, body size, intelligence, and aggressiveness. With origins of human phylogeny traced to our non-human primate ancestors (Chapais 2010), it is not clear how much of decision rules (e.g. dominance hierarchy vs. democratic poll) in humans is due to the intrinsic biology of our brains derived from evolution vs. how much is due to cultural factors. Thus, systematically investigating similarities between human groups and groups of non-human primates regarding how they make decisions appears to promise new insights into the principles that underlie decision processes in human groups.

Although group cohesion and group decision making among human as well as non-human primates are interesting in their own right, evolutionary theory would suggest that the existence of these group social systems implies that they are functional with regard to environmental factors (Caporael et al. 2005). In this respect, primatology and anthropology, on the one hand, and psychology, on the other hand, differ considerably with regard to their temporal focus and considerations of what is functionally successful and what is not. Primatology and anthropology focus on the long-term existential success of group cohesion and group decision making; that is, they ask what patterns of group cohesion and group decision making are functional for group stability and the survival of group members. In contrast, psychological research focuses more on the short-term success of group cohesion and group decision making. Social psychologists are interested in whether group processes in terms of information exchange or mutual understanding benefit from cohesion or specific types of cohesion (Cornelius and Boos 2003), and how high-quality decisions can successfully be achieved in groups (Boos 1996; Kolbe 2007). Furthermore, social psychological research on group performance is especially concerned with how group processes affect performance in a group by influencing member motivation, member capability, and/or member efforts in the group. An important finding is that as a consequence of these influences, performance in a group is not always 'successful' and can lead to process losses as well as process gains when compared to individual settings (Steiner 1972). For example, collective action in a group can lead to coordination losses among members due to the fact that their problem definitions, their goals, or their knowledge bases cannot be synchronised (Boos and Sassenberg 2001). All such human group processes examined by social psychologists affect performance consequences in the short run (e.g. anaesthesia teams' successful management of critical non-routine events; see Chap. 5), rather than a survival or selection advantage of the group in the long run.

Hence, comparative research on the consequences of group cohesion, group decision making, or – generally – group coordination and other group processes on performance criteria in human vs. non-human primate groups could offer new insights for both disciplines (cf. Wilson 1997; Wilson and Sober 1994). For example, regarding short-term consequences of group processes on performance in non-human primate groups, it is yet completely untested as to what extent the same process losses and gains that have been found in human groups also exist among non-human primates. This investigation of group-specific influences on non-human primates' task-related performance would be interesting in itself (e.g. studying capability gains among non-human primates as a function of social learning in a group), but it might also contribute significantly to our understanding of process and capability losses and gains in human group performance. Another open research question concerns motivation gains and why, under specific conditions, group members exert extra effort in a group situation: Whereas some approaches trace this behaviour back to an individualistic motive (e.g. winning the performance competition and thereby gaining status in the group), other approaches postulate a collectivistic motive (e.g. caring for the group's welfare in itself) (Semmann et al. 2003). Since most non-human primates are likely to lack collectivistic motivations, whereas individualistic motives such as striving for status can be frequently found (Silk et al. 2005), comparative studies of group vs. individual performance in tasks where performance almost exclusively depends on effort could provide interesting new evidence for this open question. Likewise, studies of human groups could take advantage of the long-term survival perspective adopted in non-human primate group research. By more extensively studying real groups in the field over extended periods of time, a more adequate picture of 'successful' human group behaviour might arise. Specifically, we might learn to what extent processes that directly impede the short-term performance of groups might nevertheless be facilitative or even essential for the performance, stability, and sustainability of a group in the long run. This would be a more consequent implementation of the principle of evolutionary selectivity within human social psychology research.

Thus, it appears that integrating research from social psychology, primatology, and anthropology harbours substantial potential benefits for investigating the main questions regarding the evolution of social coordination behaviour: The question of *how human groups coordinate* can be answered partly by means of psychological research; and the more general question of *how primates coordinate* can partly be investigated by means of research in the domain of primatology. And finally, the questions requiring anthropological research are those that consider *the differences between human and non-human primate group coordination* and *how human group coordination has evolved*. It is therefore the objective of the above-described synergistic interdisciplinary perspective to define basic aspects and evolved psychological mechanisms (Buss 2004) of group coordination and decision making and to provide foundational principles on group functioning (Caporael et al. 2005) via appropriate comparative studies of human and non-human primate groups. Specifically, this means that interdisciplinary approaches for assessing the adaptation and selection of coordination behaviour will have to be found in order to define its contribution to the general fitness of both human and non-human primate species.

We consider this an important contribution to evolutionary theory, based on the expectation that comparisons between a variety of primates should allow for determining convergent developments of social behaviour. Similarities between chimpanzee and human cultures have already been found, indicating that they share evolutionary roots (Boesch and Tomasello 1998; de Waal 2006). Furthermore, an interdisciplinary view on the evolution of social behaviour could increase our knowledge on the outlier position of human behaviour and on the importance of language and higher-order cognitive processes for group coordination such as shared mental models.

Thus, within the research objective of describing the evolution of social coordination behaviour, the following five questions can be posed:

1. Which processes and mechanisms of coordination can be found in human and non-human primate groups?
2. How do coordination processes and mechanisms differ between human and non-human primate groups?
3. What are the costs of different strategies (e.g. democratic vs. despotic) for group coordination (Conradt and Roper 2003; Larson et al. 1998)?
4. What is the role of situational adaptation of group coordination processes and mechanisms, and does it differ between human and non-human primate groups?
5. How are means of verbal and non-verbal communication used for coordination purposes in human and non-human primate groups (e.g. Clark 1991)?

These five questions will be considered in the following chapters of this book, giving a systematic overview of the research from the focal fields of primatology, social psychology, and anthropology.

1 Coordination in Human and Non-human Primate Groups: Why Compare and How?

References

Arrow H, McGrath JE, Berdahl JL (2000) Small groups as complex systems: formation, coordination, development, and adaption. Sage Publications, Thousand Oaks, CA

Baron RS, Kerr NL (2003) Group process, group decision, group action. Open University Press, Buckingham, UK

Boesch C, Tomasello M (1998) Chimpanzee and human cultures. Curr Anthropol 39:591–614

Boos M (1996) Entscheidungsfindung in Gruppen: Eine Prozessanalyse [Decision-making in groups. A process analysis]. Huber, Bern

Boos M, Sassenberg K (2001) Koordination in verteilten Arbeitsgruppen [Coordination in distributed work groups]. In: Witte EH (ed) Leistungsverbesserungen in aufgabenorientierten Kleingruppen: Beiträge des 15 Hamburger Symposiums zur Methodologie der Sozialpsychologie. Papst, Lengerich, pp 198–216 [Improvements of performance in task-oriented small groups: Contributions to the 15th Hamburger Symposium of Methodology in Social Psychology]

Boos M, Strack M (2008) The destiny of proposals in the course of group discussions. XXIX International Congress of Psychology, Berlin

Buss DM (2004) Evolutionary psychology: the new science of mind. Pearson, Boston

Caporael L, Wilson DS, Hemelrijk C, Sheldon KM (2005) Small groups from an evolutionary perspective. In: Poole MS, Hollingshead AB (eds) Theories of small groups: interdisciplinary perspectives. Sage Publications, Thousand Oaks, CA, pp 369–391

Chapais B (2010) The deep structure of human society: primate origins and evolution. In: Kappeler P, Silk JB (eds) Mind the gap. Springer, Heidelberg, pp 19–51

Clark HH (1991) Grounding in communication. In: Resnick LB, Levine JM, Teasley SD (eds) Perspectives on socially shared cognition. American Psychological Association, Washinton, DC

Conradt L, List C (2009) Group decisions in humans and animals: a survey. Philos Trans Roy Soc Lond B Biol Sci 364:719–742

Conradt L, Roper TJ (2003) Group decision-making in animals. Nature 421:155–158

Cornelius C, Boos M (2003) Enhancing mutual understanding in synchronous computer-mediated communication by training. Trade-offs in judgemental tasks. Commun Res 30:147–177

Daly M, Wilson MI (1999) Human evolutionary psychology and animal behavior. Anim Behav 57:509–519

de Waal F (2006) Der Affe in uns. Warum wir so sind, wie wir sind [in German]. Hanser, München

Festinger L (1957) A theory of cognitive dissonance. Row Peterson, Evanston, IL

Fischer J, Hammerschmidt K, Cheney DL, Seyfarth RM (2001) Acoustic features of female chacma baboon barks. J Ethol 107:33–54

Forsyth DR (2006) Group dynamics. Wadsworth, Belmont, CA

Gouran DS, Hirokawa RY, Julian KM, Leatham GB (1993) The evolution and current status of the functional perspective on communication in decision-making and problem-solving groups. In: Deetz SA (ed) Communication yearbook 16. Sage Publications, Newbury Park, CA, pp 573–600

Hackman JR, Wageman R (2005) A theory of team coaching. Acad Manage Rev 30:269–287

Hogg MA, Abrams D (1989) Social psychology: a social identity perspective. Methuen, London

Kappeler P (2006) Verhaltensbiologie [in German]. Springer, Berlin

Kerth G (2010) Group decision-making in animal societies. In: Kappeler P (ed) Animal behavior: evolution and mechanisms. Springer, Heidelberg, pp 241–265

Kolbe M (2007) Koordination von Entscheidungsprozessen in Gruppen [in German]. Die Bedeutung expliziter Koordinationsmechanismen, VDM, Saarbrücken

Larson JR, Foster-Fishman PG, Franz TM (1998) Leadership style and the discussion of shared and unshared information in decision-making groups. Pers Soc Psychol Bull 24:482–495

Orlitzky M, Hirokawa RY (2001) To err is human, to correct for it divine. A meta-analysis of research testing the functional theory of group decision-making effectiveness. Small Group Res 32:313–341

Poole MS, Hollingshead AB, McGrath JE, Moreland RL, Rohrbaugh J (2004) Interdisciplinary perspectives on small groups. Small Group Res 35:3–16

Rendall D, Seyfarth RM, Cheney DL, Owren MJ (1999) The meaning and function of grunt variants in baboons. Anim Behav 57:583–592

Schauenburg B (2004) Motivierter Informationsaustausch in Gruppen: Der Einfluss individueller Ziele und Gruppenziele [Motivated information sampling in groups: The influence of individual and group goals]. Dissertation. University of Goettingen, Goettingen. Available at http://webdoc.sub.gwdg.de/diss/2004/schauenburg/

Schulz-Hardt S, Frey D, Lüthgens C, Moscovici S (2000) Biased information search in group decision-making. J Pers Soc Psychol 78:655–669

Schweiger DM, Sandberg WR (1989) Experiential effects of dialectical inquiry, devil's advocacy and consensus approaches to strategic decision making. Acad Manage J 32:745–772

Semmann D, Krambeck HJ, Milinski M (2003) Volunteering leads to rock-paper-scissors dynamics in a public goods game. Nature 425:390–393

Silk JB, Brosnan SF, Vonk J, Henrich J, Povinelli DJ, Richardson AS, Lambeth SP, Mascaro J, Schapiro SJ (2005) Chimpanzees are indifferent to the welfare of unrelated group members. Nature 437:1357–1359

Stasser G, Titus W (1985) Pooling of unshared information in group decision making: biased information sampling during discussion. J Pers Soc Psychol 48:1467–1578

Steiner ID (1972) Group processes and productivity. Academic, New York

Stroebe W, Frey BS (1982) Self-interest and collective action: the economics and psychology of public goods. Brit J Soc Psychol 21:121–137

Trillmich J, Fichtel C, Kappeler PM (2004) Coordination of group movements in wild Verreaux's sifakas (*Propithecus verreaux*). Behaviour 141:1103–1120

Voland E (2000) Grundriss der Soziobiologie [in German]. Spektrum, Heidelberg

West MA (2004) Effective teamwork. Practical lessons from organizational research. BPS Blackwell, Oxford

Williams K, Harkins S (2003) Social performance. In: Hogg M, Cooper J (eds) The Sage handbook of social psychology. Sage Publications, London, pp 327–346

Wilson DS (1997) Incorporating group selection into the adaptionist program: a case study involving human decision making. In: Simpson JA, Kenrick DT (eds) Evolutionary social psychology. Lawrence Erlbaum, Mahwah, NJ, pp 345–386

Wilson DS, Sober E (1994) Reintroducing group selection to the human behavioral sciences. Behav Brain Sci 17:585–654

Wittenbaum GM, Hollingshead AB, Botero IC (2004) From cooperative to motivated information sharing in groups: moving beyond the hidden profile paradigm. Commun Monog 71:286–310

Chapter 2
An Inclusive Model of Group Coordination

Margarete Boos, Michaela Kolbe, and Micha Strack

Abstract The need for a cross-disciplinary inclusive model to analyse the coordination of human and non-human groups is based on observations that (1) group coordination is a fundamental and complex everyday phenomenon in both human and non-human primate groups that (2) largely impacts the functioning of these groups and (3) continues to be fragmentarily studied across disciplines. We formulate an overview of the basic group challenge (group task) of coordination and describe how the context of the group task regulates the group's *functions* (effectiveness criteria) for achieving their task. We explain the basic *entities* that have to be coordinated and therefore analysed, illustrate the concept of coordination process *mechanisms* by which the entities can be coordinated, and finally argue that these mechanisms have finite characteristics of explicitness or implicitness and can and do occur before and after the core coordination process. We then go into further detail by showing how *patterns* emerge from the various coordination dynamics, and end with a discussion of how the various coordination levels at which coordination operates also need to be analysed with a separate *IPO* (*input–process–outcome*) 'lens' that revolves around the basic analytical model, ensuring that multiple perspectives as well as levels of dissolution (macro, meso, micro) are analysed. In our final section, we review the components of contemporary small group theory and integrate these components into our inclusive functions–entities–mechanisms–patterns (FEMP[ipo]) model of human and non-human primate small group coordination.

M. Boos (✉) and M. Strack
Georg-Elias-Müller-Institute of Psychology, Georg-August-University Göttingen, Goßlerstrasse 14, 37075 Göttingen, Germany
e-mail: mboos@uni-goettingen.de; mstrack@uni-goettingen.de

M. Kolbe
Department of Management, Technology, and Economics, Organisation, Work, Technology Group, ETH Zürich, Kreuzplatz 5, KPL G 14, 8032 Zürich, Switzerland
e-mail: mkolbe@ethz.ch

M. Boos et al. (eds.), *Coordination in Human and Primate Groups*,
DOI 10.1007/978-3-642-15355-6_2, © Springer-Verlag Berlin Heidelberg 2011

2.1 Introduction

What is an inclusive model of group coordination, and why do we need it? An inclusive model of group coordination integrates, or – as the name suggests – includes, variables that determine how group coordination works. The need for such a model is based on observations that (1) group coordination is a fundamental and complex everyday phenomenon that (2) largely impacts the functioning of human and non-human primate groups and (3) continues to be fragmentarily studied.

This chapter is organised as follows. We start with a formulation of the basic group coordination challenge, that is, the task-dependent management of interdependencies of individual contributions. In the four sections that follow, we explore the many facets of the coordination challenge, such as coordination entities: the goals, meanings, and behaviours that have to be coordinated as basic psychological levels of analysis; coordination mechanisms: the means by which the entities can be coordinated; coordination dynamics: the emerging coordination patterns; and coordination levels: the levels at which coordination operates. In our final section, we use the results of this exploration of facets of the coordination challenge to integrate these components into a workable inclusive model of human and non-human primate small group coordination.

2.2 Why Coordinate? Task Types and the Coordination Challenge

We define *group coordination* as the group task-dependent management of interdependencies of individual goals, meanings, and behaviours (Arrow et al. 2000) by a hierarchically and sequentially regulated action and information flow in order to achieve a common goal (see also Chap. 1). There is a long-standing concept in small group research regarding the so-called synergistic advantage of group performance compared to the same number of persons individually performing the task (West 2004; Zysno 1998). If the task is additive, the group coordination product can be calculated as the arithmetic sum of individual contributions (e.g. Hill 1982; Shaw 1976; Steiner 1972; Williams and Sternberg 1988). For example, pulling a rope, clapping hands, or brainstorming ideas are typically additive tasks. The power of the individual rope-pullers, hand-clappers, or idea-generators equals the group's performance as a whole, and the sum of the individual ideas, for instance, defines the creativity of the group. In other words, the effectiveness of the group is measured in 'the more (pulling, clapping, ideas), the better' terms.

The consensus among primatologists regarding non-human primate groups is that group cohabitation exists because its advantages (such as consolidation of foraging efforts and strength-in-numbers defence against predators) exceed its disadvantages (feeding competition, disease transmission, mating rivalries) (see Chaps. 13–15 for thorough treatments). In contrast, there exists an argument in the literature of small group coordination that group performance is associated with a net loss in both productivity and efficiencies (Steiner 1972). However, other social scientists appear to side with the primatologists, arguing that a net poor group performance in human groups is unexpected (Caporael et al. 2005; Wilson 1997; Yeager 2001).

2.2.1 Coordination Challenge of Task Synchronisation

This debate within and across multiple disciplines shows in a salient fashion that the effectiveness of group performance – even at its most rudimentary level of additive tasks – is not so much an arithmetical problem but a sociopsychological coordination challenge. In pulling a rope, clapping hands, or generating ideas, people must coordinate their individual endeavours by pulling or clapping at exactly the same point in time; or in the case of non-human primate foraging, perform directional leading; or in human brainstorming, regulate turn-taking. Otherwise, in each of these instances, the contributions of individual group members could not be meaningfully concatenated into a group effort. This problem of synchronisation in time can be solved physically – in the human group examples at least – by pace-makers.

2.2.2 Coordination Challenge of Process Loss

The case of synchronising brainstorming is a bit more complicated, as we know from empirical research reported by Diehl and Stroebe (1987). If people come together in a real group to brainstorm ideas, the pool of ideas created by the group as a whole is smaller than the sum of ideas generated by the same number of individuals as participants of a so-called nominal group. This *productivity disadvantage* (e.g. number of ideas), also known as a *process loss*, of interactive groups compared to nominal groups is to be expected. In brainstorming, evaluation apprehension such as the fear of being evaluated negatively by other participants can hinder the creative potential and/or contribution of group members. Another potential motivational loss is social loafing (Latané 1981; Zysno 1998). One important reason for the reduced productivity of real groups compared to nominal groups is the coordination loss due to production blocking (Diehl and Stroebe 1991; Stroebe and Diehl 1994). People cannot talk at the same time, they must wait their turn in order to express their ideas, and – even more costly to productivity – they tend to forget their own ideas while listening to the contributions of the other group members. The brainstorming group coordination paradigm is a particularly useful example of a group coordination challenge because this so-called productivity loss (reduction in arithmetic sum of ideas) can also be due to a redundancy of ideas: The sum of 'group ideas' is less than the sum of ideas from individual group members if collated pre-process. In the case of brainstorming, group effectiveness is reduced if expressed *quantitatively* (number of ideas reduced due to redundancy), but the actual functional effectiveness can conceivably be increased – especially in cases of brainstorming – if expressed *qualitatively* due to the quality of ideas emerging from group interaction vs. individual members working alone (see Boos and Sassenberg 2001).

2.2.3 Coordination Challenge of Increased Requirements Based on Task Complexity

As can be seen in Table 2.1, coordination requirements increase with the complexity of the group task, and as the complexity of a group task correlates with its coordination requirement, different tasks face different functional effectiveness criteria (Boos and Sassenberg 2001). Interestingly, this coordination requirements–group complexity association can also be present in non-human primate group coordination, as alluded to in Chap. 15 in a presentation of mixed-species coordination. Generating tasks such as brainstorming only requires the coordination of individual goals or task representations. But because participants of the brainstorming process must generate ideas on the same question or problem, a preliminary group discussion on the question or problem will in all likelihood be necessary in order to jointly define the problem (group goal). However, reaching a joint problem definition and formulating a group goal or incentive for the subsequent brainstorming session is not a 'generating' task but belongs to another category of tasks, namely 'problem solving.' Group coordination tasks are categorised as 'problem solving' if there exists a potentially correct or at least optimal problem definition, and are categorised as 'decision making' if the group 'only' has to come to a consensus.

Decision-making tasks are characterised by an opaque structure and a lack of a solution that can often only be clearly perceived as the correct one after the decision has been implemented (Orlitzky and Hirokawa 2001). This task is particularly complex because (1) goals and means of goal achievement are often unclear, making their establishment an important part of the decision-making task itself, (2) they involve high information requirements, as the initial information is typically unequally distributed among group members and a final decision is only

Table 2.1 Task type, coordination requirements, and effectiveness criteria (as per Boos and Sassenberg 2001; McGrath 1984)

Task type	Coordination requirements	Effectiveness criteria
Generating ideas/plans	Problem definitions, goals	Quantity/Quality
Problem-solving	Problem definitions, goals, facts, evaluations	Validity, correctness
Decision-making	Problem definitions, goals, facts, evaluations, opinions, evaluation criteria	Validity, Group cohesion: task commitment, compliance, or consensus

2 An Inclusive Model of Group Coordination   15

possible via sharing and integrating information, and (3) they also involve high evaluation demands because the correctness of possible decision alternatives cannot be determined objectively (Kolbe and Boos 2009). Additionally, group decisions are not made in a social vacuum but involve social, affiliative, hierarchical, and agonistic aspects (Gouran and Hirokawa 1996).

2.2.4 Coordination Challenge of Other Task Complexities

Distinguishing task types as predictors of coordination requirements is useful because it shows the fundamental impact of the task on the group process. However, its limitations are obvious. In real life, few group tasks are single-faceted brainstorming or decision making in character. Instead, groups frequently face tasks consisting of different levels and qualities of complexity (see Examples 1 and 2 ahead as well as Table 2.2). Examples (and by no means an exhaustive list) of further task-defining aspects are the degree and quality of task interdependence (Grote et al. 2004; Rico et al. 2008), level of task standardisation (Grote et al. 2003), task load (Grote et al. 2010), and task routineness (Kolbe et al. under review; Rico et al. 2008). In order to meet the shortcomings of group task classifications and make more specific predictions on what has to be coordinated when and by whom, it has been suggested that performing group task analysis is helpful in sorting out predictions of task complexities and requirements (Annett 2004; Tschan 2000). For a more thorough treatment on the subject of task analysis as a means for defining group coordination requirements, see Chap. 6.

In Sect. 2.3 we will segue into a finer-grained analysis of coordination requirements, exploring different entities that are to be coordinated in groups.

> **Example 1: Family Trip**
> A family (mother, father, 13-year-old daughter, 5-year-old son, plus both sets of grandparents) spends a weekend together. The father suggests a trip to a famous modern-cuisine restaurant at a beautiful lake, which would involve a 2-hour trip together in the car. He is used to his kids' less-than-enthusiastic reactions to such suggestions but not sure how to interpret the smiling 'Sure!' from his parents and parents-in-law and even more irritated by the non-communicative facial expression of his wife.

Table 2.2 Coordination problem of Examples 1 and 2

	Example 1 "Family trip"	Example 2 "Non-human primate group"
Coordination problem	Coordination problem: This familiar group situation shows that a task envisaged as brainstorming most likely also involves classic decision-making components (and lurking problem-solving as well).	This group task includes a variety of different decision-making (e.g. where to go, when to go) and physical activities (e.g. moving both groups safely from one resource to the other).

> Example 2: Non-human Primate Group
> A mixed-species group of non-human primates moves from one feeding resource to the next (see Chap. 15).

2.3 What Is to Be Coordinated

2.3.1 Entities of Coordination: Individual Goals, Meanings, Behaviours

The coordination problem consists not only of the interdependencies of member-specific activity contributions (behaviours), but also of the coordination of terms and information (meanings), as well as special role expectations and intentions (goals) held by individual members of the group (Boos et al. 2006, 2007). Arrow et al. (2000) structured *goals*, *meanings*, and *behaviours* in an entity-levels pyramid, implying in their hierarchical design by using the label 'levels' that the coordination of individual member *goals* has an innately higher value than the coordination of individual member *meanings* (e.g. terms, information) and *behaviours* (see Fig. 2.1). We prefer not to follow this hierarchical order, as all three entities help define the coordination task itself (input) as well as the activities that will occur in the process stage of the group coordination task (process) and the functions that determine the effectiveness criteria of the group coordination task (output). For example, a case in point is coordinating spatial movements from one feeding resource to the next among non-human primate mixed-species groups (see Example 2 in Table 2.3; see also Chap. 15). Individual *goals* (satiation of hunger vs. wanting to rest), *behaviours* (some members display foraging behaviours while others nurse and care for their young), and *meanings* (some members know trail traits indicating prospective foraging grounds while other members recognise noise, odours, or other information indicating the approach of predators) are coordinated to secure a *collective action* that accomplishes *spatial cohesion* as its *function*. We therefore prefer to use an equal-lined triangle to depict a content model for the entities component of our model, implying that there is no innate hierarchical importance of individual goals, individual meanings, or individual behaviours regarding their influence on the constructs of group coordination.

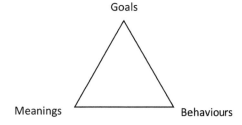

Fig. 2.1 Content model for input and output entities

Table 2.3 Coordination problem and entities of Examples 1 and 2

	Example 1 "Family trip"	Example 2 "Non-human primate group"
Coordination problem	Coordination problem: This familiar group situation shows that a task envisaged as brainstorming most likely also involves classic decision-making components (and lurking problem-solving as well).	This group task includes a variety of different decision-making (e.g. where to go, when to go) and physical activities (e.g. moving both groups safely from one resource to the other).
Coordination entities	Individual goals (satiation of hunger vs. showing off vs. having fun vs. getting it over with without quarrel), meanings (individual ideas of how to spend a day together), and behaviours (walking and driving abilities, who is sitting where in the car) have to be coordinated.	Individual goals (satiation of hunger vs. wanting to rest), meanings (some members know trail traits indicating prospective foraging grounds while other members recognise noise, scent or other information indicating approach of predators), and behaviours (some members display foraging behaviours while others care for their offspring) have to be coordinated.

2.3.2 Coordination of Goals

One of the most likely potential sources of intra-group conflict is a divergence of the goals of individual members. We all can probably recount more than one frustrating group experience where it turned out that we (1) found ourselves speculating about the hidden agendas of our group mates, or (2) had to realise conflicting individual goals within our group, or (3) found that our individual goals were not completely compatible with the group goal. Human groups seem to have an inherent preference to assume within-group goal congruence and avoid an open discussion to explicitly define individual and group goals (Hackman and Morris 1975). This seemingly pseudo-consensus is not necessarily harmful or insincere when group members actually agree on the same goals. However, diversity of interests is present in most cases, making the coordination of individual goals a necessary condition in the majority of cases for successful group functioning. In fact, it has been found that student teams working on a business simulation showed significantly better long-term performance when they made individual goals known in advance of planning their team task (Mathieu and Rapp 2009).

We argue that achieving 'consensus' regarding a group goal can be understood as the explicit or implicit convergence of individual goals to a group goal, or the setting and acceptance of a given group goal, or even a combination of these contrasting egalitarian and despotic processes – the relevant outcome being a single group goal that all group members can strive to achieve. The coordination of goals refers to a motivational process comprising the integration of goals and intentions of group members (Arrow et al. 2000). In a hypothetical example of coordinating goals, one group member might approach a group meeting with the pre-process goal/intent to convince the project manager not to include the project leader presentation, while another member might have the pre-process goal/intent to convince everyone that their former school colleague should be invited to give a talk, while the project manager him- or herself might have pre-process goals/intentions to talk about ideas for guest speakers, how to track the progress of research organisation, and exchange ideas for collaborative projects. All the above pre-process goals and intentions, no matter where the group member is placed in the organisational hierarchy, are individual goals, as they have not yet been coordinated in-process.

In the inclusive model we present, one of the important challenges in achieving effective group coordination is to set group goals that, by definition, can only be set in-process by the group (vs. despotically by the project manager) for a number of reasons, not the least of which is to enhance individual commitment to the group task.

2.3.3 Coordination of Meanings

On the level of meanings, coordination can be understood as the process of grounding and information sharing for the development of a common ground as well as the development of a shared mental model of information and the group task (see Boos and Sassenberg 2001; Poole and Hirokawa 1996; Waller and Uitdewilligen 2008; see also Chap. 10). In a recent interdisciplinary research task among some of the authors of this book, a group of researchers from primatology and psychology attempted to explore the 'Evolution of Social Behaviour.' While making efforts to reach shared mental models within this interdisciplinary research group, it soon became obvious that "meanings" in such a highly diverse group have to do with individual and discipline-specific views, perspectives, and term definitions that in a more homogeneous group can simply be assumed to be shared.

As discussed in Chaps. 10 and 11, achieving a shared mental model often requires reconciliation of the ambiguities and meanings of shared information (e.g. Poole and Hirokawa 1996; Waller and Uitdewilligen 2008). Once definitions of contributed information are settled upon, a shared mental model of evaluation criteria, with the inevitable diverse opinions, preferences, and disagreements, needs to be achieved as well (Boos and Sassenberg 2001; Orlitzky and Hirokawa 2001). Establishing a shared mental model of the group task between pre-process individual goals and in-process group goals is required to accomplish any group task.

The extent of coordination of meanings positively correlates to the extent of explicit and implicit agreement of group members regarding their shared comprehension of facts, tasks, topics, and terminology. Small group studies have shown that a shared mental model is so pivotal to the effectiveness of groups that it is positively correlated to both the risk and the complexity of the group task as well as the adaptability of the group to a dynamic task environment (e.g. Cannon-Bowers and Salas 1998; Cannon-Bowers et al. 1993; Klimoski and Mohammed 1994). A large portion of the challenge of achieving shared mental models is maximising the extent of explicitness in the consensus regarding meanings (for additional details regarding the importance of explicitness concerning meanings, see Chap. 11).

As is generally the case with coordination requirements and the complexity of group coordination, the intensity of the challenge of achieving shared mental models increases pari passu with the diversity of the group (see Table 2.1). Implicit or tacit assumptions regarding terminology are particularly disruptive to joint research efforts, as was alluded to earlier in this section regarding our interdisciplinary project. However, the appropriate degree of explicitness seems to vary among cultures (De Luque and Sommer 2000), implying that the compelling solution of 'the more goal diversity, the more explicitness, and the more effective the group' does not always apply to every setting – once again illustrating the complexity of the coordination problem.

2.3.4 Coordination of Behaviours

On the level of behaviour, coordination can be understood as the synchronisation of actions (behaviours) in time and space – the orchestration of the sequence and timing of interdependent actions (Arrow et al. 2000; Marks et al. 2001). As an example, in the operating room, anaesthesia team members each have different roles that are defined by task responsibilities as well as behavioural expectations during anaesthesia and surgery. Consequently, they have to coordinate their specific actions in a specific manner to be successful, involving measures of both explicit and implicit coordination appropriate to the individual subtasks and medical situation (Kolbe et al. 2009; Zala-Mezö et al. 2009; see also Chap. 5). One could reasonably assume that successful synchronisation of behaviours equates with the group doing the right things in the right order at the right time. For instance, groups having to work on a construction task that plan *before* they start working and *intermittently* stop to evaluate their task performance are more likely to perform well (Tschan et al. 2000). In the same vein, anaesthesia teams have been shown to perform better when their members monitored each other's performance and *subsequently either provided back-up behaviour or spoke up* (Kolbe et al. 2010). Similarly, 'closed-loop communication' involving the receiver of the message acknowledging its receipt was found to improve group performance (Salas et al. 2005). Nuclear power plant control room teams have also been shown to perform better when they exhibited *fewer, shorter*, and *less complex interaction patterns* (Stachowski et al. 2009).

An interesting example of synchronisation of behaviour via leadership in non-human primate groups has been identified in the coordinated group movements in Verreaux's sifakas, an arboreal Malagasy non-human primate living in small groups observed in a study performed by Trillmich et al. (2004). The group movement was initiated more often by female individual movements than by males – and accomplished via leadership, as observations indicated that a specific so-called grumble vocalisation was likely involved in coordinating the group movements.

As in the non-human primate Example 2 earlier, leadership can be defined as a sequence of behaviours, that is, the synchronisation of leadership and followership behaviour. As in other correlations of coordination, the effectiveness of leadership behaviour, such as initiating a group move or making a proposal, is positively correlated to how effective it is in eliciting followership behaviour. For example, the instruction to administer epinephrine is as effective in a reanimation scenario as it is followed regarding the accuracy of timing and dosage of administering. These examples in the literature of the criticality of effective synchronisation of actions illustrate the fundamental role of coordination (see Sect. 2.1).

Depending on the nature of the group task, the three entities of coordination discussed in this section (individual goals, meanings, and behaviours) have a different weight and are focused on to a variable degree. That means that the topic of coordination (both theoretical as well as practical) is complex, as it includes dynamics occurring on the goal-orientation level, the definition of terms level, and the activity (behavioural) level.

Thus, coordination is a multi-level process that references different types of entities to be coordinated and to be synchronised in one and the same process – a process we intend to elucidate further in the upcoming section.

2.4 How Entities Are Coordinated: Coordination Mechanisms

At the next level of dissolution of the coordination problem – from the atomic level of single entities such as goals, meanings, and behaviours – we can discern coordination mechanisms on the molecular level as in those of vocalisation, gesture, and odours (Conradt and Roper 2009).

Mechanisms constitute the 'toolbox' or 'processing machine' of group coordination that includes, for example, interaction and communication events such as asking questions, soliciting opinions, summing up standpoints, giving exposés on information, grumble vocalisation, and handing a scalpel to a doctor. Coordination mechanisms transform individual input entities of goals, meaning, and behaviour into group processes. As illustrated in Chap. 5, for anaesthesia teams meaning and behaviour are the two important input entities for accomplishing the group task of induction of anaesthesia. As Figs. 5.1 and 5.2 show, these two input entities – by means of coordination mechanisms – transform into the processes of information *exchanges* and *collective* actions. This is a clear example of how, depending on the task type, the emphasis on which group coordination tools are used will change, with different mechanisms occurring more (or less) often and with a different overall importance to the successful execution of the task.

For purposes of simplicity, we frame our use of the process concept of mechanisms in terms of their level of explicitness or implicitness (Entin and Serfaty 1999; Espinosa et al. 2004; Rico et al. 2008; Wittenbaum et al. 1996, 1998; Zala-Mezö et al. 2009) and their temporal occurrence (Arrow et al. 2004; Burke et al. 2006; Fiore et al. 2003; Marks et al. 2001; Tschan et al. 2000; Wittenbaum et al. 1998) (see Fig. 2.2). For a thorough discussion of these dimensions, see Chaps. 4 and 7.

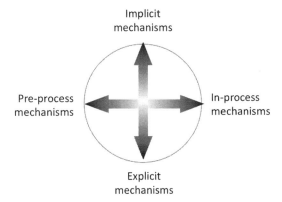

Fig. 2.2 The coordination mechanism circumplex model (CMCM) (adapted from Wittenbaum et al. 1998)

2.4.1 Explicit Versus Implicit Coordination

We regard mechanisms such as verbal or written communication as *explicit coordination* because they are used purposefully, leaving few doubts about their underlying intention. Espinosa et al. (2004) distinguish between two forms of explicit coordination: programming mechanisms (schedules, plans, procedures) and verbal communication, regarding communication itself as a coordinating mechanism. Examples of mechanisms classified as *implicit coordination* are instances when group members anticipate the actions and needs of the other group members and adjust their own behaviour accordingly, for instance, voluntarily handing a surgeon a scalpel, automatically reporting to the team where they currently stand in their group task, or synchronically targeting a flashlight when a team member is making adjustments to a piece of machinery (Rico et al. 2008; Wittenbaum et al. 1996). Contrary to explicit coordination, coordination is reached tacitly through anticipation and adjustment. As indicated in the 'Family trip' Example 1 (Table 2.4), implicit coordination can only be effective if the underlying mental models are shared as well as valid, which, not so surprisingly, is not always the case. Particularly divergent goals, unequal information distribution, and ambiguity of opinions and preferences – all characteristic of more complex and risky decision situations – require a certain amount of explicitness in order to avoid classic cases and consequences of 'miscommunication' ('I thought you got the purchase go-ahead from the boss'; 'I assumed you checked the fuel gauge before takeoff').

Given that explicit coordination as defined by many researchers (e.g. Espinosa et al. 2004) almost exclusively requires language (e.g. for defining rules, giving orders), which as far as the scientific community knows is a unique human accomplishment, one might assume that there is no explicit coordination in non-human primate groups. In fact, even though it is more difficult, it is not impossible to discern explicit versus implicit mechanisms in non-human primate groups (see Example 2, Table 2.4). In movements of non-human primate groups (see Chap. 13), if the designated silverback male in a group of mountain gorillas starts to head in his preferred direction (Watts 2000), one could conceivably construe this

Table 2.4 Coordination problem, entities, and mechanisms of Examples 1 and 2

	Example 1 "Family trip"	Example 2 "Non-human primate group"
Coordination problem	Coordination problem: This familiar group situation shows that a task envisaged as brainstorming most likely also involves classic decision-making components (and lurking problem-solving as well).	This group task includes a variety of different decision-making (e.g. where to go, when to go) and physical activities (e.g. moving both groups safely from one resource to the other).
Coordination entities	Individual goals (satiation of hunger vs. showing off vs. having fun vs. getting it over with without quarrel), meanings (individual ideas of how to spend a day together), and behaviours (walking and driving abilities, who is sitting where in the car) have to be coordinated.	Individual goals (satiation of hunger vs. wanting to rest), meanings (some members know trail traits indicating prospective foraging grounds while other members recognise noise, scent or other information indicating approach of predators), and behaviours (some members display foraging behaviours while others care for their offspring) have to be coordinated.
Coordination mechanisms	Pre-process explicit (having already talked about the trip), in-process explicit (asking the others what they would like to do, organising the trip), post-explicit (learning experience that explicit questions produce an awkward atmosphere in our family), pre-process implicit (expectations about how to spend the day, expectations about how to spend a nice day, assumptions about the expectations of the others, unspoken communication rules), in-process implicit (assuming that the others would like to make the trip and tacitly agreeing), post-process implicit (it seems that nobody wanted this trip even though they didn't say so).	Pre-process explicit (vocalisations), in-process explicit (start heading in preferred direction, vocalisations), post-process explicit (grooming of successful leader), pre-process implicit (orienting oneself in the preferred direction), in-process implicit (some individuals maintaining a particular spatial position within the moving group), post-process implicit (increased likelihood of following successful leader at next occasion).

action as 'explicit' coordination, as there is in all likelihood little doubt among any of the group members that he is initiating a group movement. Also in Chap. 13 is an unconfirmed yet conceivable example of 'implicit' coordination in non-human primate groups in South Africa in which high-ranking female baboons with dependent offspring, because of their reproductive cycle, are interpreted as compelled to stay in the centre of the group or in close vicinity of a male protector instead of taking the lead when leaving the sleeping site (Stückle and Zinner 2008). No explicit signals as such are communicated, yet their movement patterns imply a tacit 'implicit' behavioural mechanism of maintaining a physical position of protection for both themselves and their young – a behaviour that could conceivably be interpreted as a 'shared mental model' as it is not opposed (stopped or contested) by the other members of the group.

2.4.2 Pre-, In-, and Post-Process Coordination

As mentioned at the onset of this section, coordination mechanisms are also classified according to their temporal occurrence. Wittenbaum et al. (1998) were the first to add this second dimension of time, explaining that coordination can take place *before* or *during* interaction (respectively, communication). This second dimension led to a four-cell scheme known as the Coordination Mechanism Circumplex Model (CMCM; see Fig. 2.2), validated in our empirical study in Chap. 4. The CMCM describes these four cells as (1) pre-process explicit: rules, instructions, schedules, routines; (2) in-process explicit: division of labour, communication about procedures; (3) pre-process implicit: assumptions about expertise of group members and task requirements; and (4) in-process implicit: mutual adaptation of behaviour.

For the internal logic of our intended inclusive model of group coordination, we must add a third increment to the temporal dimension: post-process group coordination. In addition to pre- and in-process coordination, we can analytically and empirically identify the result of a coordination activity occurring post-process, specifically the post-coordination mechanisms that are the result of pre- and in-process coordination such as a decision, a different location of the group, or a changed mental representation of the task in the group. In an interview study we found that experienced group facilitators have a very clear grasp of post-process group coordination, perceiving their coordination mechanisms as resulting in specific consequences, which in turn impact further group processes (Kolbe and Boos 2009). For example, after a group has finished a team meeting (in-process), all members leave with explicit and/or implicit out-process tasks (task assignments/intentions, respectively). These out-process tasks will function as input into the next in-process iteration of the team's group coordination (see Fig. 2.5). An example of post-process group coordination in non-human primate groups would be when inter-specific groups go their separate ways when retiring for the evening. This action results in separate sleeping sites, which, in turn, function as group coordination input the next morning (see Fig. 2.5) when the two groups rejoin for the day as an in-process inter-specific group (see Chap. 15 for additional details).

2.5 How Coordination Evolves: Patterns of Coordination

How patterns of group coordination evolve can be exemplarily explained based on a simple micro-level behavioural sequence (see Fig. 2.3).

Patterns of group coordination can be found on all three entity levels, as described in the next sections.

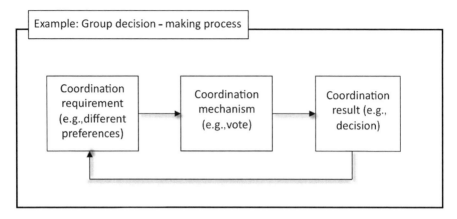

Fig. 2.3 Micro-level work model of group coordination

2.5.1 Goal-Focused Patterns

An example for a goal-focused pattern might be a case of distributed leadership, as in when somebody is presenting her information exposé to the group. During the process of her presentation, she functions as the group leader, holding the floor, steering discussion, and soliciting questions. When she gives the floor back to the project manager, the distribution of the group leadership shifts back to the project manager, where the project manager calls on the next scheduled group member to present his presentation. The leadership role then shifts to yet a third group member. This shared leadership – the dynamic group process among group members who lead one another to help reach the group goals (Pearce and Conger 2003) – has been found to be an effective coordination pattern in a variety of work groups (e.g. Avolio et al. 1996; Künzle et al. 2010b; Pearce and Sims 2002).

Another example of goal-focused patterns was described by Wittenbaum et al. (1996). They showed in an experimental study that group members supplemented others' expected recall when they anticipated a collective recall task (thus aiming to maximise the group's collective recall by remembering information that others likely would not remember), but duplicated others' expected recall when they anticipated a group decision-making task (thus facilitating the emergence of a consensus by focusing on commonly recalled information).

2.5.2 Meaning-Focused Patterns

Meaning-focused patterns can be detected where group members are funnelling idiosyncratic views into shared mental models. An example might be a design team faced with the complex non-routine situation of a creativity task where different experts (e.g. product manager, graphic artist, market statistician) must coordinate their respective expertise, design approaches, and knowledge from diverse organisational fields. That means that the group should first of all produce and differentiate a large number of design ideas in order to develop a comprehensive problem view. This differentiation has to be reduced during group interaction if the group is ever to reach a final design proposal. For that purpose, increased activity towards the integration of concepts must occur. This pattern of first divergent processes (differentiation of ideas) followed by convergent processes (integration of ideas and concepts) is typical for design processes (Boos 2006b). Another example of meaning-focused patterns has been studied by Waller and Uitdewilligen (2008) in their analysis of collective sense-making during crisis situations. They found a pattern they called 'talking to the room,' that is, undirected talk and sharing relevant information to the room at large. Talking to the room invites other group members to actively participate in effective coordination (Kolbe et al. 2010) and has been found to facilitate identifying the accurate diagnosis in medical emergency-driven groups (Tschan et al. 2009).

Meaning-focused patterns in decision-making tasks are particularly interesting. Decision making in groups is often considered a tool for exchanging and integrating their members' diverse expertise and knowledge, discussing a decision problem from different perspectives, and rationally choosing the best of the available options. However, experimental and field studies similar to the one described ahead of how decision making actually takes place often yield a different picture, namely, that initiating group action and maintaining the group's ability to act, rather than rationally elaborating the pros and cons of different alternatives, functionally underlies human group decision making [other functions] (see Kerr and Tindale 2004 for a review). For example, it has been shown that once a significant majority has emerged in the group, the group selectively searches for information supporting the alternative proposed by this majority, instead of conducting an unbiased search for advantages and disadvantages of the different alternatives [processes]. Further hindering the unbiased search for the most advantageous alternative is that dominant members of a group (e.g. those with high formal status) have the strongest impact on the group decision, irrespective of the quality of their arguments. Their proposals and their mode of argumentation turned out to be most successful [processes] (Boos and Strack 2008). This tendency of human groups to bolster an emerging dominant tendency in the group or to overestimate the performance of a member in a high position offers striking parallels to group decision making among some primates.

For example, hamadryas baboons that decide which water hole to visit appear to use similar majority rules to reach a decision about the group's behaviour. Dominance hierarchies occur in most primate species. Individuals with higher hierarchical status tend to displace those ranked below from food and mating opportunities. These hierarchies are not always fixed, however, especially among males, but instead depend on intrinsic factors such as age, body size, aggressiveness, and perhaps cognitive abilities.

2.5.3 Behaviour-Focused Patterns

For example, in a group tasked with reconciling a decision problem (*task type*), there are at least two conflicting prevalent preferences (see Table 2.1). When these distinct perspectives are defined aloud (*coordination requirement*), the group leader can then remind the group that the goal of the group is a consensus (*coordination mechanism*) and that the consequences are that the distinct perspectives, albeit conflicting, both focus on a common basis (*coordination result*). The same processes of reaching a group consensus before enacting a result hold true in groups of gorillas who will not decide about a change in their activities (e.g. leaving their resting site in order to travel to a feeding site) as long as two thirds of the adults have not uttered loud calls (Stewart and Harcourt 1994).

Such sequences can emerge into behavioural patterns, that is, a participative (majority decides) or directive (alpha male decides) style to facilitate group coordination. And yet again, the way patterns tend to evolve within the group will depend on the task focus of the group (goals, meanings, behaviour).

Behaviour-focused patterns are those instances of adaptive coordination, namely, shifting from implicit to explicit behaviour according to the requirements of the task. The adaptability of these behavioural mechanisms in response to a salient cue of the task (e.g. cardiac arrest) and team situation (e.g. resuscitation devices such as a defibrillator being available) leading to a functional outcome (e.g. regained heart activity) is shown to be a prerequisite for patient safety (Salas et al. 2007). Especially the shift from the use of implicit coordination mechanisms in routine phases of task accomplishment to the use of explicit mechanisms in complicated phases seems to be a valid predictor for group performance in anaesthesia (Künzle et al. 2010a; Risser et al. 1999). The effectiveness of adaptive coordination has been shown in a variety of studies (e.g. Grote et al. 2010; Kolbe et al. 2010; Kozlowski et al. 2009; Manser et al. 2008; Waller 1999; Waller et al. 2004).

The advantage of this sequential perspective on the coordination process lies in observing, identifying, and analysing detailed process particulars. We can discover when and under what conditions during the group process particular coordination mechanisms occur, to and from whom the mechanism shifts, and what follows these mechanisms – in other words, what mechanisms are prompted and what their dimensional characteristics are (explicit/implicit; more pre-, in- or post-process), and which coordination mechanisms are ignored (e.g. opening the floor for questions).

The work model of coordination (Fig. 2.3) allowing a micro-level-based process analysis of coordination would not make much sense if it were not embedded in the structural conditions and resources for coordination (e.g. leadership, hierarchy). As this model of the coordination process distinguishes *preceding interactions* from *coordinating actions* and also from *consequences* of the coordinating action, it zooms in on only one segment of the flow of interaction, meaning, or goal/subgoal setting. In most situations, the coordination process is part of a much larger task context or functional requirement to the group (see Example 3).

Example 3: Everyday Work-Life Decision Making in Public
Administration: A Field Study" (Boos 1994a, b, 1996)
Part and parcel to core duties of public administration is to weigh and integrate conflicting individual and public interests, for example, economic goals of extending commercial areas on the one hand and preserving ecological resources on the other hand. We found that mainly two ways of steering these heterogeneous goals and problem views were used in the organisations.

We labelled the first way of goal steering 'hierarchical decision making,' characterised by pre-process multi-department-specific criteria regarding their respective preferred decision. The final decision rests with the head of the administrative office, who is responsible for developing a workable solution, even though the departments are expected to contribute to the decision.

We also observed a goal-steering process widespread in bureaucratic organisations that we labelled 'divisional decision making,' characterised by department experts developing a pre-process solution to the problem specific to the point of view of their own department, such as an economic, ecological, or legal point of view. The head of the division was responsible for steering the decision-making process and leading the group to a consensus.

We observed group discussions about a complex decision task and found typical patterns that differentiated quite well between the two coordination strategies. In hierarchical decision-making groups, we found a recurrence of overtaxing of the group by concurrent leadership. In the divisional decision-making groups, we found that everybody had their own agendas, which, by definition, were divergent. Yet knowledge of these agendas, often quite legitimate albeit divergent, was necessary to make the appropriate decision. As small group research has established, the process of collective sharing of individual, contrasting information correlates to the quality of the group decision and can lead to a rather optimal solution (Lim and Klein 2006). The advantage of proceeding hierarchically means coming to a quick decision, mostly based on proposals of the group leader and the use of rhetorical figures of speech to get his or her point across. The disadvantages of the divisional decision-making process is that it takes longer because the success of the final decision is based not only on content but also on effectiveness of arguments related to power, status, and acceptance; additionally, this procedure requires a larger amount of coordination.

2.6 Inclusive Model of Group Coordination

2.6.1 Core Construct of Inclusive Model

From our considerations on small group coordination emerges a trimorphic pattern of components in our model (Fig. 2.5): (1) at the input level, three types of entities are coordinated: goals (*why* are we?, e.g. to safely anaesthetise a patient; to forage for food); meanings (*what* are we?, e.g. an anaesthesia team; conspecific groups foraging together); and behaviours [*who* are we?, e.g. via role-defined anaesthetist; or in a non-human primate group, some members defined as need-oriented (e.g. hungry juveniles) and some members defined as solution-oriented (e.g. food-finding lactating mothers)]. These input entities then (2) express themselves at the process 'mechanism' stage, occurring at dimensional levels of explicitness (observable and identifiable vs. often neither observable nor identifiable), and at various points on the temporal spectrum (pre-, in-, or post-process). These dimensions of process mechanisms (3) result in consequent output entities of goals, meanings, and behaviours, feeding back as input such as group-task entities (in the sense of classic functional process models such as the input–process–outcome model by Hackman and Morris 1975; Ilgen et al. 2005). These elements of input *entities*, process *mechanisms* dimensions of explicitness and temporal occurrence, and consequent output operate in an effectiveness-criteria environment *(functions)*. The environment depends on the group task, and fulfilment of functions is measured quantitatively (e.g. the more food, the better), qualitatively (e.g. the patient survives), and/or by the extent to which members either commit to, comply with, or reach consensus of the group task. In general, four basic functions of social systems are discerned (AGIL scheme; Parsons 1937), namely, (1) adaptation, (2) goal attainment, (3) integration, and (4) latent pattern maintenance. In order to manifest these social system functions, a group develops characteristic processes in coordinating their goals, meanings, and behaviours. These processes become manifold, consisting of mechanism-forming *patterns* such as democratic by majority rules, hierarchical autocratic rules, or self-organised.

2.6.2 Peripheral Input–Process–Outcome (IPO) 'Lens' for Examining Varying Levels of Dissolution

Entities, mechanisms, and process patterns can be identified as constitutive at all levels of dissolution in the analysis of group coordination, ranging from the macro- to micro- levels of perspective (see also Klein and Kozlowski 2000). Thus, within our model, the classic IPO systematic is applied as a *device of analysis* rather than as a composite element of small group coordination. We have extended the core of our model by adding an external analytical 'lens' (Fig. 2.4) device to the workings of the model that enables analysis of all levels of coordination dissolution from fine-grained atomic micro-level inter-individual interactions (e.g. initiator–follower behaviour), meso-level routines (e.g. resuscitation algorithms), to macro-level structures of small group coordination (e.g. hierarchical, egalitarian). Our resulting inclusive functions–entities–mechanisms–patterns (FEMP[ipo]) model therefore offers a practical analytical tool for both human and non-human primate group coordination that can be used at any perspective (e.g. top-down or bottom-up;

Fig. 2.4 IPO lens of group coordination

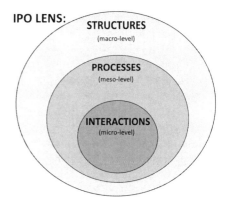

input–process–output or output–process–input) and at all levels of dissolution (micro-, meso-, and/or macro- elements).

In order to illustrate the application of the model's IPO lens (Fig. 2.4) more closely, let us return to the example of the public administration decision-making meeting from Example 3. Using this multi-dissolution analysis that the external 'lens' part of our model suggests, on the macro-level, the hierarchical and divisional structures were characteristic for every bureaucratic organisation as well as pre-determined modes of observed decision making. These structures were implemented in role instructions for the group members and the group leader in a free simulation of this public administration case (Boos 1994a). We expected different process patterns on the meso-level of dissolution under these different modes of group decision making (respectively, steering of group processes). In a recent study (Boos 2006a), we reanalysed the videos and transcripts with a combination of quantitative and qualitative methods. On the basis of interaction process coding, we identified coordination episodes in the group discussion. These episodes were interpreted according to the rules of structural hermeneutics (Oevermann 2002). Our intent was to describe the process where the two different organisational procedures (hierarchical vs. divisional) are set into action in the group. It would be naive to contend that the instructions could be implemented one by one via an intentional process such as that of the group leader in this field study, so we instead conceptualised the group process as a combination of intended individual behaviour and the unintended collective results of individual planned behaviour. As others in small group research have concluded, the structural and process levels of group coordination are intertwined and produce emergent characteristics of the group (Poole et al. 1985). In their theoretical approach, Poole and colleagues conceptualise group decision making as a 'structuration process,' meaning that the process is a pattern of interrelated events from which a structured outcome emerges. 'Structuration' in this context means that a social system produces and reproduces itself in an ongoing process via the application of generative rules (e.g. hierarchy) and resources (e.g. technical devices; routines). Applied to the example of group decision making, group decisions are not solitary events but instead more closely resemble iterative concatenations of goal settings, convergence of meanings, and synchronisation of behaviours.

This interplay of structural and process levels corresponds to two basic psychological principles: first, the constructability of a process as in our context regarding the actualisation of a coordination behaviour and its predictable outcome; and second, the spontaneity (non-predictability) of behaviour, which leads to the evolution of a pattern or 'gestalt' that can only partly be traced back to instructions or goals.

The qualitative results corresponded to results we received from detailed process coding and time-series analysis of the data (Boos and Meier 1993). The quantitative data confirmed what we hypothesised in the qualitative study: There are significant meso-level differences between these two models of group coordination (Fig. 2.5) (Boos 1996).

The design of our inclusive model with augmentations such as the embedded four-quadrant Coordination Mechanism Circumplex Model (CMCM) by Wittenbaum et al. (1998) within the process part of its IPO structure and its peripheral 'lens' to facilitate the various levels of analysis helps us to understand coordination on these different levels of dissolution as complementary notions at integrated levels. Often, coordination on the structural level is called *steering* in order to depict that there is a difference in the scope of an expectation horizon: 'Steering' in this sense means expectation-guided orientation (sense or direction) of behaviour in social systems. Coordination relates to the requirement of an ongoing, selective

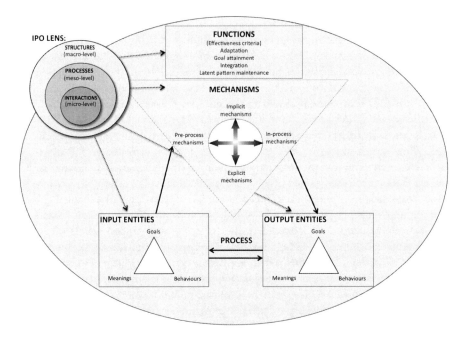

Fig. 2.5 An inclusive functions–entities–mechanisms–patterns (FEMP^{ipo}) model of group coordination

integration of events that consistently appear instantaneous, even though they can exert a long-term impact on economic and ecological structural alterations of a city in cases such as our field study that examined coordination dynamics of public administration decision making.

An example of a multi-dissolution examination of the non-human primate arena is an analysis of leadership behaviour based on maximising survival, which might be assumed to mostly occur on the structural level. Examining non-human primate leadership behaviour on a process level is whether the group – at a specific point in time – moves on the ground or in the trees. And obviously, these various levels of dissolution perspectives are usually not mutually exclusive and should therefore be analysed with assumptions of interactive emergent dynamics *and* linear, sequential cause-and-effect relationships.

2.6.3 Provisions for the Iterative Structuration Inherent in Coordination

In addition to the above differentiations of the levels of dissolution in an analysis of group coordination, depending on the nature of the group and the reasons why the group coordination's task was set up, group coordination can be either a process variable or a result variable and often times is both. As an example, we can focus the coordinated process of sharing mental models via interaction and communication or we can focus – in a specific moment in time – on a shared mental model as a result of this process. Provisions have been made for this phenomenon in our FEMP[ipo] model, with an arrow circulating from the 'output' stage of the model's core back into the 'input' stage of the model's core.

2.7 Conclusion

Here again we have the distinction between process and structure, which other models have been hard-pressed to address and thus remain in the theory stage versus the field application stage. Generally, coordination relates only to the moment where goals, meanings, and behaviours converge. And this very act of coordination is irreversible. From the structural side of coordination, which, as we mentioned earlier, is often called *steering*, individuals and groups take such moments of convergence as an opportunity to adjust their expectations and thus identify new coordination challenges. In this sense, the difference between steering and coordination corresponds to the difference in the reversibility and irreversibility of events as well as to the difference between structure and process.

It is our hope that this chapter, with its description of small group coordination theory and its consequent inclusive FEMP[ipo] model for examining coordination elements in groups, has struck a balance between conveying an appreciation for the enormous complexity of group coordination and offering a practical analytical tool for comparative studies of coordination in human and non-human primate groups.

References

Annett J (2004) Hierarchical task analysis. In: Diapor D, Stanton N (eds) The handbook of task analysis for human-computer interaction. Lawrence Erlbaum, Mahwah, NJ, pp 67–82

Arrow H, McGrath JE, Berdahl JL (2000) Small groups as complex systems: formation, coordination, development, and adaption. Sage, Thousand Oaks, CA

Arrow H, Poole MS, Henry KB, Wheelan SA, Moreland R (2004) Time, change, and development. a temporal perspective on groups. Small Group Res 35:73–105

Avolio BJ, Jung DI, Murry W, Sivasbramaniam N (1996) Building highly developed teams: focusing on shared leadership process, efficacy, trust, and performance. In: Beyerlein MM, Johnson DA, Beyerlein ST (eds) Advances in interdisciplinary studies of work teams: team leadership. JAI Press, Greenwich, CT, pp 173–209

Boos M (1994a) Entscheidungen in der öffentlichen Verwaltung. Die aufgabenorientierte Dynamik bei drei Modellen der Führung und Zusammenarbeit [in German]. Gruppendynamik 25:185–202

Boos M (1994b) The regulation of group process in decision making. German J Psych 18:207–209

Boos M (1996) Entscheidungsfindung in Gruppen: Eine Prozessanalyse [in German]. Huber, Bern

Boos M (2006a) Correlates and effects of the conversational coherence of group discussions. First Annual INGRoup Conference, Pittsburgh

Boos M (2006b) Optimal sharedness of mental models for effective group performance. CoDesign 3:21–28

Boos M, Meier F (1993) Die Regulation des Gruppenprozesses bei der Entscheidungsfindung [in German]. Z Sozialpsychol 24:3–14

Boos M, Kappeler P, Kolbe M, Strack M (2007) Coordination in human and nonhuman primate groups. Second Annual Meeting of the Interdisciplinary Network for Group Research (INGRoup). Lansing, MI.

Boos M, Kolbe M, Strack M (2006) Gruppenkoordination – Modellierung der Mechanismen [Group coordination – Modeling of the mechanisms]. 45 Kongress der Deutschen Gesellschaft für Psychologie [45th Congress of the German Society of Psychology]. Nürnberg

Boos M, Sassenberg K (2001) Koordination in verteilten Arbeitsgruppen [Coordination in distributed work groups]. In: Witte EH (ed) Leistungsverbesserungen in aufgabenorientierten Kleingruppen: Beiträge des 15 Hamburger Symposiums zur Methodologie der Sozialpsychologie [Improvements of performance in task-oriented small groups: Contributions to the 15th Hamburger Symposium of Methodology in Social Psychology]. Papst, Lengerich, pp 198–216

Boos M, Strack M (2008) The destiny of proposals in the course of group discussions. International Congress of Psychology, Berlin

Burke CS, Stagl KC, Salas E (2006) Understanding team adaptation: a conceptual analysis and model. J Appl Psychol 91:1189–1207

Cannon-Bowers JA, Salas E, Converse SA (1993) Shared mental models in expert team decision-making. In: Castellan NJJ (ed) Current issues in individual and group decision making. Lawrence Erlbaum, Hillsdale, NJ, pp 221–246

Cannon-Bowers JA, Salas E (1998) Individual and team decision making under stress: Theoretical underpinnings. In: Cannon-Bowers JA, Salas E (eds) Making decisions under stress. American Psychological Association, Washington, DC, pp 17–38

Caporael L, Wilson DS, Hemelrijk C, Sheldon KM (2005) Small groups from an evolutionary perspective. In: Poole MS, Hollingshead AB (eds) Theories of small groups interdisciplinary perspectives. Sage, Thousand Oaks, CA, pp 369–391

Conradt L, Roper TJ (2009) Conflicts of interest and the evolution of decision sharing. Philos Trans Roy Soc Lond B 364:807–819

De Luque MFS, Sommer SM (2000) The impact of culture on feedback-seeking behaviour: an integrated model and propositions. Acad Manage Rev 25:829–849

Diehl M, Stroebe W (1987) Productivity loss in brainstorming groups: toward the solution of a riddle. J Pers Soc Psychol 53:497–509

Diehl M, Stroebe W (1991) Productivity loss in idea-generating groups: tracking down the Blocking Effect. J Pers Soc Psychol 61:392–403

Entin EE, Serfaty D (1999) Adaptive team coordination. Hum Factors 41:312–325

Espinosa A, Lerch FJ, Kraut RE (2004) Explicit vs. implicit coordination mechanisms and task dependencies: one size does not fit all. In: Salas E, Fiore SM (eds) Team cognition: understanding the factors that drive process and performance. American Psychological Association, Washington, DC, pp 107–129

Fiore SM, Salas E, Cuevas HM, Bowers CA (2003) Distributed coordination space: toward a theory of distributed team process and performance. Theor Issues Ergon Sci 4:340–364

Gouran DS, Hirokawa RY (1996) Functional theory and communication in decision-making and problem-solving groups. An expanded view. In: Hirokawa RY, Poole MS (eds) Communication and group decision making. Sage, Thousand Oaks, CA, pp 55–80

Grote G, Zala-Mezö E, Grommes P (2003) Effects of standardisation on coordination and communication in high workload situations. Linguistische Berichte, Sonderheft 12:127–155

Grote G, Helmreich RL, Sträter O, Häusler R, Zala-Mezö E, Sexton JB (2004) Setting the stage: characteristics of organisations, teams and tasks influencing team processes. In: Dietrich R, Childress TM (eds) Group interaction in high risk environments. Ashgate, Aldershot, UK, pp 111–140

Grote G, Kolbe M, Zala-Mezö E, Bienefeld-Seall N, Künzle B (2010) Adaptive coordination and heedfulness make better cockpit crews. Ergonomics 52:211–228

Hackman JR, Morris CG (1975) Group tasks, group interaction process, and group performance effectiveness: a review and proposed integration. In: Berkowitz L (ed) Advances in experimental social psychology. Academic, New York, pp 45–99

Hill GW (1982) Group versus individual performance: Are $N+1$ heads better than one? Psychol Bull 91:517–539

Ilgen DR, Hollenbeck JR, Johnson M, Jundt D (2005) Teams in organisations: from input–process–output models to IMOI models. Annu Rev Psychol 56:517–543

Kerr NL, Tindale RS (2004) Group performance and decision-making. Annu Rev Psychol 55:623–655

Klein KJ, Kozlowski SWJ (2000) From micro to meso: critical steps in conceptualising and conducting multilevel research. Organ Res Methods 3:211–236

Klimoski R, Mohammed S (1994) Team mental model: Construct or metaphor? J Manage 20:403–437

Kolbe M, Boos M (2009) Facilitating group decision-making: Facilitator's subjective theories on group coordination Forum Qualitative Sozialforschung/Forum: Qualitative Social Research 10: http://nbn-resolving.de/urn:nbn:de:0114-fqs0901287

Kolbe M, Künzle B, Zala-Mezö E, Wacker J, Grote G (2009) Measuring coordination behavior in anaesthesia teams during induction of general anaesthetics. In: Flin R, Mitchell L (eds) Safer surgery analysing behavior in the operating theatre. Ashgate, Aldershot, UK, pp 203–221

Kolbe M, Künzle B, Zala-Mezö E, Wacker J, Spahn DR, Grote G (2010) Adaptive coordination makes better anaesthesia crews. 25th SIOP Annual Conference. Atlanta

Kolbe M, Künzle B, Zala-Mezö E, Burtscher MJ, Wacker J, Spahn DR, Grote G (2010) The functions of team monitoring and 'talking to the room' for performance in anesthesia teams. In Proceedings of the Human Factors and Ergonomics Society 54th Annual Meeting (pp. 857–861). Santa Monica, CA, USA: Human Factors and Ergonomics Society

Kolbe M, Künzle B, Zala-Mezö E, Wacker J, Spahn DR, Grote G (under review) Adaptive coordination during emergency situations: Is implicit or explicit more effective?

Kozlowski SWJ, Watola DJ, Jensen JM, Kim BH, Botero IC (2009) Developing adaptive teams: a theory of dynamic team leadership. In: Salas E, Goodwin GF, Burke CS (eds) Team effectiveness in complex organisations: cross-disciplinary perspectives and approaches (SIOP Frontier Series). Taylor and Francis, New York, pp 113–156

Künzle B, Kolbe M, Grote G (2010a) Ensuring patient safety through effective leadership behavior: a literature review. Safety Sci 48:1–17

Künzle B, Zala-Mezö E, Wacker J, Kolbe M, Grote G (2006) Leadership in anaesthesia teams: the most effective leadership is shared. Qual Saf Health Care online first

Latané B (1981) The psychology of social impact. Am Psychol 36:343–356

Lim BC, Klein KJ (2006) Team mental models and team performance: a field study of the effects of team mental model similarity and accuracy. J Organ Behav 27:403–418

Manser T, Howard SK, Gaba DM (2008) Adaptive coordination in cardiac anaesthesia: a study of situational changes in coordination patterns using a new observation system. Ergonomics 51:1153–1178

Marks MA, Mathieu JE, Zaccaro SJ (2001) A temporally based framework and taxonomy of team processes. Acad Manage Rev 26:356–376

Mathieu JE, Rapp TL (2009) Laying the foundation for successful team performance trajectories: the roles of team charters and performance strategies. J Appl Psychol 94:90–103

McGrath JE (1984) Groups, interaction and performance. Prentice Hall, Englewood Cliffs, NJ

Oevermann U (2002) Klinische Soziologie auf der Basis der Methodologie der objektiven Hermeneutik – Manifest der objektiv hermeneutischen Sozialforschung [Clinical sociology based on methods of objective hermeneutics]. Institut für Hermeneutische Sozial- und Kulturforschung e.V, Frankfurt aM

Orlitzky M, Hirokawa RY (2001) To err is human, to correct for it divine. A meta-analysis of research testing the functional theory of group decision-making effectiveness. Small Group Res 32:313–341

Parsons T (1937) The structure of social action. McGraw-Hill, New York

Pearce CL, Conger JA (2003) Shared leadership: reframing the how's and why's of leadership. Sage, Thousand Oaks, CA

Pearce CL, Sims HP Jr (2002) Vertical versus shared leadership as predictors of the effectiveness of change management teams: an examinations of aversive, directive, transactional, transformational, and empowering leader behaviours. Group Dyn Theory Res 6:172–197

Poole MS, Hirokawa RY (1996) Introduction. Communication and group decision making. In: Hirokawa RY, Poole MS (eds) Communication and group decision making. Sage, Thousand Oaks, CA, pp 3–18

Poole MS, Seibold DR, McPhee RD (1985) Group decision-making as a structurational process. Q J Speech 71:74–102

Rico R, Sánchez-Manzanares M, Gil F, Gibson C (2008) Team implicit coordination processes: a team knowledge-based approach. Acad Manage Rev 33:163–184

Risser DT, Rice MM, Salisbury ML, Simon R, Jay GD, Berns SD (1999) The potential for improved teamwork to reduce medical errors in the emergency department. Ann Emerg Med 34:373–383

Salas E, Sims DE, Burke CS (2005) Is there a "big five" in teamwork? Small Group Res 36:555–599

Salas E, Rosen MA, King H (2007) Managing teams managing crisis: principles of teamwork to improve patient safety in the emergency room and beyond. Theor Issues Ergon Sci 8:381–394

Shaw ME (1976) Group dynamics: the psychology of small group behaviour. McGraw-Hill, New York

Stachowski AA, Kaplan SA, Waller MJ (2009) The benefits of flexible team interaction during crisis. J Appl Psychol 94:1536–1543

Steiner ID (1972) Group processes and productivity. Academic, New York

Stewart KJ, Harcourt AH (1994) Gorillas' vocalisations during rest periods: signals of impending departure? Behaviour 130:29–40

Stroebe W, Diehl M (1994) Why groups are less effective than their members: on productivity losses in idea-generating groups. Eur Rev Soc Psychol 2:271–303

Stückle S, Zinner D (2008) To follow or not to follow: decision making and leadership during the morning departure in chacma baboons (*Papio hamadryas ursinus*). Anim Behav 75:1995–2004

Trillmich J, Fichtel C, Kappeler PM (2004) Coordination of group movements in wild Verreaux's Sifakas (*Propithecus verreauxi*). Behaviour 141:1103–1120

Tschan F (2000) Produktivität in Kleingruppen. Was machen produktive Gruppen anders und besser [in German]? Huber, Bern

Tschan F, Semmer NK, Nägele C, Gurtner A (2000) Task adaptive behaviour and performance in groups. Group Process Interg 3:367–386

Tschan F, Semmer NK, Gurtner A, Bizzari L, Spychiger M, Breuer M, Marsch SU (2009) Explicit reasoning, confirmation bias, and illusory transactive memory. A simulation study of group medical decision making. Small Group Res 40:271–300

Waller MJ (1999) The timing of adaptive group responses to nonroutine events. Acad Manage J 42:127–137

Waller MJ, Uitdewilligen S (2008) Talking to the room. Collective sensemaking during crisis situations. In: Roe RA, Waller MJ, Clegg SR (eds) Time in organisational research. Routledge, Oxford, pp 186–203

Waller MJ, Gupta N, Giambatista RC (2004) Effects of adaptive behaviours and shared mental models on control crew performance. Manage Sci 50:1534–1544

Watts D (2000) Mountain gorilla habitat use strategies and group movements. In: Boinski S, Garber P (eds) On the move: how and why animals travel in groups. University of Chicago Press, Chicago, IL, pp 351–374

West MA (2004) Effective teamwork. Practical lessons from organisational research, BPS Blackwell, Oxford

Williams WM, Sternberg RJ (1988) Group intelligence: why some groups are better than others. Intelligence 12:351–377

Wilson DS (1997) Incorporating group selection into the adaptionist program: A case study involving human decision making. In: Simpson JA, Kenrick DT (eds) Evolutionary social psychology. Lawrence Erlbaum, Mahwah, NJ, pp 345–386

Wittenbaum GM, Stasser G, Merry CJ (1996) Tacit coordination in anticipation of small group task completion. J Exp Soc Psychol 32:129–152

Wittenbaum GM, Vaughan SI, Stasser G (1998) Coordination in task-performing groups. In: Tindale RS, Heath L, Edwards J, Posavac EJ, Bryant FB, Suarez-Balcazar Y, Henderson-King E, Myers J (eds) Theory and research on small groups. Plenum, New York, pp 177–204

Yeager L (2001) Ethics as a social science: The moral philosophy of social cooperation. Edward Elgar, Cheltenham, UK

Zala-Mezö E, Wacker J, Künzle B, Brüesch M, Grote G (2009) The influence of standardisation and task load on team coordination patterns during anaesthesia inductions. Qual Saf Health Care 18:127–130

Zysno P (1998) Von Seilzug bis Brainstorming: Die Effizienz der Gruppe [Group efficiency]. In: Witte EH (ed) Sozialpsychologie der Gruppenleistung [Social psychology of group performance]. Pabst, Lengerich, pp 184–210

Chapter 3
Coordination of Group Movements in Non-human Primates

Claudia Fichtel, Lennart Pyritz, and Peter M. Kappeler

Abstract Many animals are organised into social groups. Because individuals have different preferences and diverging needs, conflicts of interests exist; these conflicts are particularly revealed and negotiated in the context of group movements. Thus, group movements provide an excellent example to study coordination processes in non-human primates. In this chapter we review several aspects related to group movements in non-human primates. We first summarise the current understanding of variation in spacing patterns, types of leadership, and decision-making processes. We then focus on methodological issues and discuss various operational definitions of group movements, and we propose an operational definition that has already been applied successfully in studies of small free-ranging groups. We conclude by discussing the possibilities and limitations of transferring concepts and methods from studies of non-human primate groups to research on human groups.

3.1 Introduction

Many animals are organised into permanent social groups. The shift from an originally solitary to a gregarious lifestyle is considered to be one of the major evolutionary transitions (Maynard Smith and Szathmáry 1995). These social groups differ enormously in size, composition, permanence, and cohesion (Parrish and Edelstein-Keshet 1999). Their members can be anonymous to each other, or they can recognise group or even individual identity. The ultimate reasons for why animals might be group-living as well as the respective optimal group size have been investigated in detail in diverse taxa (e.g. Bertram 1978; van Schaik 1983; Zemel and Lubin 1995). These evolutionary benefits include reduced individual

---

C. Fichtel (✉), L. Pyritz, and P.M. Kappeler
Department of Behavioural Ecology and Sociobiology, German Primate Center, Kellnerweg 6, 37077 Göttingen, Germany
e-mail: claudia.fichtel@gwdg.de; LennartPyritz@gmx.net; pkappel@gwdg.de

M. Boos et al. (eds.), *Coordination in Human and Primate Groups*,
DOI 10.1007/978-3-642-15355-6_3, © Springer-Verlag Berlin Heidelberg 2011

predation risk, joint resource defence, cooperative foraging, shared vigilance, and information transfer (Alexander 1974; Bertram 1978).

Living in a group also leads to interindividual conflicts and costs, such as competition over resources and mates, as well as increased pathogen transmission. These factors limit the size of groups and act as a centrifugal force on group cohesion (Alexander 1974; Bertram 1978). First and foremost, individual foraging strategies and schedules are expected to be heterogeneous and are therefore a source of conflict. Growing juveniles, pregnant or lactating females, and adult males often have divergent overall activity budgets and different dietary needs, such as types of food items eaten and time devoted to foraging for each item (see, e.g. Altmann 1980; Dunbar and Dunbar 1988). Depending on the type and distribution of particular resources, intra-group feeding competition can threaten group cohesion and influence individual and subgroup movements (van Schaik 1989; van Nordwijk et al. 1993; Pulliam and Caraco 1984). A conflict of interest may also arise between the sexes when inter-group encounters have different costs and/or benefits for males vs. females (Cheney 1987) or when mating competition interferes with foraging efforts (Alberts et al. 1996).

In order to maintain group cohesion and social stability despite these conflicts, individuals need to synchronise and coordinate their activities such as foraging, resting, social interactions, and collective movements if they want to reap the benefits of gregariousness (Conradt and Roper 2003, 2007; Rands et al. 2003; Kerth et al. 2006). How this trade-off is achieved and implemented at the behavioural level is not easily studied. That said, natural group movements among resources provide an operationally accessible and ecologically relevant context to study these fundamental mechanisms of social coordination. In the context of group movements, it is possible to quantify how members of a group achieve a communal decision about which activities will be carried out, where, and for how long (Boinski and Garber 2000).

Because group movements are characterised by dynamics operating at multiple levels, it is heuristically useful to consider group movements on four different levels: (1) normative details of the spatiotemporal patterns of space use of a group as an entity such as travel routes within the home range and their variability according to seasonal changes and climatic conditions, resource availability, predation risk, and/or the likelihood of inter-group encounters; (2) behavioural processes describing who initiates, leads, and terminates a group movement and how many members follow whom; (3) communication mechanisms that control the processes proximately, such as vocal or visual signals used to initiate a movement and to maintain group cohesion; and (4) whether leadership is distributed or monopolised. If leadership is distributed, all group members are said to contribute to a democratic decision. If a single individual leads the group and the other group members merely follow, the decisions are said to be despotic (Conradt and Roper 2003, 2005). Because information on all four aspects is not available for most species (honey bees being an exception; see, for example, Seeley and Visscher 2004), general principles are currently best inferred from inter-specific comparisons. We adopt this approach and focus on one relatively well-studied taxon with interesting variation in social organisation: non-human primates. In this chapter we review the currently available information on group movements in non-human primates with special emphasis on the four levels described above. We then raise the issue of how group movements in animals can be operationalised by human observers in the field. Final thoughts provide a current context and future outlook on inter-disciplinary research in human and non-human primates.

3.2 Group Movements in Non-human Primates

The more than 300 species of non-human primates are interesting subjects for the study of group movements for at least four reasons. First, they exhibit more variation in social organisation than most other vertebrate taxa. Primate groups range in size from two to several hundred individuals of both sexes and multiple generations (Smuts et al. 1986). Second, primates occupy a wide range of habitats, from semi-deserts to tropical rain forests and temperate mountain forests, resulting in movements that appear to be guided by these widely differing ecological needs (Eisenberg 1981). Third, non-human primates have larger brains relative to their body size than other mammals and vertebrates, suggesting that behavioural aspects of group movements may be influenced by their unusual cognitive abilities (Reader and Laland 2002; Dunbar and Shultz 2007). Finally, primates vary across species in dominance styles and predominant communication modalities (Seyfarth 1986; Zeller 1986; Sterck et al. 1997), offering interesting behavioural variation in the social component of group movements.

3.2.1 Patterns of Group Movements

Beside abiotic variables, ecological factors such as seasonal differences in resource distribution or predation risk, as well as social influences from neighbouring groups, affect daily ranging patterns of primate groups. We illustrate these effects with a few examples below.

The spatiotemporal distribution and availability of resources not only influence the size and cohesion but also the ranging patterns of primate groups (van Schaik 1983; Chapman et al. 1995). For instance, food availability has been observed to significantly affect activity profiles and habitat use of redfronted lemurs (*Eulemur fulvus rufus*) and red-bellied lemurs (*Eulemur rubriventer*) in Ranomafana National Park, a rainforest in southeastern Madagascar (Overdorff 1993, 1996). During periods of food scarcity, both species fed more and dedicated less time to travelling and resting. Redfronted lemurs in Ranomafana also conducted group movements of up to 5 km away from their usual ranges during a period of fruit scarcity in order to exploit extraordinary food abundance (a guava plantation) elsewhere. The ranging behaviour of redfronted lemurs was also affected by the differential availability of water during the dry and rainy seasons in the Kirindy Forest, a dry deciduous forest in western Madagascar. During the 8 month dry season, groups living close to ephemeral water holes made daily excursions of up to two to three home range diameters to drink, whereas groups living farther away from the river shifted their ranges nearer the water holes for several weeks or months and moved very little during this time (Scholz and Kappeler 2004). We also observed a group with permanent access to a water hole in their usual home range extending its range away from the riverbed (Pyritz et al. unpublished data). Presumably, lemurs exhibit this behaviour in order to avoid encounters with conspecifics from other groups or predators that are attracted by the lemurs gathering at the water holes in large numbers.

Resource availability has also been observed to influence travelling patterns in chacma baboons (*Papio ursinus*) (Noser and Byrne 2007a). During the dry season, the study group followed linear paths over great distances in the morning to reach sparse fruit trees and ephemeral waterholes. In the afternoons, when the baboons fed on seeds, group movements were shorter and sinuous. During the rainy season, food distribution determined the onset of group movements. The baboons left their sleeping sites earlier when visiting patchily distributed fig trees than when moving towards evenly distributed fruit resources. Therefore, these baboons seem to plan movements according to the type of feeding goal.

The presence of conspecific groups has also been observed to be an additional factor impacting the ranging behaviour of chacma baboons (Noser and Byrne 2007b). When neighbouring groups were present within a 500-m radius, the routes conducted by the focal group were less linear, the baboons travelled faster, and they covered larger distances between different resources. These changes in travelling behaviour are interpreted as measures to avoid group encounters, which can proceed quite aggressively in this species.

3.2.2 Processes and Leadership

The process of group movements depends on the species, group composition, and permanence. For example, fish swarms and bird flocks are often so large that members seem to neither know each other individually nor know which individuals possess decisive information, and they also appear to lack recruiting signals (Couzin et al. 2002). Cohesion and coordinated movements in such groups are often maintained by self-coordination such as individuals following the simple rule of 'keep a certain safe distance to the next neighbour (Parrish and Edelstein-Keshet 1999; Hemelrijk 2002; Couzin et al. 2002). In contrast, in groups where members know each other individually, such as primates, certain individuals may adopt different roles and initiate and terminate a group movement (Boinski and Garber 2000).

Studies of several primate species revealed that age, rank, or sex can be defining characteristics of group leaders. In many species, adult and therefore more experienced and knowledgeable individuals initiate and lead group movements more often than juveniles (Japanese monkeys, *Macaca fuscata*: Itani 1963; Costa Rican squirrel monkeys, *Saimiri oerstedii*: Boinski 1991; chimpanzees, *Pan troglodytes*: Boesch 1991a; white-faced capuchin monkeys, *Cebus capucinus*: Boinski and Campbell 1995; mountain gorillas, *Gorilla gorilla*: Stewart and Harcourt 1994). In some species, dominant animals rather than the most experienced lead groups more often than subordinate individuals (hamadryas baboons, *Papio hamadryas*: Kummer 1968; mountain gorillas: Watts 1994; white-faced capuchin monkeys: Boinski 1993; ringtailed lemurs, *Lemur catta:* Sauther and Sussman 1993). However, rank is often confounded with age or sex, which handicaps untangling the relative importance of these variables in structuring group leadership.

Many studies showed females to lead groups more often than males (see Table 3.1; Neville 1968; Rowell 1969; Struhsaker 1967a; Dunbar and Dunbar 1975; Oates 1977; van Nordwijk and van Schaik 1987; Boinski 1988; Mitchell et al. 1991; Erhart and Overdorff 1999; Leca et al. 2003; Trillmich et al. 2004). This sex difference is usually attributed to higher nutritional needs of females due to the energetic costs of gestation and lactation (Boinski 1988, 1991; Erhart and Overdorff 1999; Trillmich et al. 2004).

Reasons for male leadership are surmised to include dominance or mating competition (Table 3.1). For example, male mountain gorillas initiate group movements after contact with a rival (Watts 1994), and in spider monkeys (*Ateles geoffroyi*), males frequently lead their group to the edge of the home range presumably to make contact with females from other groups (Chapman 1990). Sex differences in leadership of groups have also been explained by sex-specific patterns of residency and dispersal and a corresponding improved information status of the philopatric sex regarding the distribution and availability of different resources (Struhsaker 1967b; Goodall 1968; Sigg and Stolba 1981; van Nordwijk and van Schaik 1987; Watts 1994; Trillmich et al. 2004).

Animals were identified as leaders when they had been observed initiating movement and were therefore at the forefront of collective movements. However, the initiating individual did not always remain in the leading position during the entire movement, meaning that either changes in their forefront positioning occurred (hamadryas baboons: Kummer 1968; guinea baboons, *Papio papio*: Byrne 1981, 2000) or the movement was terminated by an individual different from the initiator (indris, *Indri indri*: Pollock 1997). There are also reports of distributed leadership where all group members equally initiated and led movements (Leca et al. 2003; Meunier et al. 2006; Jacobs et al. 2008).

How and why leadership and followership evolved and how such a system can be stable have been the subject of a number of recent studies (e.g. Conradt and Roper 2005; Couzin et al. 2005; van Vugt 2006; Rands et al. 2008; Sueur and Petit 2008a). On the one hand, leadership is interpreted as a byproduct of dominance and submission in animal groups (e.g. Alexander 1987). Several other studies that mainly focused on non-primate species with no clear dominance hierarchy identified correlates of leaders, including intrinsic factors such as size or physiological

Table 3.1 Predominant sex of leaders, initiation signals and decision-making processes described in non-human primates

Species	Predominantly leading sex	Initiation signals			Decision making process		Data[a]	References
		Visual displays	Vocal displays	Combination of visual and vocal displays	Shared	Unshared		
Redfronted lemur, *Eulemur fulvus rufus*	Females	No	unknown (maybe grunt frequency)	no	Yes (partially)	No	S Q	Erhart and Overdorff (1999), Pyritz et al. unpublished data
Verreaux's sifaka, *Propithecus verreauxi*	Females	No	No	No	Yes (partially)	No	S	Trillmich et al. (2004)
Diademed sifaka, *Propithecus diadema edwardsii*	Females	Unknown	Unknown	Unknown	Yes (partially)	No	S	Erhart and Overdorff (1999)
Pygmy marmoset, *Callithrix pymaea*	Unknown	Yes	Yes	Unknown (probably no)	Unknown	Unknown	A	Soini (1981)
Golden lion tamarin, *Leontopithecus rosalia*	Unknown	Unknown (probably no)	Yes	Unknown (probably no)	Unknown	Unknown	S	Boinski et al. (1994)
Costa Rican squirrel monkey, *Saimiri oerstedii*	Females	Unknown (probably no)	Yes	Unknown (probably no)	Unknown	Unknown	S S S	Boinski (1988), Boinski (1991), Mitchell et al. (1991)
White-faced capuchin monkey, *Cebus capucinus*	Females	Unknown (probably no)	Yes	Unknown (probably no)	Yes	No	S S S S	Boinski (1993), Boinski and Campbell (1995), Leca et al. (2003), Meunier et al. (2006)

C. Fichtel et al.

Spider monkey, *Ateles geoffroyi*	Males	Unknown	Unknown	Unknown	Unknown	Unknown	A	Chapman (1990)
Mantled howler monkey, *Alouatta palliata*	Unknown	Yes	Yes	Unknown (probably no)	Unknown	Unknown	A A	Milton (1980), Whitehead (1989)
Dusky titi monkey, *Callicebus moloch*	Unknown	Yes	Unknown (probably no)	Unknown (probably no)	Unknown	Unknown	S	Menzel (1993)
Guinea baboon, *Papio papio*	Unknown	Unknown (probably no)	Yes	Unknown (probably no)	Unknown	Unknown	S	Byrne (1981)
Olive baboon, *Papio anubis*	Females	Yes	Unknown (probably no)	Unknown (probably no)	Unknown	Unknown	A A	Rowell (1969), Rowell (1972b)
Chacma baboon, *Papio ursinus*	Unknown	Unknown	Unknown	Unknown	Yes (departure off sleeping site)	Yes (movements in foraging experiment)	S S	King et al. (2008), Stueckle and Zinner (2008)
Hamadryas baboon, *Papio hamadryas*	Unknown	Yes	Unknown (probably no)	Unknown (probably no)	Yes (partially)	No	S	Kummer (1995)
Yellow baboon, *Papio cynocephalus*	Unknown	Yes	Unknown (probably no)	Unknown (probably no)	Unknown	Unknown	A	Norton (1986)
Gelada baboon, *Theropithecus gelada*	Females	Unknown	Unknown	Unknown	Unknown	Unknown	A	Dunbar and Dunbar (1975)
Hanuman langur, *Semnopithecus entellus*	Unknown	Yes	Yes	Unknown (probably no)	Unknown	Unknown	A	Vogel (1973)
Capped leaf monkey, *Trachypithecus pileata*	Unknown	Yes	Unknown (probably no)	Unknown (probably no)	Unknown	Unknown	Q	Stanford (1990)

(*continued*)

Table 3.1 (continued)

Species	Predominantly leading sex	Initiation signals			Decision making process		Data[a]	References
		Visual displays	Vocal displays	Combination of visual and vocal displays	Shared	Unshared		
Red colobus, *Procolobus badius*	Unknown	Unknown (probably no)	Yes	Unknown (probably no)	Unknown	Unknown	A	Struhsaker (1975)
Guereza, *Colobus guereza*	Females	Yes	Unknown (probably no)	Unknown (probably no)	Unknown	Unknown	A A	Oates (1977), Marler (1965)
Drill, *Mandrillus leucophaeus*	Unknown	Unknown (probably no)	Yes	Unknown (probably no)	Unknown	Unknown	A	Jouventin (1975)
Mandrill, *Mandrillus sphinx*	Unknown	Unknown (probably no)	Yes	Unknown (probably no)	Unknown	Unknown	A	Kudo (1987)
Barbary macaque, *Macaca sylvanus*	Unknown	Unknown (probably no)	Unknown (probably no)	Yes	Unknown	Unknown	Q	Mehlman (1996)
Long-tailed macaque, *Macaca fascicularis*	Females	Unknown	Unknown	Unknown	Unknown	Unknown	Q	van Nordwijk and van Schaik (1987)
Rhesus macaque, *Macaca mulatta*	Females	Unknown	Unknown	Unknown	Yes (partially)	No	A S	Neville (1968), Sueur and Petit (2008a)
Tonkean macaque, *Macaca tonkeana*	Unknown	Unknown	Unknown	Unknown	Yes	No	S	Sueur and Petit (2008a)
Vervet monkey, *Cercopithecus aethiops*	Females	Unknown	Unknown	Unknown	Unknown	Unknown	A	Struhsaker (1967a)

Species							References
Mountain gorilla, *Gorilla gorilla berengei*	Males	Yes	Yes	Unknown (probably no)	No	Yes	A, S	Schaller (1963), Stewart and Harcourt (1994)
Chimpanzee, *Pan troglodytes*	Unknown	Yes	Unknown	Unknown	Unknown	Unknown	Q	Boesch (1991b)
Bonobo, *Pan paniscus*	Unknown (probably no)	Unknown (probably no)	Unknown (probably no)	Yes	Unknown	Unknown	Q	Ingmanson (1996)

[a] Indicates type of data: *A* anecdotal, *Q* quantified but without robust statistical analyses, *S* robust statistical analyses

state (Krause et al. 1998; Rands et al. 2003; Fischhoff et al. 2007), personality characteristics such as activity (Beauchamp 2000) and boldness (Ward et al. 2004; Leblond and Reebs 2006), positive social feedback between group members (Harcourt et al. 2009), and asymmetries in information or knowledge (Reebs 2000, 2001; Dyer et al. 2009 as an example for human groups). Because in most non-human primates, several individuals of a group may act as leaders, a combination of dominance, physiological state, personality characteristics, and also knowledge may explain why several individuals emerge as principal leaders of a group.

Although leadership also involves costs such as reduced attention (Piyapong et al. 2007), individuals that lead group movements have the advantage of promoting their own interests compared to followers. Hence, conflicts over the leading position would seem likely to arise (reviewed in Conradt and Roper 2005), but they are, in fact, rarely observed. Leading and following animals may simply differ in the degree of their incentives (Erhart and Overdorff 1998), or the long-term fitness benefits related to social ties or kinship could compensate for the short-term costs of following a leader in a given situation (Silk et al. 2003; Cheney and Seyfarth 2007; King et al. 2008; Sueur and Petit 2008a). Alternatively, following may simply not be costly in each and every case, so that these conflicts do not arise permanently.

3.2.3 Mechanisms of Group Coordination

Visual or acoustical displays are obvious signals to initiate group movements. Visual displays such as staring or intentional movements in the direction of the adopted course have been reported in several primate species (Table 3.1; reviewed by Boinski 2000). For example, the dominant male in mountain gorillas usually uses a simple characteristic gesture to initiate a movement: He walks stiff-leggedly and rapidly in a certain direction (Schaller 1963). Acoustical displays used to coordinate group movements, so-called travel calls (Boinski 1991), have also been reported for a number of primate species primarily from the New World (Boinski 1991, 1993; Boinski et al. 1994; Boinski and Campbell 1995; Boinski and Cropp 1999; Leca et al. 2003). In squirrel and capuchin monkeys, travelling is initiated when an individual (occasionally two or three) moves to the edge of the group and produces a specific travel call. Byrne (1981) observed the use of vocalisations in the context of group movements in Guinea baboons (*P. papio*). The individuals exchange barks to stay cohesive as a group in areas of poor visibility (dense grass, thickets), as well as to coordinate themselves before the group splits up into subgroups or fusions.

Some species combine visual and acoustic displays. Barbary macaques (*Macaca sylvanus*) shake twigs or drum on dead wood (Mehlman 1996), and bonobos (*Pan paniscus*) have been observed dragging branches behind them to make their conspecifics move (Ingmanson 1996). However, in other species such as sifakas, the initiation of group movements is not accompanied by any acoustical or visual displays (Trillmich et al. 2004). The above description of inter-specific variation in the existence and type of initiation signals is extremely abbreviated. It is conceivable that future studies of additional species may reveal that the existence of initiation signals is a function of group size and cohesion, with species living in larger groups exhibiting specific calls to initiate travel, and that the existence of multiple signals is related to habitat characteristics that influence the propagation of certain signals.

3.2.4 Decision Types

Group decisions can be defined as 'when the members of a group choose between two or more mutually exclusive actions with the aim of reaching a consensus' (Table 3.1; see also Conradt and Roper 2005). Decisions can principally be shared, unshared, or based on self-organised processes (Hemelrijk 2002; Conradt and Roper 2003, 2007). In all cases, decisions to perform a certain activity or to travel in a certain direction appear to ultimately be made by single individuals, but their consequences are manifested on the level of the group in the form of a communal decision. Only a few studies to date have described decision-making processes in non-human primates. In capuchin monkeys and Tonkean macaques (*Macaca tonkeana*), each individual can principally influence the travel direction, resulting in a shared-consensus decision-making process; whereas in Rhesus macaques, dominant and older group members take a prominent role, resulting in only partially shared consensus decisions (Leca et al. 2003; Meunier et al. 2006; Sueur and Petit 2008a, b).

A despotic decision-making process has been described in mountain gorillas. In this species, the entire daily routine – the time of rising, the distance and direction of travel, as well as the place and time of nest building – is determined by the silverback male. When he starts moving in a certain direction, the whole group, which seems to be constantly aware of the location and activity of the dominant male, follows (Schaller 1963). Conflicting results have been reported regarding decision-making processes in baboons. In one population, King et al. (2008) conducted a foraging experiment with two wild chacma baboon groups and found that the dominant male of the group consistently led all foraging movements to experimental feeding sites. Social ties are held responsible for subordinate individuals following the despotic leader. In contrast, Stueckle and Zinner (2008) observed in another population of chacma baboons a democratic decision-making process during their departure from the sleeping site, with adult males contributing more to the decision outcome than adult females. Thus, differences in decision-making processes either might be related to taxonomic differences or may vary according to the decision that has to be made: It was observed that going to a feeding site that can be monopolised by the dominant male resulted in a despotic decision, whereas the departure from the sleeping site at dawn, which probably all group members want to leave to move on to forage, resulted in a democratic decision. Divergent or common context-dependent interests of group members may therefore result in different decision processes.

3.3 Operationalisation of Group Movements in the Field

Because coordination processes are mostly studied in the context of group movements, we would like to raise the issue of how human observers can identify and operationally define such a movement. In fact, group movements do not always proceed in a coordinated manner and, therefore, cannot always be easily captured by a single definition. For example, several or all animals of a group sometimes travel during foraging activities ('feed-as-you-go'), resulting in amoeboid-like movements that do not necessarily require an initiator or coordination among group members (e.g. bonobos, *Pan paniscus*: Wrangham 2000). Therefore, it is important to separate those movements from directed movements between sleeping and feeding sites or movements to patrol the border of the home range which require a certain degree of coordination among group members (Boinski and Garber 2000; Kappeler 2000; Pyritz et al. 2010).

Early studies addressing questions about leadership, coordination processes, and communication mechanisms in collective movements employed rather basic and unspecific definitions. One of the first definitions was provided by Altmann (1979), who defined a group movement simply as 'a displacement of the centre of the group.' Because a displacement of the centre also occurs during the rather amoeboid-like foraging movements, this definition does not allow differentiating between the latter and more coordinated movements. In other studies, group movements were defined by the departure of the group from a resting or feeding site (Schaller 1963; Stewart and Harcourt 1994; King et al. 2008), but this definition may not capture all movements. According to definitions of more recent studies that specifically address questions about the coordination of group movements, a group movement starts when an individual moves a certain distance towards the edge of the troop in a defined time period, e.g. 10 m within 40 s (Leca et al. 2003; Sueur and Petit 2008a, b; Stueckle and Zinner 2008), and is followed by at least one conspecific. Although these definitions have the advantages of being more precise, less presumptuous regarding resting and foraging motives, and inter-subjectively comprehensible, the distance that had to be travelled in a certain timeframe to initiate a group movement was not established empirically with regard to the species-specific travel pattern. Boinski (1991, 1993, 2000) defined group movements in a number of New World monkeys by a specific travel call uttered by the initiating individual, but because not all species produce specific travel calls, this definition is not generally applicable. Some researchers therefore use combinations of the definitions described above (e.g. Erhart and Overdorff 1999).

In general, a definition of group movements has to include a number of different travelling types: Primates do not only move between feeding and resting sites, but also to patrol home range boundaries and/or to search for or to avoid neighbouring groups. Because groups of different primate taxa vary widely in size, composition, and cohesion, which has consequences for home range size and travel distances, the minimum meaningful distance an individual has to cover to initiate a movement as well as the definition of the corresponding followers behaviour have to be species-specific (Pyritz et al. 2010). Below we suggest a procedure to generate an operational definition of group movements for different taxa built upon empirical data collected during a pilot study (Pyritz et al. 2010).

In order to define objective rules for directed vs. amoeboid-like movements, we suggest observing a number of randomly chosen focal animals for a period of several days. During such a pilot study, any movement of more than a body length can be recorded and estimated to the nearest metre. In addition, the latency between two movements, the total distance covered, as well as the distance to the nearest neighbours after the end of the movement can be noted. Based on meaningful breaks in the corresponding frequency distributions, a movement can be defined as follows:

Start: An individual has been stationary for at least x minutes and then moves a minimum distance of x metres in a directed manner without pausing.

Initiator: The individual that started the movement is the initiator.

Leadership: The individual at the forefront of the moving group is considered to lead the group movement.

Takeover: An individual overtakes the leader by more than several body lengths without diverging more than 45° from the initial trajectory of travel.

Followers: Group members moving behind the leader are termed followers unless their movements diverge more than 45° from the leader's trajectory. If they differ by more, the individual's movement is regarded as a separate movement. Followers have to arrive within an x-metre radius around the terminator, no later than x minutes after termination of the movement.

Termination: The end of the movement occurs when the leader is stationary again for at least x minutes (see above definition for 'Start').

Regarding these definitions, it is important to keep in mind that the initiator does not always remain the leader during the entire movement (hamadryas baboons: Kummer 1968; Guinea baboons: Byrne 1981, 2000) and that the terminator can differ from the initiating individual (indris: Pollock 1997). In the Kirindy Forest, we recorded group movements of redfronted lemurs according to the above definition with two observers: one following the initiator, the other following the leader in case a change of leadership occurred (Pyritz et al. unpublished data). A number of times the overtaking animal was only followed by a portion of group mates. This subgroup later returned to the other individuals, which were grouped around the original initiator, who had continued leading the rest of the group. Hence, the initiator still functioned as the pace-maker of the movement, even after being temporarily overtaken by a new leader. Hidden leadership such as this has to be taken into account when defining the decision type of a certain species. Furthermore, it highlights the importance of at least two observers following groups on the move.

We also studied group movements in Verreaux's sifakas using the method introduced above. The two species are syntopic but differ in both group and home range size. The virtually exclusively arboreal Verreaux's sifakas in the Kirindy live in multi-male, multi-female groups, with an average of 4.1 adult individuals per group that occupy home ranges averaging 7.3 ha (Benadi et al. 2008; Kappeler and Schäffler 2008). The cathemeral redfronted lemurs also live in multi-male, multi-female groups composed of on average 5.6 adult individuals (Kappeler and Port 2008). Average home range size of this species is 18 ha in the Kirindy (Pyritz et al. unpublished data). Redfronted lemurs spend a significant proportion of their time on the ground, especially during long group movements.

In Verreaux's sifakas, we employed the following group movement definition: A start attempt is made when an individual is stationary for at least 4 min, then moves at least 5 m, and is followed by at least one group mate. Other group members were termed followers unless their movement diverged more than 45° from the trajectory of the movement of the initiator. A movement was considered terminated when the leading individual was again stationary for at least 4 min (Trillmich et al. 2004). By applying this definition, we found that both sexes initiated group movements but that females did so more often, led groups over greater distances, and enlisted more followers than males. Presumably, this more active role enables females to positively influence their individual foraging efficiency and nutritional intake, especially during gestation and lactation (see Boinski 1991; Erhart and Overdorff 1999). However, the sex of the leader had no effect on the probability that a group would feed or rest after a successful movement. A certain vocalisation, the so-called grumble, was emitted by both leaders and followers at high rates, both before and during group progressions, but grumbles uttered just before an individual moved were characterised by a significantly steeper frequency modulation at the beginning of the call and higher call frequencies in both females and males (Trillmich et al. 2004). The results of this study indicate that sifakas converge with many other group-living primates in several fundamental proximate aspects of group coordination and cohesion. In contrast to many other primates, however, sifakas do not use a particular call or other signals to initiate or control group movements.

Our earlier pilot study suggested a group movement definition similar to the one employed for sifakas for the ongoing study on coordination of group movements in redfronted lemurs: A movement is initiated when an individual is stationary for at least 4 min, then moves at least 15 m, and is followed by at least one group mate. A movement was considered terminated when the leading individual was again stationary for at least 4 min. Followers are defined as individuals moving behind the initiator without diverging more than 45° from the trajectory, arriving within 6 metres proximity to the terminator, and no later than 10 min after termination. Preliminary results suggest that adults of both sexes initiated movements but that females do so significantly more often, both during the day and at night. Socially powerful males, so-called central males (Ostner and Kappeler 1999), did not initiate or lead group movements more often than other males. Female prevalence concerning the initiation of group movements may be due to higher and more complex nutritional needs during times of reproduction or female philopatry, but this should be true for most primates and other mammals. No specific initiation movements or travel calls have been observed thus far.

Comparing the operational definitions used in these two studies revealed that only the criteria for the start of a group movement and for the followers varied between sifakas and redfronted lemurs: The distance travelled for initiation of a group movement varied between 5 m in sifakas to 15 m in redfronted lemurs, and time intervals used to determine followers at the end of a group movement varied between 4 min in sifakas to 10 min in redfronted lemurs. Both sets of criteria clearly reflect the difference in daily path length as well as home range size [daily path length for sifakas: 1.1 km, home range size: 4.5 ha (Trillmich et al. 2004); daily path length for redfronted lemurs: 2 km, home range size: 18 ha (Pyritz et al. unpublished data)] and group size between species in the Kirindy, indicating that the use of such an operational group definition indeed helps to develop an appropriate way to quantify species-specific group movements. We therefore hope that future studies of primate group movements will continue to use, and eventually converge upon, similar criteria, increasing the potential for meaningful inter-specific comparisons.

3.4 Interdisciplinary Outlook

Although group cohesion and group decision making, both among humans as well as in non-human primates, are interesting in their own right, evolutionary theory suggests that both have to be functional with regard to environmental factors. In this respect, primatology and anthropology, on the one hand, and social psychology, on the other hand, differ considerably in their approaches. Primatology and anthropology focus on the long-term success of group cohesion and group decision making; that is, they ask what patterns are functional for group stability and the survival of group members. In contrast, psychological research focuses more on the short-term success of group cohesion and the mechanisms and processes underlying group decision making. For example, social psychologists are interested in whether group processes in terms of information exchange or mutual understanding benefit from cohesion or specific types of cohesion (Cornelius and Boos 2003), or how high-quality decisions can be achieved in groups (Schulz-Hardt et al. 2006).

Hence, comparative research on the consequences of group cohesion, group decision making, and other group processes on performance criteria in human vs. non-human primate groups could offer new insights for both disciplines. For instance, the short-term consequences of group processes on performance could be investigated in non-human primate groups. For example, it remains unknown to what extent the same process losses and gains that have been found in human groups also exist among non-human primates. Such an investigation of group-specific influences on non-human primates' task-related performance would be interesting in itself (e.g. studying capability gains among non-human primates as a function of social learning in a group), and might also significantly contribute to our understanding of human group performance. An open question in research on motivation gains in groups is why group members exert extra effort in a group situation under specific conditions. Whereas some approaches trace this behaviour back to a selfish motive (e.g. winning the performance competition and thereby gaining status in the group), other approaches postulate a more prosocial motive (e.g. caring for the group's welfare).

Since most primate species most likely lack collectivistic motivations or prosocial tendencies, whereas individualistic motives such as striving for status can be frequently found among them, comparative studies of group vs. individual performance in tasks such as predator mobbing or inter-group encounters where performance almost exclusively depends on effort could provide interesting new evidence for this open question. It also seems feasible that studies of human groups could take advantage of the long-term perspective adopted in non-human primate group research. By more extensively studying real groups in the field over extended periods of time, a more adequate picture of 'successful' human group behaviour might arise. Specifically, we might learn to what extent processes that directly impede the short-term performance of groups might nevertheless be facilitative or even essential for the stability and survival of a group in the long run.

References

Alberts SC, Altmann J, Wilson ML (1996) Mate guarding constrains foraging activity of male baboons. Anim Behav 51:1269–1277

Alexander RD (1974) The evolution of social behavior. Annu Rev Ecol Syst 5:325–383

Alexander RD (1987) The biology of moral systems. Aldine, London

Altmann J (1980) Baboon mothers and infants. Harvard University Press, Cambridge, MA

Altmann SA (1979) Baboon progressions: order or chaos? A study of one dimensional group geometry. Anim Behav 27:46–80

Beauchamp G (2000) Individual differences in activity and exploration influence leadership in pairs of foraging zebra finches. Behaviour 137:301–314

Benadi G, Fichtel C, Kappeler P (2008) Intergroup relations and home range use in Verreaux's sifaka (Propithecus verreauxi). Am J Primatol 70:956–965

Bertram BCR (1978) Living in groups: predators and prey. In: Krebs JR, Davies JB (eds) Behavioural ecology, 3rd edn. Blackwell Scientific, Oxford, pp 64–96

Boesch C (1991a) The effects of leopard predation on grouping patterns in forest chimpanzees. Behaviour 117:220–242

Boesch C (1991b) Symbolic communication in wild chimpanzees? Hum Evol 6:81–90

Boinski S (1988) Sex differences in the foraging behavior of squirrel monkeys in a seasonal habitat. Behav Ecol Sociobiol 23:177–186

Boinski S (1991) The coordination of spatial position: a field study of the vocal behavior of adult female squirrel monkeys. Anim Behav 41:89–102

Boinski S (1993) Vocal coordination of group movement among white-faced capuchin monkeys, Cebus capucinus. Am J Primatol 30:85–100

Boinski S (2000) Social manipulation within and between troops mediates primate group movement. In: Boinski S, Garber PA (eds) On the move: how and why animals travel in groups. University of Chicago Press, Chicago, pp 421–469

Boinski S, Campbell AF (1995) Use of trill vocalisations to coordinate troop movement among white-faced capuchins: a second field test. Behaviour 132:875–901

3 Coordination of Group Movements in Non-human Primates

Boinski S, Cropp S (1999) Disparate data sets resolve squirrel monkey (*Saimiri*) taxonomy: Implications for behavioral ecology and biomedical usage. Int J Primatol 20:237–256

Boinski S, Garber PA (eds) (2000) On the move: How and why animals travel in groups. University of Chicago Press, Chicago

Boinski S, Moraes E, Kleiman D, Dietz J, Baker A (1994) Intra-group vocal behavior in wild golden lion tamarins, *Leontopithecus rosalia*: Honest communication of individual activity. Behaviour 130:53–76

Byrne RW (1981) Distance vocalization of Guinea baboons (*Papio papio*): an analysis of function. Behaviour 78:283–312

Byrne RW (2000) How monkeys find their way: leadership, coordination, and cognitive maps of African baboons. In: Boinski S, Garber P (eds) On the move. University of Chicago Press, Chicago, pp 491–518

Chapman CA (1990) Association patterns of spider monkeys: the influence of ecology and sex on social organization. Behav Ecol Sociobiol 26:409–414

Chapman CA, Wrangham R, Chapman L (1995) Ecological constraints on group size: an analysis of spider monkey and chimpanzee subgroups. Behav Ecol Sociobiol 36:59–70

Cheney DL (1987) Interactions and relationships between groups. In: Smuts BB, Cheney DL, Seyfarth RM, Wrangham RW, Struhsaker TT (eds) Primate societies. University of Chicago Press, Chicago, pp 267–281

Cheney DL, Seyfarth RM (2007) Baboon metaphysics: the evolution of a social mind. University of Chicago Press, Chicago

Conradt L, Roper TJ (2003) Group decision-making in animals. Nature 421:155–158

Conradt L, Roper TJ (2005) Consensus decision making in animals. Trends Ecol Evol 20:449–456

Conradt L, Roper TJ (2007) Democracy in animals: the evolution of shared group decisions. Proc Roy Soc Lond B 274:2317–2326

Cornelius C, Boos M (2003) Enhancing mutual understanding in synchronous computer-mediated communication by training: trade-offs in judgmental tasks. Commun Res 30:147–177

Couzin ID, Krause J, James R, Ruxton GD, Franks NR (2002) Collective memory and spatial sorting in animal groups. J Theoret Biol 218:1–11

Couzin ID, Krause J, Franks NR, Levin SA (2005) Effective leadership and decision-making in animal groups on the move. Nature 43:513–516

Dunbar RIM, Dunbar EP (1975) Social dynamics of gelada baboons. Contrib Primatol 6:1–157

Dunbar RIM, Dunbar EP (1988) Maternal time budgets of gelada baboons. Anim Behav 36:970–980

Dunbar RIM, Shultz S (2007) Evolution in the social brain. Science 317:1344–1347

Dyer JRG, Johansson A, Helbing D, Couzin ID, Krause J (2009) Leadership, consensus decision making and collective behavior in humans. Phil Trans Roy Soc Lond B 364:781–789

Eisenberg JF (1981) The mammalian radiations. University of Chicago Press, Chicago

Erhart EM, Overdorff DJ (1998) Group leadership and feeding priority in wild *Propithecus diadema edwardsi* and *Eulemur fulvus rufus*. Am J Primatol 45:178–179

Erhart EM, Overdorff DJ (1999) Female coordination of group travel in wild *Propithecus* and *Eulemur*. Int J Primatol 20:927–941

Fischhoff IR, Sundaresan SR, Cordingley J, Larkin HM, Sellier M-J, Rubenstein DI (2007) Social relationships and reproductive state influence leadership roles in movements of plains zebra, *Equus burchellii*. Anim Behav 73:825–831

Goodall J (1968) The behavior of free-living chimpanzees in the Gombe Stream Reserve. Anim Behav Monogr 1:161–311

Harcourt J, Ang T, Sweetman G, Johnstone R, Manica A (2009) Social feedback and the emergence of leaders and followers. Curr Biol 19:248–252

Hemelrijk CK (2002) Understanding social behavior with the help of complexity science. Ethology 108:655–671

Ingmanson EJ (1996) Tool using behavior in wild *Pan paniscus*: social and ecological considerations. In: Russon AE, Bard K, Taylor S (eds) Reaching into thought: the minds of the great apes. Cambridge University Press, Cambridge, pp 190–210

Itani J (1963) Vocal communication of the wild Japanese monkeys. Primates 4:11–66

Jacobs A, Maumy M, Petit O (2008) The influence of social organisation on leadership in brown lemurs (*Eulemur fulvus fulvus*) in a controlled environment. Behav Process 79:111–113

Jouventin P (1975) Observations sur la socioecologie du mandrill [in French]. Terre et Vie 29:493–532

Kappeler PM (2000) Grouping and movement patterns in Malagasy primates. In: Boinski S, Garber PA (eds) On the move: how and why animals travel in groups. University of Chicago Press, Chicago, pp 470–490

Kappeler P, Port M (2008) Mutual tolerance or reproductive competition? Patterns of reproductive skew among male redfronted lemurs (*Eulemur fulvus rufus*). Behav Ecol Sociobiol 62:1477–1488

Kappeler PM, Schäffler L (2008) The lemur syndrome unresolved: Extreme male reproductive sew in sifakas (*Propithecus verreauxi*), a sexually monomorphic primate with female dominance. Behav Ecol Sociobiol 62:1007–1015

Kerth G, Ebert C, Schmidtke C (2006) Group decision making in fission-fusion societies: evidence from two field-experiments in Bechstein's bats. Proc Roy Soc Lond B 273:2785–2790

King AJ, Douglas CMS, Huchard E, Isaac NJB, Cowlishaw G (2008) Dominance and affiliation mediate despotism in a social primate. Curr Biol 18:1833–1838

Krause J, Reeves P, Hoare D (1998) Positioning behavior in roach shoals: the role of body length and nutritional state. Behaviour 135:1031–1039

Kudo H (1987) The study of vocal communication of wild mandrills in Cameroon in relation to their social structure. Primates 28:289–308

Kummer H (1968) Social organisation of hamadryas baboons. University of Chicago Press, Chicago

Kummer H (1995) In quest of the sacred baboon: a scientist's journey. Princeton University Press, Princeton, NJ

Leblond C, Reebs SG (2006) Individual leadership and boldness in shoals of golden shiners (*Notemigonus crysoleucas*). Behaviour 143:1263–1280

Leca JB, Gunst N, Thierry B, Petit O (2003) Distributed leadership in semi-free ranging white-faced capuchin monkeys. Anim Behav 66:1045–1052

Marler P (1965) Communication in monkeys and apes. In: DeVore I (ed) Primate behavior. Holt Rinehart Winston, New York, pp 544–584

Maynard Smith J, Szathmáry E (1995) The major transitions in evolution. WH Freeman/Spektrum, Oxford

Mehlman PT (1996) Branch shaking and related displays in wild Barbary macaques. In: Fa JE, Lindburg DG (eds) Evolution and ecology of macaque societies. Cambridge University Press, Cambridge, pp 503–526

Menzel CR (1993) Coordination and conflict in *Callicebus* social groups. In: Mason WA, Mendoza SP (eds) Primate social conflict. State University of New York Press, Albany, pp 253–290

Meunier H, Leca JB, Deneubourg JL, Petit O (2006) Group movement decisions in capuchin monkeys: the utility of an experimental study and a mathematical model to explore the relationship between individual and collective behaviors. Behaviour 143:1511–1527

Milton K (1980) The foraging strategy of howler monkeys: a study in primate economics. Columbia University Press, New York

Mitchell CL, Boinski S, van Schaik CP (1991) Competitive regimes and females bonding in two species of squirrel monkey (*Saimiri oerstedi* and *S. sciureus*). Behav Ecol Sociobiol 28:55–60

Neville MK (1968) Ecology and activity of Himalayan foothill rhesus monkeys (*Macaca mulatta*). Ecology 49:110–123

Norton GW (1986) Leadership decision processes of group movement in yellow baboons. In: Else JG, Lee PC (eds) Primate ecology and conservation. Cambridge University Press, Cambridge, pp 145–156

Noser R, Byrne RW (2007a) Travel routes and planning of visits to out-of-sight resources in wild chacma baboons, *Papio ursinus*. Anim Behav 73:257–266

Noser R, Byrne RW (2007b) Mental maps in chacma baboons (*Papio ursinus*): Using inter-group encounters as a natural experiment. Anim Cogn 10:331–340

Oates JF (1977) The guereza and its food. In: Clutton-Brock TH (ed) Primate ecology. Academic, London, pp 276–321

Ostner J, Kappeler PM (1999) Central males instead of multiple pairs in redfronted lemurs, *Eulemur fulvus rufus* (Primates, Lemuridae)? Anim Behav 58:1069–1078

Overdorff DJ (1993) Similarities, differences, and seasonal patterns in diets of *Eulemur rubriventer* and *Eulemur fulvus rufus* in the Ranomafana National Park, Madagascar. Int J Primatol 14:721–753

Overdorff DJ (1996) Ecological correlates to activity and habitat use of two prosimian primates: *Eulemur rubriventer* and *Eulemur fulvus rufus* in Madagascar. Am J Primatol 40:327–342

Parrish JK, Edelstein-Keshet L (1999) Complexity, pattern, and evolutionary trade-offs in animal aggregation. Science 284:99–101

Piyapong C, Morrell LJ, Croft DP, Dyer JRG, Ioannou CC, Krause J (2007) A cost of leadership in human groups. Ethology 113:821–824

Pollock IJ (1997) Female dominance in *Indri indri*. Folia Primatol 31:143–164

Pulliam HR, Caraco T (1984) Living in groups: is there an optimal group size? In: Krebs JR, Davies MNO (eds) Behavioural ecology. Blackwell, Oxford, pp 122–147

Pyritz L, Fichtel C, Kappeler P (2010) Conceptual and methodological issues in the comparative study of collective group movements. Behav Process 84:681–684

Rands SA, Cowlishaw G, Pettifor RA, Rowcliffe JM, Johnstone RA (2003) Spontaneous emergence of leaders and followers in foraging pairs. Nature 423:432–434

Rands SA, Cowlishaw G, Pettifor RA, Rowcliffe JM, Johnstone RA (2008) The emergence of leaders and followers in foraging pairs when the qualities of individuals differ. BMC Evol Biol 8:51

Reader SM, Laland KN (2002) Social intelligence, innovation, and enhanced brain size in primates. Proc Natl Acad Sci USA 99:4436–4441

Reebs SG (2000) Can a minority of informed leaders determine the foraging movements of a fish shoal? Anim Behav 59:403–409

Reebs SG (2001) Influence of body size on leadership in shoals of Golden Shiners, *Notemigonus crysoleucas*. Behaviour 138:797–809

Rowell TE (1969) Intra-sexual behavior and female reproductive cycles of baboons (*Papio anubis*). Anim Behav 17:159–167

Rowell TE (1972) Female reproductive cycles and social behavior in primates. Adv Stud Behav 4:69–105

Sauther ML, Sussman RW (1993) A new interpretation of the social organization and mating system of the ring-tailed lemur *(Lemur catta*). In: Kappeler PM, Ganzhorn JU (eds) Lemur social systems and their ecological basis. Plenum, New York, pp 111–121

Schaller G (1963) The mountain gorilla: Ecology and behavior. University of Chicago Press, Chicago

Scholz F, Kappeler PM (2004) Effects of seasonal water scarcity on the ranging behavior of *Eulemur fulvus rufus*. Int J Primatol 25:599–613

Schulz-Hardt S, Brodbeck FC, Mojzisch A, Kerschreiter R, Frey D (2006) Group decision making in hidden profile situations: dissent as a facilitator for decision quality. J Pers Soc Psychol 91:1080–1091

Seeley TD, Visscher PK (2004) Quorum sensing during nest-site selection by honeybee swarms. Behav Ecol Sociobiol 56:594–601

Seyfarth RM (1986) Vocal communication and its relation to language. In: Smuts BB, Cheney DL, Seyfarth RM, Wrangham RW, Struhsaker TT (eds) Primate societies. Chicago University Press, Chicago, pp 440–451

Sigg H, Stolba A (1981) Home range and daily march in a hamadryas baboon troop. Folia Primatol 36:40–75

Silk JB, Alberts SC, Altmann J (2003) Social bonds of female baboons enhance infant survival. Science 302:1231–1234

Smuts BB, Cheney DL, Seyfarth RM, Wrangham RW, Struhsaker TT (eds) (1986) Primate societies. University of Chicago Press, Chicago

Soini P (1981) The pygmy marmoset, genus *Cebuella*. In: Mittermeier RA, Rylands AB, Coimbra-Filho AF (eds) Ecology and behavior of Neotropical primates. Academia Brasileira de Ciencias, Rio de Janeiro, pp 79–129

Stanford CB (1990) Colobine socioecology and female-bonded models of primate social structure. Kroeber Anthropol Soc Pap 71–72:21–28

Sterck EHM, Watts DP, van Schaik CP (1997) The evolution of female social relationships in nonhuman primates. Behav Ecol Sociobiol 41:291–309

Stewart KJ, Harcourt AH (1994) Gorillas' vocalisations during rest periods: signals of impending departure? Behaviour 130:29–40

Struhsaker TT (1967a) Auditory communication among vervet monkeys (*Cercopithecus aethiops*). In: Altmann SA (ed) Social communication among primates. University of Chicago Press, Chicago, pp 281–324

Struhsaker TT (1967b) Ecology of vervet monkeys (*Cercopithecus aethiops*) in the Masai-Amboseli Game Reserve, Kenya. Ecology 48:891–904

Struhsaker TT (1975) The red colobus monkey. University of Chicago Press, Chicago

Stueckle S, Zinner D (2008) To follow or not to follow: decision making and leadership during the morning departure in chacma baboons. Anim Behav 75:1995–2004

Sueur C, Petit O (2008a) Shared or unshared consensus decision in macaques? Behav Process 78:84–92

Sueur C, Petit O (2008b) Organization of group members at departure is driven by social structure in Macaca. Int J Primatol 29:1085–1098

Trillmich J, Fichtel C, Kappeler PM (2004) Coordination of group movements in wild Verreux's sifakas (*Propithecus verreauxi*). Behaviour 141:1103–1120

van Nordwijk MA, van Schaik CP (1987) Competition among female long-tailed macaques, *Macaca fascicularis*. Anim Behav 35:577–589

van Nordwijk MA, Hemelrijk C, Herremans L, Sterck E (1993) Spatial position and behavioral sex differences in juvenile long-tailed macaques. In: Pereira M, Fairbanks L (eds) Juvenile primates. Oxford University Press, New York, pp 77–85

van Schaik CP (1983) Why are diurnal primates living in groups? Behaviour 87:120–144

van Schaik CP (1989) The ecology of social relationships amongst female primates. In: Standen V, Foley RA (eds) Comparative socioecology. The behavioral ecology of humans and other mammals. Blackwell, Oxford, pp 195–218

van Vugt M (2006) Evolutionary origins of leadership and followership. Pers Soc Psychol Rev 10:354–371

Vogel C (1973) Acoustical communication among free-ranging common Indian langurs (*Presbytis entellus*) in two different habitats of North India. Am J Phys Anthropol 38:469–480

Ward AJW, Thomas P, Hart PJB, Krause J (2004) Correlates of boldness in three-spined stickle-backs (*Gasterosteus aculeatus*). Behav Ecol Sociobiol 55:561–568

Watts DP (1994) The influence of male mating tactics on habitat use in mountain gorillas (*Gorilla gorilla berengei*). Primates 35:35–47

Whitehead JM (1989) The effect of the location of a simulated intruder on responses to long-distance vocalizations of mantled howling monkeys, *Alouatta palliata palliata*. Anim Behav 108:73–103

Wrangham RW (2000) Why are male chimpanzees more gregarious than mothers? A scramble competition hypothesis. In: Kappeler PM (ed) Primate males. Causes and consequences of variation in group composition. Cambridge University Press, Cambridge, pp 248–258

Zeller AC (1986) Communication by sight and smell. In: Smuts BB, Cheney DL, Seyfarth RM, Wrangham RW, Struhsaker TT (eds) Primate societies. University of Chicago Press, Chicago, pp 433–439

Zemel A, Lubin Y (1995) Inter-group competition and stable group sizes. Anim Behav 50:485–488

Chapter 4
Dimensions of Group Coordination: Applicability Test of the Coordination Mechanism Circumplex Model

Micha Strack, Michaela Kolbe, and Margarete Boos

Abstract This chapter discusses the Coordination Mechanism Circumplex Model, a content model of group coordination mechanisms that proposes the dimension of explicitness and the dimension of timing (Wittenbaum et al. 1998). It aims at solving confounds in former taxonomies of coordination mechanisms. We first critique these two dimension definitions. We then report on our coder agreement study of the intelligibility of the two dimensions. As hypothesised, empirical agreement among the coders in our study varies with the built-in difficulty of the mechanism sets (macro-, meso-, and micro- level of coordination), and the expertise level of the coders (experts vs. novices) compensates for this mechanism set difficulty. Plots of mechanisms in the Coordination Mechanism Circumplex Model accomplish the extensional definition of its two dimensions of explicitness and timing. We close by discussing next steps in theory building, including the elimination of the intentionality construct and the consideration of the perspective of producers and targets of coordination mechanisms.

4.1 The Coordination Circumplex

As stated in the inclusive group coordination model described in Chap. 2, the elements of coordination in a group (e.g. the group's task and functions as well as its mechanisms and processes) need to be as well specified as possible in order to describe and explain the coordination of a particular group. In this chapter we

---

M. Strack (✉) and M. Boos
Georg-Elias-Müller-Institute of Psychology, Georg-August-University Göttingen, Goßlerstrasse 14, 37075 Göttingen, Germany
e-mail: mstrack@uni-goettingen.de; mboos@uni-goettingen.de

M. Kolbe
Department of Management, Technology, and Economics, Organisation, Work, Technology Group, ETH Zürich, Kreuzplatz 5, KPL G 14, 8032 Zürich, Switzerland
e-mail: mkolbe@ethz.ch

M. Boos et al. (eds.), *Coordination in Human and Primate Groups*,
DOI 10.1007/978-3-642-15355-6_4, © Springer-Verlag Berlin Heidelberg 2011

concentrate on specifying *mechanisms* of successful coordination, implying a plurality of mechanisms, many of which can be neutralised, moderated, or substituted. Prior to the development of a full coordination model, the coordination mechanisms themselves must be described and structured in order to explain their efficiency. Literature from different disciplines suggests lists of mechanisms providing for the same or similar functions and entities of coordination in social systems. Some authors of theoretical papers have attempted to categorise the mechanisms into two categories.

The dichotomies in Table 4.1 are intuitively arranged and are not meant in all cases to match the other mechanisms category of the same column (examples and comparisons are given later in the text). Two-category systems often confound attributes of exemplars. To resolve this, Wittenbaum et al. (1998) proposed a model with *two* dimensions intended to disentangle confounds of group coordination attributes. This Coordination Mechanism Circumplex Model (CMCM) (Fig. 4.1) structures coordination mechanisms according to their explicitness (implicit/explicit)

Table 4.1 Some two-category systems of coordination mechanisms

	Coordination category 1	Coordination category 2
March and Simon (1958)	Plans and prespecified programmes	Feedback and mutual adjustment.
Burns and Stalker (1961)	Mechanic	Organic
Van de Ven et al. (1976), Raven (1999)	Impersonal	Personal
Andersen et al. (2000)	Artefact-based	Oral
Argote (1982)	Programmed means	Non-programmed means
Mintzberg (1979)	Standardisation of processes, inputs, outputs and norms	Direct supervision and mutual adjustment
Entin and Serfaty (1999), Entin et al. (2005), MacMillan et al. (2004)	Explicit, verbal	Implicit, cognitive
Espinosa et al. (2004)	Explicit, intended	Implicit, unintended
Faraj and Xiao (2006)	Expertise coordination practise	Dialogic coordination

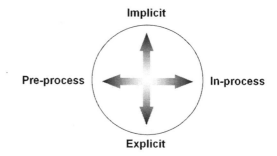

Fig. 4.1 The coordination mechanism circumplex model (CMCM), (adapted from Wittenbaum et al. 1998)

4 Dimensions of Group Coordination 59

and according to the temporal phase (pre-process/in-process) when their coordination impact is accomplished. Our first aim is to discuss the advantages and weaknesses of this model.

Wittenbaum et al. (1998) explained the dimensions by giving examples for the four quadrants: *Explicit in-process* coordination sums up leadership, facilitation, negotiation with verbal agreements, and other overt forms of communication between group members during their interaction. According to their explicitness and their temporal occurrence within the actual group process, these mechanisms are easily observed by a third party and therefore dominate small group research. Coordination 'by feedback', 'personal coordination', 'direct supervision', 'dialogic', and 'oral coordination' from Table 4.1 are classified as explicit in-process coordination mechanisms.

In contrast, *explicit pre-process* coordination mechanisms are realised and perceived prior to group interaction. 'Predefined plans', 'programmed means', 'mechanistic coordination', 'standardisations' documented on hardcopy or within software systems are examples of explicit pre-process coordination from Table 4.1. In small group research guided by the input–process–outcome model (Hackman and Morris 1975), pre-process coordination mechanisms such as agendas or legal rules are commonly grouped as the *task* or as mere *context factors*.

Implicit pre-process coordination mechanisms begin to take effect before the exchange of group members occurs, but they are less salient, less intentionally constructed, less appellative, and therefore less observable. Wittenbaum et al. (1998) specified expectations and shared scripts regarding the task, the other members, and context factors as examples of implicit pre-process coordination mechanisms. Constructs such as culture, common knowledge, shared mental models, transactive memory, pre-knowledge, internalised conventions, expertise, and professionalism subsume to this mechanism type (e.g. Evans et al. 2004; Ramon et al. 2008). Small group research frequently incorporates implicit pre-process coordination indirectly by considering input variables such as group combination, homogeneity–heterogeneity, and group history.

Implicit in-process coordination includes mechanisms of tacit coordination (Wittenbaum et al. 1996), mutual adjustment, and local self-organisation (Fichtel et al. offer the term 'self-coordination' in Chap. 3 of this book to help explain tacit coordination). It might be the most challenging quadrant for empirical research, as these mechanisms are nearly impossible to observe in overt behaviour. Nevertheless, in some sense they also embody the core of social psychology mechanisms: The informational social influence as demonstrated by Sherif (1935) and the normative impact of a consensual majority (Asch 1952) are both important prototypes for implicit in-process coordination.

Bearing in mind Carnap's (1947) distinction of intensional versus extensional definitions of concepts, describing the dimensions of the Coordination Mechanism Circumplex Model merely through examples leaves the intensions of the terms implicit and explicit insufficiently defined. That said, even concrete examples become difficult to categorise: Is coordination by rituals, such as the greeting cycle of a telephone call, or by a behaviour setting (Barker 1968) implicit or explicit? Do other approaches outlined in Table 4.1 offer answers? Does explicitness require a persistent (verbal) code? Kim and Kim (2008) defined implicit/explicit coordination tautologically by implicit/ explicit communication, never venturing outside the in-process phase. With a similar in-process focus, explicitness for the Aptima research group (Entin and Serfaty 1999, Entin et al. 2005; MacMillan et al. 2004) means verbalisation (e.g. requests) and that implicitness works via silent but elaborated cognitions (e.g. expectations and perspective taking). Wittenbaum et al. contrasted 'unspoken' versus 'verbalised' coordination mechanisms (1998, p. 5). Taking a different perspective, Grote et al. (2003) argued that implicitness is related to automatic processes, eliciting psychological compliance without cognitive control and conscious effort, whereas Andersen et al. (2000) contrarily proposed artefact-based coordination (see Table 4.1) for its automation. Godart et al. (2001) associated explicitness with extra processes and implicitness with mutual awareness, seemingly the main diagonal of Fig. 4.1. Espinosa et al. (2004) related their implicit/explicit distinction with the concept of intention: Explicit coordination mechanisms are realised, grasped, or used with the intention to coordinate a group. However, a study on subjective coordination theories reveals that implicitness can also be used intentionally (Kolbe and Boos 2009). A necessity to check for intentionality further challenges the level of precision of the CMCM. There is a long history of debate on scientific concepts of intentionality and its related subject of perspective. Coordination mechanisms can be compared to signs studied by semiotics. The relation of a sign and intentionality is connected here to the distinction of sender- versus receiver-theories of meaning (e.g. Nöth 1995, p. 109). Concerning biosemiotics, whose subject is communication among living systems not endowed with speech, the intentionality concept was replaced by the notion of semantisation and semantic specialisation: A proper sign is produced in order to signal, with an end result of conveying meaning. An object solely interpreted by perceivers as standing for something (e.g. smoke for fire) lacks this semantisation and semantic specialisation. Then there is the development of natural communicative signs such as body structures (e.g. colours in peacocks). Such signal structures are not at all intended by any organism: They evolved and changed their function from a pragmatic one to a semantic one without personal will, but rather by the mutual communicative benefit of receivers and senders. With this semiotic background in mind, the question arises as to whether implicitness should be defined from the perspective of the producers of a coordination mechanism (if there even is a producer), or from the perspective of the targets of that mechanism. The automatic processes that Grote et al. (2003) presented seem to be defined from the targets' perspective; the intentionality of Espinosa et al. (2004) and others might point toward the senders' perspective.

Clearly, there are a lot of tangents to the simple distinctions of the Coordination Mechanism Circumplex Model. Yet, in the context of group coordination mechanisms, the CMCM represents a marked progress from the two-category taxonomies in Table 4.1, which sometime confound a dimension such as explicitness with the pre-process phase, and implicitness with the in-process phase. Additionally, the model allows for a continuous distribution within each dimension and therefore offers at the very least an ordinal scaling of mechanisms within any given mechanism set. Although the dimension definitions must be articulated more precisely as research

4 Dimensions of Group Coordination

progresses, we maintain that the Coordination Mechanism Circumplex Model (CMCM, Fig.4.1) is a viable framework for coordination theory and research.

To establish the construct validity of the model, we conducted an empirical study to test the applicability of the dimensions. Although the intensional definitions remain unclear, the intelligibility of the proposed dimensions was hypothesised to be distinct enough to apply them for comparisons and to categorise observed mechanisms in the CMCM (Fig. 4.1). The main hypothesis of our study therefore proposed coder agreement for different coordination mechanisms. If different coders agreed on the relative explicitness/implicitness and on the relative pre-process/in-process status of the various mechanisms tested, the intelligibility of the Coordination Mechanism Circumplex Model would be validated.

4.2 Empirical Applicability

In the study of intercoder agreement on the explicit/implicit and on the pre-process/in-process position of a coordination mechanism, the construct validity of the model is reflected in the dependency of agreement from relevant factors. The design of the study therefore took into consideration different levels of task difficulty (macro-, meso-, and micro- levels of human coordination) and different expertise levels of the coders (expert vs. novice). The latter factor was considered a compensating factor for the former. This meant that if not only the ratings of the novices resembled those of the experts on the easier tasks, but the experts reached more agreement than novices on the more difficult tasks, then the two proposed dimensions of the model (explicitness and timing) would reflect greater construct validity than a mere agreement score for all coders.

4.2.1 Study Design

The objects for the coding task were drawn from three sets of coordination mechanisms with varying levels of task difficulty (Table 4.2). The simplest task was the coordination of time and space in road traffic, potentially due to random

Table 4.2 The difficulty (3) × expertise (2) design of the coder-agreement study

Coder expertise	Low difficulty (macro-level): coordination of road traffic	Medium difficulty (meso level): group coordination by leadership substitutes	High difficulty (micro level): verbal interacts in group discussions
Experts (the three authors)			
Novices (sets of students)			

everyday occurrence and its broad, macro-level categories. The meso-level of complexity was the coordination of groups by leadership substitutes. The most difficult set was the process analysis of micro-level verbal interactions in group discussions.

We depicted road traffic coordination using the following six mechanisms (alphabetically): 'eye contact', 'road traffic laws', 'speed bump', 'stop signs' 'traffic lights', and 'yield-to-the-right'.

The theory of substitutes for leadership (Kerr and Jermier 1978) was utilised for the medium level of coding difficulty. As touched upon earlier, this theory proposes that certain attributes of the task, the group members, the group, and/or the organisation can serve as neutralisers or substitutes for actions of group leaders. From the list of substitutes we chose eight mechanisms: 'group cohesion', 'competencies of the members', 'expert roles in the group', 'information technologies', 'prescriptions, plans, and formalisms', 'professionalism of actors', 'task-inherent feedback', and 'task structure'. The 'executive manager' was added as the ninth mechanism in this set.

As the domain with high difficulty, we chose the category system designed for micro-level process analysis of verbal coordination in decision-making groups by Kolbe et al. (MICRO-CO; see Chap. 11 and Kolbe 2007). The categories are ordered hierarchically (see Fig. 11.1). On the subcategory level, seven content-related acts (verbally conveying information, opinions, etc.) and 23 coordination acts are distinguished. Are the dimensions of implicit/explicit and pre-/in-process coordination applicable to non-coordinating content acts of communication? We decided to retain these content categories in the analysis because we were curious about their plotted location in the CMCM (Fig. 4.1). Second, we anticipated it to be difficult to utilise the pre-process time dimension pole (pre- vs. in-process) for interacts that were generally all expected to take place during the discussion. Third, the intended difficulty in coding the mechanisms of MICRO-CO (Fig. 11.1) was based on the richness in details of such a micro-level system. For example, MICRO-CO distinguishes seven types of questions. We were curious to see whether they would cluster in a small region or disperse all over the CMCM.

4.2.2 The Coding Task

An absolute coding judgment (e.g. "This is a pre-process mechanism") seemed unreasonable for dimensions lacking an intensional definition and socially shared anchoring points. Additionally, a simple cognitive anchoring of said judgments contradicts the notion of continuity of a circumplex. For example, a traffic sign restricting the speed limit to 30 km/h affects traffic participants more in-process than a prior learned rule to slow down in small villages. But do traffic signs act more pre-process than police stopping cars appearing unexpectedly at that location? With the history of psychological measurement in mind, we chose a pair-comparison task. The coders were instructed to consider a specific pair of mechanisms and

4 Dimensions of Group Coordination

decide (1) which of the two respective mechanisms was more explicit than the other and (2) which one was relatively more pre-process than the other. Because some paired mechanisms work equally well on either of the two dimensions, we allowed for equality judgments. In those cases, however, we asked the coder to specify the tendency of both mechanisms on the axis in question. Figure 4.2 shows a section of instructions for the first coding task on road traffic (simplest task level). Three of the 15 pairs of road traffic mechanisms were used as examples in the instructions and therefore omitted from the raw data results.

The $m^*(m - 1)/2$ pairs in the medium-difficulty set of $m = 9$ leader substitute mechanisms resulted in 36 trials per dimension.

Pairs of all categories in MICRO-CO (Fig. 11.1) would lead to too many trials for an expected mean motivation of a novice participant. We therefore divided the category system of MICRO-CO into three subsets: Subset A encompassed the seven content-related subcategories and the five remaining second-level categories (from 'addressings' to 'interruption'). These 12 mechanisms formed 66 pairs. Subset B contained the two subcategories of 'addressings' and the six subcategories of 'instructions' and the four remaining second-level categories (from 'structurings' to 'content-related statements'; see Fig. 11.1), also resulting in 66 pairs. Subset C included the six subcategories of 'structurings' and the seven 'questions' plus the four remaining second-level categories, resulting in 136 pairs for each dimension.

Instruction: Please compare two (vertically arranged) mechanisms: Write down in each of the four cells per dimension one alphabetic character per row (see the first examples). Sometimes you might be unable to differentiate the two mechanisms on one dimension, as the example of the road traffic law and the traffic light in the explicit/implicit dimension. In that case, the same character: here e & e for 'both equally explicit' could be used. But try to differentiate when possible. Please: Compare each pair of mechanisms separately: Which one coordinates traffic participants more explicitly, which one more implicitly than the other one? Which one coordinates the participants more likely pre-process, which one likely not until in-process?

... coordinates traffic participants	explicit	implicit	pre process means before process	actual, in-process
road traffic law	e		p	
traffic light	e			a

Fig. 4.2 The last section of the instructions for the first set of mechanisms

4.2.3 Coders and Procedures

According to the design of the study, agreement among the expert coders was to be compared with agreement among the novices. The three authors of this chapter served as the expert coders. We recruited nine university students at a psychology lecture on group coordination to function as the novice coders. The novice coders were divided into three groups of three members in order to ensure that the agreement scores for the expert and novice coders were statistically comparable.

Novices were instructed on how to categorise coordination mechanisms in the models dimensions with a page and a half of instructions (including Fig. 4.1 and ending with Fig. 4.2). Each novice coder worked on all three levels of task difficulty – always in the fixed order given in Table 4.2. This meant that the road traffic coding task had to additionally function as experience for subsequent coding of the leadership substitute mechanisms. With this accumulated experience, the novice coders were then assigned to code subsets of the verbal MICRO-CO interact categories. With this procedure, the first-level road traffic set and second-level leader substitutes set were coded by all nine student novice coders and analysed in three triads. This procedure also meant that each of the three subsets of MICRO-CO interacts (see Sect. 2.2) was coded by only three of the nine students.

The expert coders worked in the same sequence on the same material, but, unlike the novice coders, they answered all three subsets of the MICRO-CO task.

It is noteworthy that without being surveyed, all coders – experts and novices – reported the task to be very difficult, minimally indicating that they took their task seriously.

4.2.4 Dependent Measures and Statistics

As pair comparisons yield nominal data (see the first pair comparison in Fig. 4.2), we computed kappa coefficients utilising the formula of Fleiss (1971). Computations for the three levels of coding difficulty, for each dimension of the model, and for each mechanism (sub)set resulted in 28 kappa coefficients.

Additionally, each mechanism was plotted on the circumplex model axes by aggregation of all $(m - 1)$ codes per mechanism received by one participant (with $m =$ number of exemplars per set). The sum of codes for explicitness was subtracted from the sum of codes for implicitness, and the sum of codes for pre-process was subtracted from the sum of codes for in-process coordination. Location of plotted scores ranged between $\pm(m - 1)$ per set of mechanisms and was regarded as interval scaled. The agreement within a three-subject group per dimension was estimated by Cronbach's α, an agreement score for interval scaled data again resulting in 28 coefficients.

4.2.5 Results

As visible in Figs. 4.3 and 4.4, coding agreement in both dependent variables (kappa for the raw data binary decisions and Cronbach's α for the scaled position on the axes) was higher for the base-level set of traffic coordination mechanisms than for the meso-level set of leadership substitutes, and lowest of all for the most difficult coding level of MICRO-CO verbal interacts.

Although according to strict statistical logic, an agreement score (kappa or alpha) is not additive, we regressed the 28 scores on the dummy-coded design factors based on the difficulty level of the set, the binary expertise of each three-person coding group, and then on the interaction of these two factors to test for compensation between the expertise level of the coders and the difficulty level of the tasks. The results confirmed the expected compensatory interaction (Fig. 4.5).

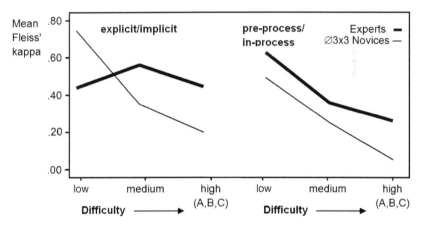

Fig. 4.3 Agreement in the coding of each mechanism in each pair comparison (kappa)

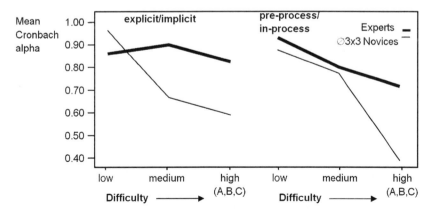

Fig. 4.4 Agreement in the dimension location of each mechanism (Cronbach's α)

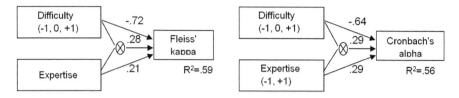

Fig. 4.5 Agreement regressed on difficulty of the set, coder status, and their interaction

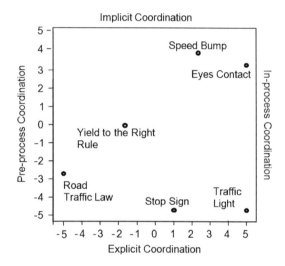

Fig. 4.6 Location of the traffic road mechanisms (coded by three experts)

For raw decisions (kappa): $\beta = +0.28$, $t = 2.08$ $p < 0.05$; for locations of the dimensions (alpha): $\beta = +0.29$, $t = 2.02$ $p = 0.05$; confirming that the expertise of coders compensated for the difficulty of the coordination domains.

Coding among the novice group was in as high agreement as that of the expert coders for the road traffic mechanisms (mean kappa = 0.59, mean alpha = 0.91). As expected, novices failed to reach agreement on the verbal interacts of MICRO-CO (Fig. 11.1): they reached mean kappa = 0.13 in raw data, and mean alpha = 0.49 on dimensions, whereas experts reached mean kappa = 0.35 in raw data, and mean alpha = 0.77 on dimensions. Therefore, the plotted location results of the expert triad are valid for reporting.

The mean location of the road traffic coordination mechanisms (as coded by the experts) is depicted in Fig. 4.6. Each of the six mechanisms was involved in five pairs, the axis ranging from −5 to +5. The explicit mechanisms of 'traffic light' and 'stop sign' reached perfect agreement among the experts. The 'yield to the right' rule evoked the highest relative level of disagreement (Euclidian distances between its locations) found among the expert coders: One of the expert coders considered the 'yield to the right' rule as explicit and pre-process functioning, another expert regarded it as explicit and in-process functioning, and the third perceived it as an implicit and pre-process functioning mechanism.

Fig. 4.7 Location of the leadership substitutes (coded by three experts)

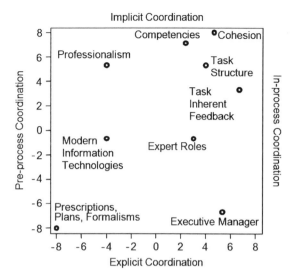

Figure 4.7 shows the experts' mean plotted location of the leadership substitutes – the meso-level coding task. Only the substitution by 'prescriptions, plans, and formalisms' obtained perfect agreement as a maximum pre-process and explicit coordination mechanism. However, 'professionalism of persons' (on both dimensions) and 'task-inherent feedback' (on the implicit/explicit dimension) had the least agreement.

To reintegrate the three subsets of the verbal interact categories of MICRO-CO, a main component analysis with pairwise data inclusion was calculated in order to estimate the standardised positions for the six mechanisms intersecting two of the three subsets. A regression of the mechanisms of each subset and dimension on these main component scores standardised all mean plotted locations of the various verbal coordination interacts. Therefore, the axes of these means (Fig. 4.8) appear as z-transformed scores.

The categories of verbal interacts were widely distributed over the circumplex rather than clustered in any concentrated region (see Fig. 4.8). This result demonstrates the relativity or reference-system dependency of the axes and, for the coders, a rather deep understanding of the CMCM dimensions. At a macro-level perspective, all the micro-level categories of verbal interaction can potentially be coded as explicit and in-process. Within the reference system of micro-level interacts, the three expert coders agreed most consistently on some pre-process explicit mechanisms such as 'giving instructions' (a second-level category of the category system; see Chap. 11) and on the first-level category of 'defining a goal', a structuring activity. It also was strongly agreed that 'interruptions' function plainly as in-process and that 'comments' and other content utterances coordinate the group discussion implicitly.

Taken together, the three difficulty levels of coordination mechanisms – the most difficult at least by the experts – were understood in terms of the dimensions of

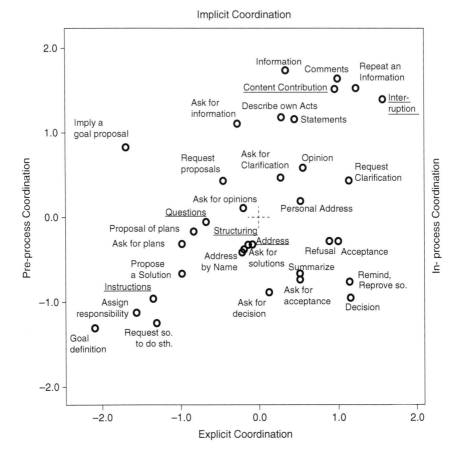

Fig. 4.8 Location of categories of verbal interacts (see Fig. 11.1; second-level categories underlined; coded by three experts, standardised by six exemplars intersecting the subsets)

the model. Intelligibility as a first validity criterion allowed a grounding plot in the quadrants of the CMCM. One of the validated features of this CMCM test is that the plots allow a scientific communication of the multi-dimensional and circumplex nature of coordination mechanisms.

4.2.6 Discussion and Outlook

In this chapter we examined the applicability of the two coordination mechanism dimensions of explicitness and timing adapted from Wittenbaum et al. (1998). Releasing the restriction of absolute judgments and allowing for relativity due to reference system dependence using pair comparisons, the three expert coders

4 Dimensions of Group Coordination

reached acceptable agreement for assigning coordination mechanisms to these two dimensions (kappa $= 0.53$, 0.46, 0.35 on decision's raw data; and Cronbach's $\alpha = 0.90$, 0.85, 0.77 on dimension locations for the least difficult, medium, and most difficult set, respectively). The novice student coders, on the other hand, matched the experts' agreement level on the easier set but failed to agree on the coordination mechanisms of the more difficult levels. Nevertheless, the hypothesised compensatory interaction of task difficulty and coding expertise was statistically established as illustrated in Fig. 4.6. Thus, despite of the lack of intensional (and therewith producer/target) definitions of the dimensions discussed in the Introduction, experts acquainted with the observation of coordination mechanisms at different levels of social systems (e.g. in organisations from a macro-level point of view), in groups from a medium level, and within a single discussion in a micro-level attitude managed to cope rather well with the model. Experts in group coordination were able to decide the relative pre-process versus in-process influence as well as the relative amount of implicitness versus explicitness of two given mechanisms within the context of a mechanism set reference system. However, our results also illustrate the necessity to code mechanisms by more than one expert in order to achieve the desired level of reliability and robustness of results. This rather cumbersome and time-intensive procedure needs to be maintained until clear and intelligible intensional definitions are formulated.

Our main observation is a reconfirmation that coordination is executed on different levels of interaction: the macro-, meso-, and micro- level, respectively. But because we also have learned that switching cognitively among these levels can lead to qualitative changes in the meaning of pre-process and in-process or implicitness and explicitness, questions remain regarding a characterisation process for intensional and perspective aspects of coordination mechanisms. Our tentative solution is to adhere to the pair-comparison approach within a reference set of mechanisms until these questions are resolved.

Secondly, our test helped to illustrate the unsolved question that perspective of coordination mechanisms (producer vs. target) was ignored in former literature on the CMCM. To pique a discussion of this problem, our injection of the component of varying levels of coordination complexity as an attribute of the reference set of coordination mechanisms seemed to have helped. For coordination of large-scale human social systems (macro-level), the usual research focus is the so-called architecture of control (Lockton 2005), where the intention of the producer becomes the salient position. Pre-process and explicit versus in-process and implicit mechanisms seem the obvious prototypes, being well understood from the producer's perspective, perhaps because it's easy to identify with Lockton's *controller* when analysing macro-coordination. However, even in the road traffic set, we chose not to apply the perspective of the producer. Speed bumps, a typical design artefact explicitly intended by their producer to slow down traffic, were coded as an implicit in-process mechanism (see Fig. 4.6). Speed bumps appear to be implicit and in-process functioning if viewed from the perspective of the target of the mechanism. In terms of intentionality, targets (in this case: drivers) adaptively slow down mainly to secure their cars and their comfort. In other words, low motivation to comply with traffic road laws would not change their actual driving behaviour when faced with a speed bump due to the overriding risk to their car and their comfort. This helps explain why explicit coordination mechanisms from the perspective of the target generally need the target's compliance (a weaker form of intentionality) in order for the mechanism to be executed in the first place. Is it this freedom not to comply that makes a coordination mechanism explicit (and in our speed bump scenario, 'implicit' because the targets' reduced speed has nothing to do with intended compliance to the speed limit) and also explains why some explicit mechanisms might work less than perfectly? But the effectiveness of a mechanism does not necessarily bias its positioning in the coordination circumplex (Fig. 4.1). From the perspective of the target, implicit mechanisms can carry a higher compliance risk (and, in some cases, a correlating compliance motivation) because the target is free to overlook them, but at the peril of their car or worse in our example.

We also observed that both task structure and task-inherent feedback as leadership substitutes in teams also function implicitly and in-process (see Fig. 4.7). Similarly to the speed bump coordination mechanism, they both seem related to the perspective of the target. The implicit in-process quadrant of the coordination mechanism circle looking at the micro-level (Fig. 4.8) is filled with content contributions. Content does not convey normative information, but implicitly changes the micro-level knowledge environment and task for thoughts and acts. Content contributions may function as a 'neighbour thought', evoking 'self-coordination' in discussions. No explicit intentions are needed to evoke the changes these content contributions make to the ongoing group process.

Harkening back to the landmark of Jones and Gerard (1967) behaviouristic model of three types of interaction patterns, pseudo-contingent behaviour (rooted in a third information source) is caused by such in-process implicit mechanisms as in the rhythm of music for dance movements, speed bumps for car drivers, and task structures and/or actual content of an ongoing discussion. The implicitness of these mechanisms is unrelated to producer intention even though the music may have been chosen by a disc jockey, the speed bumps planned by city traffic management, and even the task structure of an ongoing discussion carefully designed by symbolic leadership (Schein 1992). It is even conceivable that interruptions and repetitions are sometimes produced intentionally to control the discussion (Kolbe and Boos 2009). But from the perspective of the coping individual (the target's perspective), their adaptations to affordances of the mechanisms in all three scenarios are uncorrelated to the existence of manipulation (producer) intentions. The discussion of mechanism intentionality and perspective seems like a bottomless pit. In the discipline of semiotics, objects with a *major pragmatic* function (judged from external perspective), even if accomplishing a *minor semantic* function (from target and external perspective), are distinguished from signs with a *major semantic* function (semantic specialisation, e.g. Nöth 1995, pp. 156, 441), as biosemiotics by definition excludes external attribute intentions of non-human animals and plants. Analogously, according to the mechanism's functional specialisation, explicit mechanisms realise coordination as their major function (external perspective), whereas implicit mechanisms lack this functional speciali-sation. Task-inherent feedback and content contributions from the targets' perspec-tive seem functionally unspecialised regarding their coordination realisation. But what about speed bumps, with an explicit functional intent from the producers' perspective but an implicit coordination realisation from the targets' perspective? Even though the CMCM is neutral regarding perspective considerations, we con-tend that the targets' perspective as the reference point for the external coordination realisation is more reflective of the actual affects of the coordination mechanisms (see Figs. 4.6–4.8). Following these considerations, explicitness means *accomplish-ing a major coordination function*: An explicit coordination mechanism possesses specialisation; an implicit coordination mechanism does not.

Additionally, the functions of coordination mechanisms can change over time: Some mechanisms can continuously specialise themselves for different functions, and therefore change their temporal (pre-/in-/post-process) dimension location as well as their explicit/implicit dimension location in the CMCM quadrants (Fig. 4.1).

This lack of clarity regarding the theoretical status of the 'intention' construct and how it affects coordination realisation, especially in the context of non-human primate group coordination, has generated several questions for further theory building and empirical research. Small group research should continue to develop a convention of the implicit/explicit and pre-process/in-process mechanisms for different complexity levels and forms of coordination processes. This is absolutely essential if we are ever to hope for consensus among researchers from different backgrounds and disciplines regarding a fully functional characterisation model of coordination mechanisms of both human and non-human primate groups. It is also important that questions regarding producer vs. target perspective relative to the two coordination circumplex dimensions are further researched and eventually accounted for in such a model. For instance, if a coordination mechanism is defined as explicit due to the intention of a producer, but as implicit due to going unnoticed as such by the target, yet nevertheless as a successful coordination mechanism due to its asserted effect on the target's behaviour (e.g. our speed bump scenario), its plotting on the CMCM becomes split. Two circles would be needed in order for the CMCM to accommodate the plotting of this scenario: one for the controller, one for the target. In such scenarios we prefer the perspective of the target because this perspective represents the actual realisation of the mechanism.

Then there are questions evoked by temporality that need to be addressed. The CMCM distinguishes pre-process and in-process phases. Two interpretations are potentially applicable: (1) the onset timing of a coordination mechanism and (2) the durability of its effectiveness, or 'power'. Consider the basis of power (Raven 1965) as a set of coordination mechanisms on a meso- or macro-level. Raven (1999) later considered the durability of power based on 'information' as longer lasting and having more sustainable effects without in-process surveillance compared to other power bases such as assertion of authority. In our study of verbal acts, content contributions (statements, comments, information), if identified as coordination mechanisms, were coded as relatively implicit and in-process coordination (Fig. 4.7). But requests, goal definitions, and implications of goals were plotted on the pre-process section of the micro-level CMCM because their effects are sustained over longer durations. Both aspects of temporality (onset and durability) seem to fit, depending on the macro- or micro- level of the coordination reference set.

We show in this chapter some data on the intelligibility and therefore applicability of the Coordination Mechanism Circumplex Model. But assessment of the overall *effectiveness* of the model is another matter. Yet even with all the above-mentioned caveats requiring additional clarification and study, we nevertheless believe in the applicability of the CMCM as a helpful model when analysing the timing and explicit/implicit descriptions of coordination in both human and, potentially, non-human primates.

References

Andersen PB, Carstensen PH, Nielsen M (2000) Dimensions of coordination. LAP 2000. In: The Fifth International Workshop on the Language-Action Perspective on Communication Modelling. Available at http://www.cs.aau.dk/

Argote L (1982) Input uncertainty and organizational coordination in hospital emergency units. Admin Sci Quart 27:420–434

Asch SE (1952) Social psychology. Prentice Hall, Englewood Cliffs, NJ

Barker RG (1968) Ecological psychology: concepts and methods for studying the environment of human behaviour. Stanford University Press, Palo Alto, CA

Burns T, Stalker GM (1961) The management of innovation. Tavistock, London

Carnap R (1947) Meaning and necessity. A study in semantics and modal logic. University of Chicago Press, Chicago, IL

Entin EE, Serfaty D (1999) Adaptive team coordination. Hum Factors 41:312–325

Entin EE, Diedrich FJ, Weil SA, See KA, Serfaty D (2005) Understanding team adaptation via team communication. In: Proceedings of the Human Systems Integration Conference, Washington, DC. Available at http://www.aptima.com/

Espinosa A, Lerch FJ, Kraut RE (2004) Explicit vs. implicit coordination mechanisms and task dependencies: one size does not fit all. In: Salas E, Fiore SM (eds) Team cognition: Understanding the factors that drive process and performance. American Psychological Association, Washington, DC, pp 107–129

Evans AW, Harper ME, Jentsch F (2004) I know what you're thinking: eliciting mental models about familiar teammates. In: Cañas AJ, Novak JD, González FM (Eds) Concept maps: Theory, methodology, technology. Proceedings of the First International Conference on Concept Mapping. Pamplona, Spain 2004. Available at http://www.cmc.ihmc.us

Faraj S, Xiao Y (2006) Coordination in fast-response organizations. Manage Sci 55:1155–1169

Fleiss JL (1971) Measuring nominal scale agreement among many raters. Psychol Bull 76:378–382

Godart C, Halin G, Bignon JC, Bouthier C, Malcurat O, Molli P (2001) Implicit or explicit coordination of virtual teams in building design. CAADRIA01 Sydney, Australia, pp 429–434. Available at http://www.crai.archi.fr

Grote G, Zala-Mezö E, Grommes P (2003) Effects of standardization on coordination and communication in high workload situations. In: Dietrich R (ed) Communication in high risk environment. Helmut Buske, Hamburg, pp 127–154

Hackman JR, Morris CG (1975) Group tasks, group interaction process, and group performance effectiveness: a review and proposed integration. In: Berkowitz L (ed) Advances in experimental social psychology, vol 8. Academic, New York, pp 45–99

Jones EE, Gerard HB (1967) Foundations of social psychology. Wiley, New York

Kerr S, Jermier JM (1978) Substitutes for leadership: their meaning and measurement. Organ Behav Hum Perf 22:375–403

Kim H, Kim D (2008) The effects of the coordination support on shared mental models and coordinated action. Brit J Educ Technol 39:522–537

Kolbe M (2007) Explicit process coordination of decision-making groups (German: Explizite Prozesskoordination von Entscheidungsfindungsgruppen). SUB, University of Goettingen. Available at http://webdoc.sub.gwdg.de/diss/2007/kolbe/

Kolbe M, Boos M (2009) Facilitating group decision-making: facilitator's subjective theories on group coordination. Forum: Qual Soc Res 10(1):29

Lockton D (2005) Architectures of control in consumer product design. Master's thesis, Judge Institute of Management, University of Cambridge. Available at http://www.danlockton.co.uk/research/Architectures_of_Control_v1_01.pdf

MacMillan J, Entin EE, Serfaty D (2004) Communication overhead: The hidden cost of team cognition. In: Salas E, Fiore SM (Eds) Team cognition: Process and performance at the inter- and intra-individual level. American Psychological Association, Washington, DC. Available at http://www.aptima.com/publications/2004_MacMillan_EntinEE_Serfaty.pdf

March J, Simon HA (1958) Organizations. Wiley, New York

Mintzberg H (1979) The structuring of organizations. Prentice Hall, Englewood Cliffs, NJ

Nöth W (1995) Handbook of semiotics. Indiana University Press, Bloomington, IN

Ramon R, Sanchez-Manzanarez M, Gil F, Gibson C (2008) Team implicit coordination processes: a team knowledge-based approach. Acad Manage Rev 33:163–184

Raven BH (1965) Social influence and power. In: Steiner ID, Fishbein M (eds) Current studies in social psychology. Holt, Rinehart and Winston, New York, pp 371–382

Raven BH (1999) Influence, power, religion, and the mechanism of social control. J Soc Issues 55:161–186

Schein EH (1992) Organizational culture and leadership, 2nd edn. Jossey-Bass, San Francisco, CA

Sherif M (1935) The psychology of social norms. Harper and Row, New York

Van de Ven AH, Delbecq AL, Koenig RJ (1976) Determinants of coordination modes within organizations. Am Sociol Rev 41:322–338

Wittenbaum GM, Stasser G, Merry CJ (1996) Tacit coordination in anticipation of small group task completion. J Exp Soc Psychol 32:129–152

Wittenbaum GM, Stasser G, Vaughan SI (1998) Coordination in task-performing groups. In: Tindale RS, Heath L, Edwards J, Posavac EJ, Bryant FB, Suarez-Balcazar Y, Henderson-King E, Myers J (eds) Theory and research on small groups. Plenum, New York

Chapter 5
The Role of Coordination in Preventing Harm in Healthcare Groups: Research Examples from Anaesthesia and an Integrated Model of Coordination for Action Teams in Health Care

Michaela Kolbe, Michael Burtscher, Tanja Manser, Barbara Künzle, and Gudela Grote

Abstract In this chapter we discuss the role of group coordination as a means of preventing iatrogenic harm in health care using anaesthesia teams as our forum of research. Applying the inclusive model of group coordination in Chap. 2 (see Fig. 2.5), we outline that (1) clinical performance and patient safety are functions of group coordination, (2) information and actions are key input entities of group coordination, (3) adaptation to situational demands serves as a critical coordination process, and (4) explicit and implicit coordination are essential coordination mechanisms. We will present recent findings regarding the role of each of these concepts for teamwork in health care. Combining theoretical considerations and empirical results, we will offer an integrated model of coordination for action teams in health care. The core idea of this model is that coordination can be classified along two independent dimensions (1) mechanisms such as explicit vs. implicit coordination, and (2) input entities such as behaviours (e.g. actions) and meanings (e.g. information). We suggest that the usefulness of team coordination should hence be considered with regard to this distinction.

5.1 Introduction

Group work plays a vital role in health care. In fact, many medical procedures such as surgery, emergency medicine, and anaesthesia can only be performed by groups. In these group procedures, clinical performance and patient safety are key functions

---

M. Kolbe (✉), M. Burtscher, B. Künzle, and G. Grote
Department of Management, Technology, and Economics, Organisation, Work, Technology Group, ETH Zürich, Kreuzplatz 5, KPL G 14, 8032 Zürich, Switzerland
e-mail: mkolbe@ethz.ch, mburtscher@ethz.ch, bkuenzle@ethz.ch, ggrote@ethz.ch

T. Manser
Industrial Psychology Research Centre, School of Psychology, King's College, University of Aberdeen, G32 William Guild Building, Aberdeen AB24 2UB, UK
e-mail: t.manser@abdn.ac.uk

M. Boos et al. (eds.), *Coordination in Human and Primate Groups*,
DOI 10.1007/978-3-642-15355-6_5, © Springer-Verlag Berlin Heidelberg 2011

of group coordination. Anaesthesia in particular involves a variety of risks, and the consequences of potential failures may be life-threatening to the patient. Therefore, safety is valued as a prime goal, and today anaesthesia is safer than ever and acknowledged as the leading specialty in considering patient safety (Gaba 2000). Recent studies, however, have shown that incidents (e.g. breathing-circuit disconnections; Cooper et al. 2002) resulting in harm or death still occur (Catchpole et al. 2008; Gravenstein 2002).

Such iatrogenic injuries – inadvertent injuries caused by or resulting from medical treatment – involve human error in about 82% of cases (Cooper et al. 2002) and are often related to breakdowns in the quality of group work, such as in communication (Arbous et al. 2001; Currie et al. 1993; Lingard et al. 2004), particularly information loss (Christian et al. 2006) and difficulties in discussing errors (Sexton et al. 2000).

In this chapter we use the example of team administration of anaesthesia to discuss the role of group coordination as a means of preventing iatrogenic harm in health care. We apply the inclusive model of group coordination presented in Chap. 2 (see Fig. 2.5) that distinguishes functions, entities, and mechanisms of group coordination and specify it for the specific situations of healthcare action teams. Within the context of human factors in high-risk healthcare domains, we analyse coordination as a critical success factor and outline that (see Fig. 5.1)

- Clinical performance and patient safety are functions of group coordination
- Information and actions are key input entities of group coordination that respectively become information exchange and collective actions during the process of group coordination
- Explicit and implicit coordination are essential coordination mechanisms
- Adaptation to situational demands serves as a critical coordination process

This chapter is organized as follows: We begin by outlining the functions of group coordination in anaesthesia by referring to studies showing that breakdowns

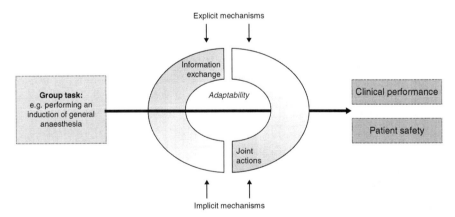

Fig. 5.1 Coordination in healthcare action teams as an interplay of entities, mechanisms, process, and functions

in the quality of coordination are responsible for impaired clinical performance. We then illustrate the main input entities, processes, and mechanisms of medical group coordination, giving examples from our research in anaesthesia. Combining these considerations, we propose an integrated model of coordination for action teams in health care. We conclude by discussing respective further research needs as well as implications for training programmes and clinical practice.

5.2 Groups in Anaesthesia

Anaesthesia can be regarded as a classic small-group performance situation. In the analysis of teamwork, common functional models suggest three levels of focus: individual, group, and context (Ilgen et al. 2005). We will analyse characteristics of teamwork in anaesthesia groups with regard to these three levels.

In routine cases, groups in anaesthesia typically consist of an anaesthesia nurse and an anaesthesia resident physician. An attending physician is on standby and available during the course of the anaesthetic procedure. In a non-routine case, more members – for instance, the attending physician on standby and/or additional nurses – can join the group.[1] One of the most significant characteristics of anaesthesia groups is their structure as a crew (Arrow et al. 2000; Gaba 1994; Tschan et al. 2006; Webber and Klimoski 2004) or action team (Manser 2009). Action teams are "highly skilled specialist teams cooperating in brief performance events that require improvisation in unpredictable circumstances" (Sundstrom et al. 1990, p. 121). Thus, anaesthesia teams[2] often have no previous experience of working together and almost no formal training in teamwork, two circumstances that greatly challenge effective coordination. Further characteristics of anaesthesia teams refer to the individual team members as well as the context level in which the team operates, which will be specified subsequently.

At the level of the individual team member, anaesthesia team members are characterized by technical and non-technical competence (Fletcher et al. 2003), heterogeneous knowledge (Rosen et al. 2008), and high work commitment (Nyssen et al. 2003). Individual differences regarding these factors, especially between physicians and nurses who receive different training, are very likely to affect communication culture and approaches to team coordination.

At the context level, two significant factors influence teamwork in anaesthesia: First, the teams operate in hospitals – highly structured, high-risk organization. In such environments, errors may have serious consequences and therefore make safety one of the top priorities (e.g. Baker et al. 2006). This challenging structural aspect is exacerbated by the fact that anaesthesia teams are embedded in the context of an

---

[1]Anaesthesia group composition depends on the surgical procedure and can vary across countries.

[2]For ease of reading, we will use the term "anaesthesia team" instead of "anaesthesia action team" throughout this chapter.

overall operating room team. The operating room team consists of several sub-teams that need to coordinate their joint work: nursing, surgery, and anaesthesia (Gaba 1994). It can be considered a multi-system team, which is defined as consisting of "two or more teams that interface directly and interdependently in response to environmental contingencies towards the accomplishment of collective goals" (Mathieu et al. 2001, p. 290). As a consequence, team coordination does not only occur within the anaesthesia team, but also within and among the multi-system teams, giving rise to potential inter-team conflicts and errors. A second important context factor lies in the task of anaesthesia itself. As Gaba (1994, pp. 198–199) points out: "The dominant features of the anaesthetist's environment include a combination of extreme dynamism, intense time pressure, high complexity, frequent uncertainty, and palpable risk. This combination is considerably different from that encountered in most medical fields". Anaesthesia involves constant attention to a range of tasks (Leedal and Smith 2007; Weinger and Englund 1990), which hold planning (e.g. planning the correct amount of a certain medication), intellectual (e.g. finding the correct reason for suddenly rising blood pressure), decision making (e.g. deciding on the right time for intubation), psycho-motor performance (e.g. laryngoscopy), and mixed-motive components (e.g. choosing between the surgeon's request to proceed with the operation and the need to further stabilize the patient). Additionally, anaesthesia is influenced by less than completely predictable patients as well as by a variety of external (e.g. noise, temperature, lighting, and other workplace constraints) and human (interpersonal relations, fatigue and sleep deprivation, boredom, workload, and task characteristics) factors (Weinger and Englund 1990). The dynamism of anaesthesia is reflected in rapid shifts from executing routine procedures to handling critical situations. This shift is defined in the literature by the occurrence of non-routine events – events perceived by care providers or skilled observers to be unusual (Weinger and Slagle 2002) – and that pose unusual challenges to anaesthesia teams. These events refer not only to critical incidents but also to a broad range of events that might not lead to immediate adverse outcomes but nevertheless could be early indicators of later adverse incidents (Oken et al. 2007; Wacker et al. 2008).

We will now outline two functions of group coordination in health care (see Fig. 5.1), referring to studies showing how breakdowns in the quality of coordination are responsible for impaired clinical performance.

5.3 Functions of Group Coordination in Anaesthesia

As suggested by the functional perspective of teamwork (Hackman and Morris 1975; Marks et al. 2001; Wittenbaum et al. 2004), the characteristics of anaesthesia teams described above (e.g. action team structure, role-specific team member training, being embedded in complex OR teams, dynamic and risky coordination task) influence clinical performance via the interaction process of the anaesthesia team members. This interaction requires coordination in order to perform safe and effective patient treatment (Dickinson and McIntyre 1997; Rosen et al. 2008;

Tschan et al. 2006). From an action regulation perspective, coordination mechanisms are considered important for regulating task-related collaborative behaviours and team performance within task execution (Rousseau et al. 2006). As Steiner (1972) pointed out, teams rarely achieve their potential performance, and as Stroebe and Frey (1982) pointed out, they frequently suffer from process losses due to a lack in team member motivation and coordination. Given that anaesthesia is usually performed by teams that are in turn embedded in a complex multi-team system (i.e. operating room teams), the risk of iatrogenic errors occurring at the team level is increased (e.g. failures to communicate, communication misunderstandings, nonshared terminology or procedural definitions). For instance, team communication failure was found to be a factor contributing to drug error (Abeysekera et al. 2005), surgical injuries (Greenberg et al. 2007; Zingg et al. 2008), and operating room team functioning (Lingard et al. 2004). In an interview study, Gawande and colleagues found that one of "the most commonly cited system factors contributing to errors were [...] communication breakdowns among personnel (43%)" (Gawande et al. 2003, p. 614). The importance of non-technical skills such as communication and teamwork for good anaesthetic practice has been highlighted by other research as well (Fletcher et al. 2002). Giving further credence to this point is the increasing number of studies on human factors in various operating room teams that have stressed the role of team coordination for maintaining patient safety (see Manser 2009 for a review). Thus, managing errors and preventing iatrogenic harm are the main coordination functions of team coordination in anaesthesia. They are classified as coordination functions (see Fig. 2.5) because they define the overall performance objectives of the anaesthesia process. By applying the functional perspective on teamwork (Hackman and Morris 1975; Healey et al. 2004; Marks et al. 2001; Wittenbaum et al. 2004), we consider patient safety and clinical performance to be the most significant outcomes of the anaesthesia procedure, influenced by the variety of interrelated factors outlined above. In the following section we discuss two entities to be coordinated while performing anaesthesia that are pivotal for meeting these objections of patient safety and clinical performance.

5.4 Information Exchange and Joint Actions Within Anaesthesia Groups

Coordination is a constituent component of team work (Brannick and Prince 1997) and has been defined as "orchestrating the sequence and the timing of interdependent actions" (Marks et al. 2001, p. 363). However, it is not only the coordination of the input entity actions but also the coordination of the input entity information (e.g. sharing information regarding the patient or the drug that has just been administered) that is essential for clinical performance (Arrow et al. 2000). Thus, joint actions and information exchange are key objectives that have to be achieved by coordination (see Figs. 5.1 and 5.2). The exchange of information and the joining of actions occur in the process stage of the inclusive IPO (input–process–output)

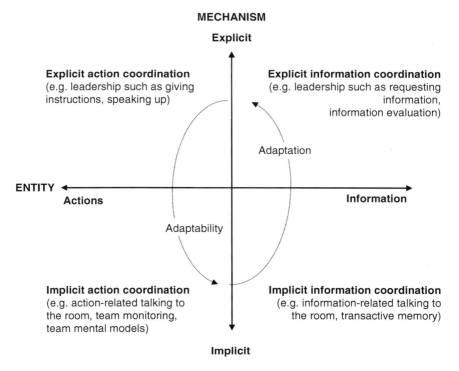

Fig. 5.2 Integrated model of coordination for action teams in health care

model of group coordination (Fig. 2.5). As sharing information and discussing its implications for the current task are pivotal for collective sense-making during crisis situations (Waller and Uitdewilligen 2008), coordination of these subtasks is necessary because important information is often unshared before initiating anaesthesia and has to be obtained in-process from various sources, including the patient, other team members, written notes, and the different monitors in the operating room. Also, physicians and nurses differ in their information-gathering behaviour (Thuilliez et al. 2005). Unfortunately, as known from research on team information processing, teams are often less than perfect at sharing relevant information (Mesmer-Magnus and DeChurch 2009; Stasser and Titus 1985), confirmed by the frequently reported incidents in anaesthesia of failure to appropriately communicate relevant information (e.g. patient allergies) to all team members at a time (pre-process) and in a fashion (explicitly) when problems could be avoided (Catchpole et al. 2008). Bogenstätter et al. (2009) showed for resuscitation teams that information transmitted within the team was only partly reliable and concluded that physicians and nurses should be trained in explicit coordination in the sense of standardized communication procedures.

Action coordination refers to the management of joint actions, which are defined activities in which two or more participants have to coordinate with each other in order to succeed. Action coordination involves the timing and sequencing of interdependent activities. For example, a cardiac arrest requires the team members to quickly decide which drug(s), by whom, and when to be administered, who will perform a precordial thump, who will call for additional help, who will take the lead, and who will monitor the progress. This involves team members acting in and sometimes outside their respective roles (e.g. resident as responsible leader and nurse as assistant), backing each other up when necessary (Clark 1999).

The distinction between information exchange and joint actions as two discrete coordination entities has already been acknowledged in taxonomies of coordination behaviour in anaesthesia teams. For example, the taxonomy for explicit and implicit team coordination and heedful interrelating behaviour (Kolbe et al. 2009) includes two main categories: (1) coordination of information exchange and (2) coordination of action within each explicit and implicit mechanism. Similarly, Manser and her colleagues' observation system for coordination behaviour in anaesthesia crews distinguishes between information management and task management (Manser et al. 2008; 2009a, b). While the former category includes activities related to information sharing (e.g. "information request"), the latter refers to the coordination of team actions (e.g. "prioritizing").

To coordinate joint actions and information exchange, teams can use two basic mechanisms: explicit and implicit coordination (see Fig. 4.1 as well as the mechanism section in Fig. 2.5). This common dichotomy (e.g. Chaps. 4 and 6) will be outlined in the following section.

5.5 Explicit and Implicit Coordination Mechanisms

In this section we will discuss explicit and implicit coordination mechanisms, including their related concepts of leadership and shared mental models (see Fig. 5.2). Depending on characteristics of both the team and the situation, the timing and nature of these mechanisms can vary in how beneficial they are to performance.

5.5.1 Explicit Coordination

Coordination behaviour can be regarded as explicit when it is intentionally used for coordination purposes and expressed in an unequivocal manner that is usually plain and easy to understand (Espinosa et al. 2004; Kolbe and Boos 2009; see also Chaps. 4 and 11; Serfaty and Kleinman 1990; Wittenbaum et al. 1998). Typical examples of explicit coordination include defining rules to standardize behaviour in advance (e.g. in case of emergency, call a staff anaesthetist), requesting information (e.g. information on patient allergies), or giving instructions (e.g. instructions regarding drug administration). Another classic example of explicitness is closed-loop communication – the acknowledged exchange of information between a sender and a receiver. For example, a physician's instructions regarding administering a specific amount of a specific drug are frequently acknowledged by the respective nurse (e.g. "Okay") or affirmed when executed (e.g. "0.2 Fentanyl given"). As such, explicitness can serve as a double-check, preventing failures in information transfer and facilitating doing the right thing in the right situation. Related research has shown that explicit coordination in terms of strategic planning has been triggered by resource inadequacy or time pressure (Xiao et al. 2004) and that it has proven to be a successful strategy during non-routine events for healthcare teams treating cardiac arrest (Marsch et al. 2004; see also Chap. 6).

Leadership is another example of personal and mostly explicit coordination, with team leaders being responsible for effective coordination and the creation of shared team cognition (Salas et al. 2001). Teams operating in critical care need directive, task-oriented, as well as relationship-oriented leadership in order to effectively complete the task and to keep members both calm and functioning (Künzle et al. 2010). In healthcare action teams, however, directive, task-based leadership behaviour – as opposed to interpersonal or developing behaviour – is most critical for task fulfilment (Künzle 2003; Zala-Mezö et al. 2004). These teams inherently share motivation and have a clear common goal – to save patients' lives – and therefore do not require long-term functions of leadership such as developing strategies or building trust (Klein et al. 2006; Zala-Mezö et al. 2004). This view is in line with the substitutes for leadership theory suggesting that a pressing and important task might substitute or eliminate the need for a motivating leader (Kerr and Jermier 1978). This has indeed been found for anaesthesia teams: The amount of leadership increased significantly during non-routine, high-taskload situations, while it was significantly reduced if the level of standardization was high (e.g. Grote et al. 2003; Klein et al. 2006; Künzle et al. in press; Zala-Mezö et al. 2009).

The main drawback of explicit coordination is that it is costly in terms of communicative effort and time – both limitedly available during non-routine situations. Given that non-routine events are ill-defined situations with high levels of uncertainty and time pressure, they might require but not allow for explicitness. Alternatively, explicit coordination can be used early during the team process to establish a shared understanding of the team and the task and thus facilitate later implicitness (Orasanu 1993, see also Chap. 10). Implicit coordination, with its potential for efficiency, and its main prerequisite – team mental models – are the focus of the following section.

5.5.2 Implicit Coordination

In contrast to the directive and verbal nature of explicitness, implicitness relies on the anticipation of actions and needs of the other team members and the subsequent adjustment of their own behaviour accordingly (Entin and Serfaty 1999; Grote et al. 2003; MacMillan et al. 2004; Rico et al. 2008; Toups and Kerne 2007; Wittenbaum et al. 1996, 1998). In fact, some researchers equate the level of implicit coordination with the level of team effectiveness (Stachowski et al. 2009). Although we consider implicit coordination highly relevant for team performance, we still regard it as a process rather than an outcome factor.

In anaesthesia some procedures, such as intubation, are highly standardized and allow for team members to predict the behaviour of their colleagues. For example, nurses can directly observe the behaviour of the physician, anticipate when their assistance is needed, and then provide unsolicited information or action accordingly (Kolbe et al. 2009). This tacit management is frequently regarded as the ideal way of coordination (Zala-Mezö et al. 2004), particularly when the patient is still awake (Hindmarsh and Pilnick 2002). It can, however, only be successful if both physician and nurse have an accurate and shared team mental model (TMM). That is, they need a common understanding of their task and a mutual agreement regarding the steps that are to be taken in a certain situation. In the context of anaesthesia, that means, for instance, that both nurse and resident have a shared understanding of which drugs are to be administered in the course of the induction, who has the leadership role, and what is to be done in case of emergency.

The importance of an TMM for team performance is emphasized in many current theoretical approaches to teamwork (for a review, see Chap. 9). In general, a positive correlation between the degree of "sharedness" amongst team members and the team's performance is assumed. Current studies in other high-risk industries such as aviation and the military support this view (Lim and Klein 2006; Smith-Jentsch et al. 2005). Because TMMs are related to team communication and coordination (Gurtner et al. 2007; Marks et al. 2002), this means that if both the anaesthesia resident and the nurse have an TMM, they will not have to discuss drug administration at length and will therefore save the time and effort costs distracted from the task at hand. Without an TMM, they would have to compensate by explicitly coordinating the medication process, which, due to its costs, could conceivably heighten the risk of iatrogenic injuries.

That said, thus far there is only little empirical evidence for postulated relationships among TMM, implicit coordination, and performance in medical teams (Burtscher and Manser submitted). However, the findings of Michinov et al. (2008) showed that transactive memory systems, a concept related to TMM in terms of knowing who knows what, predicted anaesthesia team member perceptions of team effectiveness and also affective outcomes such as team identification and job satisfaction. More recent work in anaesthesia has also addressed the function of implicit coordination processes as a mediator of the relationship between TMM and objective medical team performance by proposing a anaesthesia-specific measurement tool based on concept mapping (Burtscher et al. 2009).

In view of the above distinctions – explicit vs. implicit and information exchange vs. joint actions – anaesthesia teams face the challenge of prioritizing which entity and mechanism is most important in a given situation. Since time and resources are limited, when team members are not able to perform all entities and mechanisms simultaneously, they instead need to constantly adapt their coordination behaviours to the demands of the situation. The importance of adaptation as team process will be outlined in the following section.

5.6 Adaptation as a Key Coordination Process

The relevance of coordination entities and the appropriateness of coordination mechanisms are not stable and change relative to the situation – an idea similar to contingency models of human leadership (e.g. Burke et al. 2006; Entin and Serfaty 1999; Fiedler 1964; Grote et al. 2003, 2010; Rico et al. 2008; Salas et al. 2007a, b). Thus, to provide safe and efficient patient care, medical teams need to adapt quickly to changing situational demands such as the occurrence of non-routine events (see Fig. 5.2). Adaptation as such is generated by the team's "response to actual or anticipated changes in the embedding contexts" (McGrath and Tschan 2004, p. 6). In particular, task uncertainty, standardization, and time pressure are situational constraints stimulating dynamic changes in team coordination behaviour (Grote et al. 2010; Serfaty and Entin 2002). As Ballard et al. (2008, p. 339) noted that "group processes may be divided through the occurrence of critical events", coordination mechanisms that may be effective during routine work may hence not be effective during non-routine events (Gersick and Hackman 1990; Waller 1999). Even within a single non-routine situation, certain behaviours might only be relevant during a very specific sub-phase. For example, in the study on leadership during the management of a cardiac arrest, Tschan et al. (2006) showed that the amount of leadership of the incoming resident mattered only during the first 30 s. In contrast to long-term, quasi-permanent adjustment of performance strategies, adaptation in action teams involves very quick changes in coordination behaviour within a single episodic or task cycle vs. from one episode to another (Marks et al. 2001). These quick changes require that the team members constantly know what is going on during the task cycle and regularly evaluate and adapt to the current situation.

Several studies have shown that anaesthesia teams effectively adapt to situational demands by adjusting the time they spend on task and information management (Burtscher et al. 2010; Manser et al. 2008, 2009a, b) and by adjusting the appropriate level of explicitness (Kolbe et al. under review; Zala-Mezö et al. 2009). These findings indicate that adaptation is fundamental to establishing safety (Salas et al. 2007a, b). However, there is also an ongoing discussion that adaptability is still a rather elusive and ill-defined concept within the psychological literature (e.g. Pulakos et al. 2000). Recent studies are attempting to deal with this issue by descriptively depicting the dynamics of adaptive interaction patterns (Grote et al. 2010; Stachowski et al. 2009).

5.7 An Integrated Model of Coordination for Action Teams in Health Care

Having reported various empirical results regarding team coordination and its effects on safety and performance in health care, we need to relate these various findings to each other in order to get a more concise picture of team coordination in health care and to enable us to make predictions regarding the appropriateness of coordination mechanisms in specific task circumstances. Although existing models of team coordination point out that the degree of explicitness should adjust according to task demands and that TMMs are a basis for implicit coordination (e.g. Rico et al. 2008; Wittenbaum et al. 1998), they do not specifically differentiate coordination entities nor the particular characteristics of action teams. Applying the inclusive model of group coordination (Fig. 2.5), we attempt to make an initial step in filling this gap by proposing a specific model for systematizing coordination mechanisms in healthcare action teams and interlinking the coordination input entities (what is coordinated?), coordination mechanisms (how is it coordinated?), and coordination process (adaptation).

When discussing the role of team coordination as a means of preventing iatrogenic harm, we showed that clinical performance and patient safety are key functions of team coordination; information and actions are the entities of team coordination; explicit coordination (including leadership) and implicit coordination (including TMM) are essential coordination mechanisms; and adaptation to situational demands serves as a critical coordination process. Yet the question remains: How are these concepts of coordination interrelated? Assuming that information exchange and collective actions, adaptation, and explicit and implicit mechanisms operate in an interrelated manner to establish safe performance as the overall function, we propose an integrated model that regards entities and mechanisms as two independent dimensions and adaptation as the interlinking process (Fig. 5.2).

Within this model, each coordination act can be classified along two independent dimensions: entity (information vs. action) and mechanism (explicit vs. implicit). That is, these entities refer to the question of what has to be coordinated, while the mechanisms describe how it can be coordinated. The resulting two-by-two matrix with the four quadrants[3] represents different coordination styles as systematized by the Coordination Mechanism Circumplex Model (see Chap. 4): Explicit action coordination includes explicit coordination acts that aim at coordinating joint actions. Explicit information coordination, on the other hand, includes communicative acts that explicitly aim at managing the information processing within the team. Both "styles" can be associated with leadership behaviours. On the other side of the spectrum, team cognition – team mental models and transactive memory – provides the basis for implicit coordination. Thereby, implicit information coordination includes acts for tacitly managing team information processing. Finally, implicit action coordination includes acts that facilitate action coordination via mutual anticipation.

Adaptation refers to the process of regulating these different coordination styles, for example, by switching from one style to another. As we and others have illustrated, in many settings it is crucial to adapt to changing situational and task demands; different situations require different styles of coordination. As a

---

[3]Although these dimensions represent continua instead of dichotomies, for ease of presentation we will discuss four styles of coordination located at the extreme of the continua.

consequence, in order to realize the key functions of team coordination – clinical performance and patient safety – teams have to balance their coordination style via the process of adaptation.

This integrated model specific to healthcare groups (and potentially to those in other high-risk industries) can be used to analyse the interplay of explicit and implicit coordination which should be considered with regard to the respective coordination entity. It allows us to make predictions regarding the usefulness of coordination styles in a given situation where either information exchange or joint actions (or both) are pivotal. We will illustrate this with an example of a routine induction of general anaesthesia with an unexpected occurrence of a cardiac arrest as a non-routine event, which was part of a simulator-based study on adaptive team coordination (Kolbe et al. under review; see example below).

Example: Adaptive Coordination During the Management of a Cardiac Arrest

Anaesthesia induction is the first step in all operations requiring general anaesthesia and – compared to other tasks involved in the anaesthetic process – is very coordination-demanding (Phipps et al. 2008). An induction starts with preparing the patient and equipment and injecting drugs to anaesthetize and paralyze the patient. Once the patient shows no voluntary muscle movement, a tube is inserted into the patient's trachea using direct laryngoscopy.[4] As these steps are highly standardized, we assumed that during this non-routine phase, team members (Hypothesis 1)

- Would require only limited explicit action coordination and could be managed by relying on team mental models and thus on implicit action coordination,
- Would still require higher levels of explicit information coordination in order to effectively process relevant information about the status of the patient and his or her reaction to the administered drugs.

In the study mentioned above (Kolbe et al. under review), the simulation continued with an asystole (cardiac arrest) during laryngoscopy. A sudden cardiac arrest is a rarely occurring but time-critical and life-threatening non-routine situation in anaesthesia. It can, however, be unambiguously diagnosed (visible flat line on monitor with electrodes properly fixed) and handled (administering atropine, chest compressions). For this non-routine emergency situation, we assumed that (Hypothesis 2)

(continued)

---

[4]During laryngoscopy, vocal cords are directly visualized with the laryngoscope blade. A light beam from the blade tip facilitates the introduction of a plastic tube through the mouth into the trachea under visual control (orotracheal intubation).

- Explicit information coordination such as verbally requesting, providing, and verifying information might not be the most effective coordination method because it requires time and effort, which are not available,
- Team members in a high-risk non-routine event such as this better coordinate their information exchange using the implicit information coordination style such as gathering and sharing information without being asked,
- The actual management of action steps – who is doing what at what time – would still require higher levels of explicitness because of the unfamiliarity of the situation. Here, the explicit action coordination style might be most appropriate.

Indeed, an analysis of coordination behaviour showed that both residents and nurses showed significantly higher levels of explicit action coordination during the management of the cardiac arrest than during the initial routine situation. It was also found, as predicted, that explicit information coordination decreased in the transition from the routine to the non-routine situation. The level of implicit action coordination was high in both phases but significantly increased further during the management of the cardiac arrest, indicating that anaesthesia teams use means of explicit as well as implicit action coordination during such a non-routine situation. Levels of implicit information coordination were generally low and did not change in relation to the cardiac arrest.

As shown in the example, the integrated model of coordination for action teams in health care can be used to make differential predictions regarding the usefulness of explicit and implicit coordination mechanisms with respect to the management of joint actions and processing of information and thus help to clarify previous inconsistent findings on their respective efficiency in non-routine situations (e.g. Entin and Serfaty 1999; Tschan et al. 2006).

5.8 Directions for Future Research

Using the example of anaesthesia, in this chapter we discussed the role of team coordination as a means of preventing iatrogenic harm to patients. We applied the inclusive model of group coordination presented in Chap. 2 (Fig. 2.5) to the specific situation of healthcare action teams. We specified that in healthcare teams, clinical performance and patient safety are key functions of group coordination; information and actions are entities of group coordination; explicit coordination (including leadership) and implicit coordination (including team mental models) are essential coordination mechanisms; and adaptation to situational demands serves as a critical coordination process. We outlined the relationships among these concepts and proposed an integrated model of coordination for action teams in health care (Fig. 5.2). The core idea of this model is that coordination can be classified along two independent dimensions: entity (actions vs. information) and mechanism (explicit vs. implicit coordination). The usefulness of team coordination should hence be considered with regard to this distinction. Different coordination tools such as TMM or leadership can be assigned accordingly and their functionality can be judged with regard to task requirements and constraints. For example, an unexpected crisis situation may call for explicit coordination of information (Bogenstätter et al. 2009), and thus leadership behaviour will likely be more appropriate than implicit coordination based on an TMM. We are convinced that theoretical models of coordination in healthcare teams are necessary not only to guide future research but also to enable the development of training curricula that will improve team performance and patient safety, perhaps even rendering helpful coordination paradigms transferable to other high-risk industries.

References

Abeysekera A, Bergman IJ, Kluger MT, Short TG (2005) Drug error in anaesthetic practice: a review of 896 reports from the Australian Incident Monitoring Study database. Anaesthesia 60:220–227

Arbous MS, Grobee DE, van Kleef JW, de Lange JJ, Spoormans HHAJM, Touw P, Werner FM, Meursing AEE (2001) Mortality associated with anaesthesia: a qualitative analysis to identify risk factors. Anaesthesia 56:1141–1153

Arrow H, McGrath JE, Berdahl JL (2000) Small groups as complex systems: formation, coordination, development, and adaption. Sage, Thousand Oaks, CA

Baker DP, Days R, Salas E (2006) Teamwork as an essential component of high-reliability organizations. Health Serv Res 41:1576–1598

Ballard DI, Tschan F, Waller MJ (2008) All in the timing: considering time at multiple stages of group research. Small Group Res 39:328–351

Bogenstätter Y, Tschan F, Semmer NK, Spychiger M, Breuer M, Marsch SU (2009) How accurate is information transmitted to medical professionals joining a medical emergency? A simulator study. Hum Factors 51:115–125

Brannick MT, Prince C (1997) An overview of team performance measurement. In: Brannick MT, Salas E, Prince C (eds) Team performance assessment and measurement theory, methods, and applications. Lawrence Erlbaum, Mahwah, NJ, pp 3–16

Burke CS, Stagl KC, Salas E (2006) Understanding team adaptation: a conceptual analysis and model. J Appl Psychol 91:1189–1207

Burtscher MJ, Manser T (under revision) Assessment of shared mental models in different teamwork settings: a review and implications for future research

Burtscher MJ, Wacker J, Sevdalis N, Manser T (2009) Shared mental models in healthcare – Development and testing of a measurement tool. 15th Annual Meeting of Society in Europe for Simulation Applied to Medicine (SESAM), Mainz

Burtscher MJ, Wacker J, Grote G, Manser T (2010) Managing non-routine events in anesthesia: the role of adaptive coordination. Hum Factors. Human Factors Advance online publication

Catchpole K, Bell MDD, Johnson S (2008) Safety on anaesthesia: a study of 12606 reported incidents from the UK National Reporting and Learning System. Anaesthesia 63:340–346

5 The Role of Coordination in Preventing Harm in Healthcare Groups 89

Christian CK, Gustafson ML, Roth EM, Sheridan TB, Gandhi TK, Dwyer K, Zinner MJ, Dierks MM (2006) A prospective study of patient safety in the operating room. Surgery 139:159–173

Clark HH (1999) On the origins of conversation. Verbum 21:147–161

Cooper JB, Newbower RS, Long CD, McPeek B (2002) Preventable anesthesia mishaps: a study of human factors. Qual Saf Health Care 11:277–283

Currie M, Mackay P, Morgen C, Runciman WB, Russell WJ, Sellen A, Webb RK, Williamson JA (1993) The "wrong drug" problem in anaesthesia: an analysis of 2000 incident reports. Anaesth Intensive Care 21:596–601

Dickinson TL, McIntyre RM (1997) A conceptual framework for teamwork measurement. In: Brannick MT, Salas E, Prince C (eds) Team performance assessment and measurement. Lawrence Erlbaum, Mahwah, NJ, pp 19–43

Entin EE, Serfaty D (1999) Adaptive team coordination. Hum Factors 41:312–325

Espinosa A, Lerch FJ, Kraut RE (2004) Explicit vs. implicit coordination mechanisms and task dependencies: one size does not fit all. In: Salas E, Fiore SM (eds) Team cognition: understanding the factors that drive process and performance. American Psychological Association, Washington, DC, pp 107–129

Fiedler FE (1964) A contingency model of leadership effectiveness. In: Berkowitz L (ed) Advances in experimental social psychology. Academic, New York, pp 149–190

Fletcher GCL, McGeorge P, Flin RH, Glavin RJ, Maran NJ (2002) The role of non-technical skills in anaesthesia: a review of current literature. Brit J Anaesth 88:418–429

Fletcher G, Flin R, McGeorge P, Glavin R, Maran N, Patey R (2003) Anaesthetists' non-technical skills (ANTS): evaluation of a behavioural marker system. Brit J Anaesth 90:580–588

Gaba DM (1994) Human error in dynamic medical domains. In: Bogner MS (ed) Human error in medicine. Lawrence Erlbaum, Hillsdale, NJ, pp 197–224

Gaba DM (2000) Anaesthesiology as a model for patient safety in health care. Brit Med J 320:785–788

Gawande AA, Zinner MJ, Studdert DM, Brennan TA (2003) Analysis of errors reported by surgeons at three teaching hospitals. Surgery 133:614–621

Gersick C, Hackman JR (1990) Habitual routines in task-performing groups. Organ Behav Hum Dec 47:65–97

Gravenstein JS (2002) Safety in anesthesia. Anaesthesist 51:754–759

Greenberg CC, Regenbogen SE, Studdert DM, Lipsitz SR, Rogers SO, Zinner MJ, Gawande AA (2007) Patterns of communication breakdowns resulting in injury to surgical patients. J Am Coll Surgeons 204:533–540

Grote G, Zala-Mezö E, Grommes P (2003) Effects of standardization on coordination and communication in high workload situations. Linguistische Berichte, Sonderheft 12:127–155

Grote G, Kolbe M, Zala-Mezö E, Bienefeld-Seall N, Künzle B (2010) Adaptive coordination and heedfulness make better cockpit crews. Ergonomics 52:211–228

Gurtner A, Tschan F, Semmer NK, Nägele C (2007) Getting groups to develop good strategies: effects of reflexivity interventions on team process, team performance, and shared mental models. Organ Behav Hum Dec 102:127–142

Hackman JR, Morris CG (1975) Group tasks, group interaction process, and group performance effectiveness: a review and proposed integration. In: Berkowitz L (ed) Advances in experimental social psychology. Academic, New York, pp 45–99

Healey AN, Undre S, Vincent CA (2004) Developing observational measures of performance in surgical teams. Qual Saf Health Care 13:33–40

Hindmarsh J, Pilnick A (2002) The tacit order of teamwork: collaboration and embodied conduct in anaesthesia. Sociol Quart 43:139–164

Ilgen DR, Hollenbeck JR, Johnson M, Jundt D (2005) Teams in organizations: from input-process-output models to IMOI models. Annu Rev Psychol 56:517–543

Kerr S, Jermier JM (1978) Substitutes for leadership: their meaning and measurement. Organ Behav Hum Perf 22:375–403

Klein KJ, Ziegert JC, Knight AP, Xiao Y (2006) Dynamic delegation: shared, hierarchical, and deindividualized leadership in extreme action teams. Admin Sci Quart 51:590–621

Kolbe M, Boos M (2009) Facilitating group decision-making: Facilitator's subjective theories on group coordination Forum Qualitative Sozialforschung/Forum: Qualitative Social Research 10: http://nbn-resolving.de/urn:nbn:de:0114-fqs0901287

Kolbe M, Künzle B, Zala-Mezö E, Wacker J, Grote G (2009) Measuring coordination behaviour in anaesthesia teams during induction of general anaesthetics. In: Flin R, Mitchell L (eds) Safer surgery: analysing behaviour in the operating theatre. Ashgate, Aldershot, UK, pp 203–221

Kolbe M, Künzle B, Zala-Mezö E, Wacker J, Spahn DR, Grote G (under review) Adaptive coordination during emergency situations: Is implicit or explicit more effective?

Künzle B (2003) Teamprozesse und Koordination im Anästhesieteam [Team processes and coordination in the anaesthesia team]. University of Zurich, Lizenziatsarbeit

Künzle B, Kolbe M, Grote G (2010) Ensuring patient safety through effective leadership behaviour: a literature review. Safety Sci 48:1–17

Künzle B, Zala-Mezö E, Kolbe M, Wacker J, Grote G (in press) Substitutes for leadership in anaesthesia teams and their impact on leadership effectiveness. Eur J Work Organ Psy

Leedal JM, Smith AF (2007) Methodological approaches to anaesthetists' workload in the operating theatre. Brit J Anaesth 94:702–709

Lim B-C, Klein K-J (2006) Team mental models and team performance: a field study of the effects of team mental model similarity and accuracy. J Organ Behav 27:403–418

Lingard L, Espin S, Whyte S, Regehr G, Baker GR, Reznick R, Bohnen J, Orser B, Doran D, Grober E (2004) Communication failures in the operating room: an observational classification of recurrent types and effects. Qual Saf Health Care 13:330–334

MacMillan J, Entin EE, Serfaty D (2004) Communication overhead: the hidden cost of team cognition. In: Salas E, Fiore SM (eds) Team cognition: understanding the factors that drive process and performance. American Psychological Association, Washington, DC, pp 61–82

Manser T (2009) Teamwork and patient safety in dynamic domains of healthcare: a review of the literature. Acta Anaesth Scand 53:143–151

Manser T, Howard SK, Gaba DM (2008) Adaptive coordination in cardiac anaesthesia: a study of situational changes in coordination patterns using a new observation system. Ergonomics 51:1153–1178

Manser T, Harrison TK, Gaba DM, Howard SK (2009a) Coordination patterns related to high clinical performance in a simulated anesthetic crisis. Anesth Analg 108:1606–1615

Manser T, Howard SK, Gaba DM (2009b) Identifying characteristics of effective teamwork in complex medical work environments: adaptive crew coordination in anaesthesia. In: Flin R, Mitchell L (eds) Safer surgery: analysing behaviour in the operating theatre. Ashgate, Aldershot, UK, pp 223–239

Marks MA, Mathieu JE, Zaccaro SJ (2001) A temporally based framework and taxonomy of team processes. Acad Manage Rev 26:356–376

Marks MA, Sabella MJ, Burke CS, Zaccaro SJ (2002) The impact of cross-training on team effectiveness. J Appl Psychol 87:3–13

Marsch SCU, Müller C, Marquardt K, Conrad G, Tschan F, Hunziker PR (2004) Human factors affect the quality of cardiopulmonary resuscitation in simulated cardiac arrests. Resuscitation 60:51–56

Mathieu JE, Marks MA, Zaccaro S-J (2001) Multi-team systems. In: Anderson N, Ones D, Sinangil HK, Viswesvaran C (eds) International handbook of work and organizational psychology. Sage, London, pp 289–313

McGrath JE, Tschan F (2004) Temporal matters in social psychology. Examining the role of time in the lives of groups and individuals. American Psychological Association, Washington, DC

Mesmer-Magnus JR, DeChurch LA (2009) Information sharing and team performance: a meta-analysis. J Appl Psychol 94:535–546

Michinov E, Olivier-Chiron E, Rusch E, Chiron B (2008) Influence of transactive memory on perceived performance, job satisfaction and identification in anaesthesia teams. Brit J Anaesth 100:327–332

Nyssen AS, Hansez I, Baele P, Lamy M, De Keyser V (2003) Occupational stress and burnout in anaesthesia. Brit J Anaesth 90:333–337

Oken A, Rasmusson MD, Slagle JM, Jain S, Kuykendall T, Ordonez N, Weinger MB (2007) A facilitated survey instrument captures significantly more anesthesia events than does traditional voluntary event reporting. Anesthesiology 107:909–922

Orasanu JM (1993) Decision-making in the cockpit. In: Wiener EL, Kanki BG, Helmreich RL (eds) Cockpit ressource management. Academic, San Diego, pp 137–172

Phipps D, Meakin GH, Beatty PCW, Nsoedo C, Parker D (2008) Human factors in anaesthetic practice: insights from a task analysis. Brit J Anaesth 100:333–343

Pulakos ED, Arad S, Donovan MA, Plamondon KE (2000) Adaptability in the workplace: development of a taxonomy of adaptive performance. J Appl Psychol 85:612–624

Rico R, Sánchez-Manzanares M, Gil F, Gibson C (2008) Team implicit coordination processes: a team knowledge-based approach. Acad Manage Rev 33:163–184

Rosen MA, Salas E, Wilson KA, King HB, Salisbury ML, Augenstein JS, Robinson DW, Birnbach DJ (2008) Measuring team performance in simulation-based training: adopting best practices for healthcare. Simul Healthcare 3:33–41

Rousseau V, Aubé C, Savoie A (2006) Teamwork behaviors: a review and an integration of frameworks. Small Group Res 37:540–570

Salas E, Bowers C, Edens E (2001) Improving teamwork in organizations: application of resource management training. Laurence Erlbaum, Mahwah, NJ

Salas E, Nichols DR, Driskell JE (2007a) Testing three team training strategies in intact teams. A meta-analysis. Small Group Res 38:471–488

Salas E, Rosen MA, King H (2007b) Managing teams managing crisis: principles of teamwork to improve patient safety in the emergency room and beyond. Theor Issues Ergon Sci 8:381–394

Serfaty D, Entin EE (2002) Team adaptation and coordination training. In: Flin R, Salas E, Strub M, Martin C (eds) Decision making under stress: emerging themes and applications. Ashgate, Alderhot, UK, pp 170–184

Serfaty D, Kleinman DL (1990) Adaptation processes in team decisionmaking and coordination. In: Proceedings of the IEEE International Conference on Systems, Man and Cybernetics. IEEE, Los Angeles, pp 394–395

Sexton JB, Thomas EJ, Helmreich RL (2000) Error, stress, and teamwork in medicine and aviation: cross sectional surveys. Brit Med J 320:745–749

Smith-Jentsch KA, Mathieu JE, Kraiger K (2005) Investigating linear and interactive effects of shared mental models on safety and efficiency in a field setting. J Appl Psychol 90:523–535

Stachowski AA, Kaplan SA, Waller MJ (2009) The benefits of flexible team interaction during crisis. J Appl Psychol 94:1536–1543

Stasser G, Titus W (1985) Pooling of unshared information in group decision making: biased information sampling during discussion. J Pers Soc Psychol 48:1467–1578

Steiner ID (1972) Group processes and productivity. Academic, New York

Stroebe W, Frey BS (1982) Self-interest and collective action: the economics and psychology of public goods. Brit J Soc Psychol 21:121–137

Sundstrom E, de Meuse KP, Futrell D (1990) Work teams. Applications and effectiveness. Am Psychol 45:120–133

Thuilliez H, Anceaux F, Hoc J-M (2005) Rôle de l'opérateur et du statut fonctionnel des informations lors de la prise d'informations en anesthésie [in French]. Le Travail Humain 68:225–252

Toups ZO, Kerne A (2007) Implicit coordination in firefighting practice: design implications for teaching fire emergency responders. In: Proceedings of the SIGCHI Conference on Human Factors in Computing Systems. ACM Press, New York, pp 707–716

Tschan F, Semmer NK, Gautschi D, Hunziker P, Spychiger M, Marsch SU (2006) Leading to recovery: group performance and coordinative activities in medical emergency driven groups. Hum Perform 19:277–304

Wacker J, Kleeb B, Künzle B, Leisinger E, Kobler A, Kolbe M, Manser T, Burtscher M, Spahn DR, Grote G (2008) High incidence of non-routine events during standard anesthesia inductions: a prospective analysis of synchronized video and vital parameter recordings. Annual Conference of the Swiss Society of Anaesthesiology and Reanimation, Fribourg, Switzerland

Waller MJ (1999) The timing of adaptive group responses to nonroutine events. Acad Manage J 42:127–137

Waller MJ, Uitdewilligen S (2008) Talking to the room. Collective sensemaking during crisis situations. In: Roe RA, Waller MJ, Clegg SR (eds) Time in organizational research. Oxford, Routledge, pp 186–203

Webber SS, Klimoski RJ (2004) Crews: a distinct type of work team. J Bus Psychol 18:261–279

Weinger MB, Englund CE (1990) Ergonomic and human factors affecting anesthetic vigilance and monitoring performance in the operating room environment. Anesthesiology 73:995–1021

Weinger MB, Slagle J (2002) Human factors research in anesthesia patient safety: techniques to elucidate factors affecting clinical task performance and decision making. J Amer Med Assoc 9:58–63

Wittenbaum GM, Stasser G, Merry CJ (1996) Tacit coordination in anticipation of small group task completion. J Exp Soc Psychol 32:129–152

Wittenbaum GM, Vaughan SI, Stasser G (1998) Coordination in task-performing groups. In: Tindale RS, Heath L, Edwards J, Posavac EJ, Bryant FB, Suarez-Balcazar Y, Henderson-King E, Myers J (eds) Theory and research on small groups. Plenum, New York, pp 177–204

Wittenbaum GM, Hollingshead AB, Paulus PB, Hirokawa RY, Ancona DG, Peterson RS, Jehn KA, Yoon K (2004) The functional perspective as a lens for understanding groups. Small Group Res 35:17–43

Xiao Y, Seagull FJ, Mackenzie CF, Klein K (2004) Adaptive leadership in trauma resuscitation teams: a grounded theory approach to video analysis. Cogn Tech Work 6:158–164

Zala-Mezö E, Künzle B, Wacker J, Grote G (2004) Zusammenarbeit in Anästhesieteams aus Sicht der Teammitglieder (Cooperation in anesthesia teams as seen by team members). Z Arbeitswiss 58:199–207

Zala-Mezö E, Wacker J, Künzle B, Brüesch M, Grote G (2009) The influence of standardisation and task load on team coordination patterns during anaesthesia inductions. Qual Saf Health Care 18:127–130

Zingg U, Zala-Mezö E, Künzle B, Licht A, Metzger U, Grote G, Platz A (2008) Evaluation of critical incidents in general surgery. Brit J Surg 95:1420–1425

Chapter 6
Developing Observational Categories for Group Process Research Based on Task and Coordination Requirement Analysis: Examples from Research on Medical Emergency-Driven Teams

Franziska Tschan, Norbert K. Semmer, Maria Vetterli, Andrea Gurtner, Sabina Hunziker, and Stephan U. Marsch

Abstract In this chapter, we argue that the task is an important influence for teams and that task aspects should be more explicitly, and more specifically, included in the study of team processes and team performance. Using a cardiopulmonary resuscitation task as an example, we show how an adaptation of hierarchical task analysis that assesses task requirements (taskwork) and coordination requirements (teamwork) can be useful in identifying a task's goals and sub-goals, defining qualifiers of good goal attainment, identifying coordination requirements, and developing hypotheses about which teamwork and coordination behaviour should specifically be related to the performance of different aspects of complex tasks. Our argument is based on concepts that extend the general input–process–output model of groups.

F. Tschan (✉) and M. Vetterli
University of Neuchâtel, Institut de Psychologie du Travail et des Organisations, Rue Emile Argand 11, 2000 Neuchâtel, Switzerland
e-mail: franziska.tschan@unine.ch; maria.vetterli@unine.ch

N.K. Semmer
University of Berne, Institute of Psychology, Muesmattstrasse 45, 3000 Bern 9, Switzerland
e-mail: norbert.semmer@psy.unibe.ch

A. Gurtner
Applied University of Berne, Berner Fachhochschule, Fachbereich Wirtschaft und Verwaltung, Morgartenstrasse 2c, 3014 Bern, Switzerland
e-mail: andrea.gurtner@bfh.ch

S. Hunziker and S.U. Marsch
Departement für Innere Medizin, University Hospital of Basel, Abteilung für Intensivmedizin, Kantonsspital, 4031 Basel, Switzerland
e-mail: smarsch@uhbs.ch

M. Boos et al. (eds.), *Coordination in Human and Primate Groups*,
DOI 10.1007/978-3-642-15355-6_6, © Springer-Verlag Berlin Heidelberg 2011

6.1 Introduction

Imagine the following situation: A physician talks with a patient resting in the recovery room after a small surgical intervention related to his heart condition. The surgery went well. The doctor controls the patient's vital signs and chats with him about an upcoming soccer game they both want to watch on television. Suddenly, the patient states that he is feeling dizzy; immediately thereafter, the patient suffers a sudden cardiac arrest, clearly visible on the surveillance monitor. The physician sounds the alarm, and by the time two other physicians rush into the room, she has already started cardiopulmonary resuscitation. She informs her colleagues that this is a cardiac arrest situation, and the three of them continue resuscitation, a complex task that is ideally performed in groups of three or four people. It is an emergency situation that has to be carried out under a lot of time pressure, as every minute of untreated cardiac arrest diminishes survival chances of a patient by 7–10% (von Planta 2004).

Although all physicians undergo regular resuscitation training, previous research has revealed important performance shortcomings of cardiopulmonary resuscitation, even when performed by well-trained hospital staff (Abella et al. 2005; Ravakhah et al. 1998). These shortcomings are often related to coordination and collaboration problems (Marsch et al. 2004a, b). Thus, it seems important to analyse what hinders or enhances the performance of these teams. The authors of this chapter are an interdisciplinary team of researchers (psychologists and physicians) who collaborate in studying teams of physicians and medical students confronted with complex medical problems. For this research, we use a high-fidelity patient simulator and video-tape processes of the medical teams as they perform the tasks. The overall goal of our research is to evaluate what influences team performance in emergency medical situations in order to help craft better training methods for such teams (Hunziker et al. 2010). To achieve this goal, we need good methods to assess group performance, and we need to identify teamwork and coordination behaviour that influence performance on the tasks we study. As we study complex tasks that are performed in groups of medical specialists, we need methods applicable for the analysis of performance as well as of coordination requirements of complex tasks. In this chapter we will show how and why task analysis can be an important help in studying groups.

There are a few instruments for group process analysis that are conceived as generic instruments applicable to a wide range of tasks. Their main advantage is that they suggest common categories for coding behaviour of a wide range of groups, permitting comparisons across groups and tasks. Their main disadvantages are that they either contain many categories, that may not all be of interest for specific research questions of categories for many applications (e.g. Kauffeld et al. 2009) or that the behavioural categories observed are defined in very general a way (Bales 1950; Futoran et al. 1989), limiting their application to specific tasks. Given this dilemma, in publications on current practices of group observation and analysis methods, method specialists emphasize that there is no agreed set of categories for

6 Developing Observational Categories for Group Process Research Based on Task 95

team behaviour observed in groups and suggest that researchers should choose observational categories according to their specific research question (e.g. Brett et al. 2004; McGrath and Altermatt 2001; Weingart 1997). The choice of behavioural categories may be difficult, however, especially for complex tasks. In this chapter we contribute to this issue by emphasizing the utility and importance of assessing task requirements through detailed task analysis in order to develop observational categories and to assess group performance and group coordination, especially for complex tasks.

The chapter is structured as follows. First, we present the most important extensions of the general input–process–output (IPO) model of groups. Researchers who have suggested refining the IPO model advocate division of the general group process into smaller phases, episodes, or cycles (see Chap. 2). After presenting their arguments, we will show that knowledge about the tasks involved can be particularly helpful in identifying such phases. We then present hierarchical task analysis (HTA) (e.g. Shepherd 1998) as a way of disentangling and describing sub-tasks of the defining team task and its coordination requirements. Referring back to our example of the resuscitation task described above,[1] we will demonstrate a simplified HTA of this task. In the fourth section, we will argue that, for many tasks, performance should not solely be defined in terms of results or output, but in terms of process performance markers. Again, we will show how task analysis can help to define process performance markers, and we will illustrate the usefulness of performance markers in our own research. In the fifth section, we will show the usefulness of HTA for deciding which teamwork behaviours may be particularly important at a particular moment or phase in the group process. On this basis, we suggest that it is possible to develop hypotheses for *predicting* group performance. The chapter ends with a general conclusion.

6.2 Extensions of the General Input–Process–Output Model: Phases, Episodes, and Cycles

The well-known input–process–output (IPO) model of group performance suggests that *input* factors, such as group composition or the group's environment, influence the *group process* (taskwork and teamwork/cooperation) and that input as well as process variables influences the group's *outcome* (e.g. performance) (Hackman and Morris 1975; McGrath 1964). The generic IPO model thus distinguishes among "before the group has started working" as the input, "while the group

---

[1]The task-focused approach to group process analysis presented here is obviously not restricted to the cardiopulmonary resuscitation task. We will nevertheless use this task to illustrate how process performance markers can be developed and how the analysis of task requirements can be helpful in relating specific behaviour to performance. Examples referring to other tasks will also be mentioned.

interacts" as the process, and "after the group is done" as the output (note: the inclusive model presented in Chap. 2 overcomes this restrictive before- vs. during-process design. At first glance, this model seem to imply that a researcher who is interested in group performance can first measure or even manipulate input variables (for example, group composition, member experience, etc.). He or she will then assess the process of the group (for example, by observing the group members' behaviours) and then relate input and/or process variables (as mediators, moderators, or both) to the result, measured as the product after the group has finished working. One of the main advantages of the IPO model is indeed that it draws attention to and therefore emphasizes the group process and its relationship with productivity. Small group research has tended, and still tends, to neglect the analysis of the group process itself (McGrath and Altermatt 2001; Moreland et al. 2010; Weingart 1997).

Although focusing on the group process is very important, a number of questions arise with regard to its assessment. Many studies focus on process variables in general, such as the amount of communication, planning, and reflecting. Such an approach makes the assumption, at least implicitly, that these group processes are important throughout the whole time the group works on a task, and for all aspects of the task. This is a questionable assumption even for well-circumscribed tasks. Some process aspects, such as planning, typically are important at the beginning and at specific "turning points" (Hackman and Wageman 2005; Marks et al. 2001; Waller 1999); some, such as evaluation, are important throughout, but only after some action has been carried out (Tschan 1995, 2002); some are important only for specific sub-tasks (e.g. in an air traffic control task, planes identified as foes require different observation behaviour than do planes identified as friends; Tschan et al. 2000). Assessing behaviours in terms of frequencies over the entire process may therefore in many cases not capture the important aspects and may even yield misleading results.

The problems of overall process measures are even more pronounced in teams. Teams have a longer existence than the ad hoc groups often studied in group research. Multiple tasks and thus multiple processes, dynamic changes in tasks, as well as changes in input factors (e.g. in team membership, member competences, etc.) are very common for teams. Furthermore, a team may begin a new task while still continuing another one, and at the same time may terminate yet another task. Teams may thus have to coordinate between different tasks, and progress on one task may influence the work on another task. As different processes occur at the same time, it is difficult to analyse "the" process. Furthermore, as a team progresses on one task, team members may learn or change their attitudes, or members may leave or join a team, so that input factors may change as well. For teams, the boundaries of input, process, and output may be especially difficult. It is therefore not surprising that it is the domain of team research in which authors have critically discussed limitations of a general IPO model and have suggested important extensions (Antoni and Hertel 2009; Arrow et al. 2000; Ilgen et al. 2005; Marks et al. 2001; McGrath 1991). For these authors, the realities of teamwork – multiple tasks and changes in input factors – are the background to the argument that it may not be easy, or even inappropriate, to study teams according to an overall input–process–output framework. Thus, many researchers have recommended approaching the overall group process with more fine-grained analyses (Arrow et al. 2000; Marks et al. 2001; McGrath 1991; Tschan 1995; Weingart 1992, 1997).

Researchers who have extended the IPO models have suggested different conceptualizations of the overall group process. For example, Ilgen et al. (2005) adopted a temporal structure and distinguish several phases of the group's increasing experience (forming, functioning, finishing). They then related core topics to each of the stages. In the forming stage, trust, planning, and structuring are primordial, whereas in the functioning stage, bonding, adapting, and learning may be core concerns. Marks et al. (2001) suggested dividing the overall group process into temporal cycles of goal-directed activities they call *episodes*. Episodes are "IPO-type" micro-cycles and are defined as action-feedback cycles that are preceded and followed by periods of transitions between tasks, similarly to cycles described by Tschan (1995). Finally, McGrath and Tschan (2004) distinguished three hierarchical-temporal levels related to the overall group process: (1) at the purpose or project level, where the group selects, accepts, or modifies the group's projects; (2) at the planning level, where the group structures the process (what will be done, when, by whom and how); and (3) at the action level, where the process consists of a series of interrelated "orient–enact–monitor– modify" cycles. The cycles are related to the different goals or sub-tasks the group has to carry out. The cycle concept is similar to the concept of episodes and transitions suggested by Tschan (1995), and by Marks et al. (2001), described above. The three hierarchical levels (project, planning, and action) constitute a general temporal pattern similar to the one described by Ilgen et al. (2005). Thus, project choice and planning are more likely in the earlier stages (i.e. the forming stage), and the action level corresponds to the functioning stage.

All of these refinements of the general IPO model implicitly or explicitly assume that group goals (or group tasks that are defined by goals; see below) are an important influence on the overall group process: Episodes (Marks et al. 2001) and cycles (McGrath and Tschan 2004) are both defined as related to the group's tasks and goals; and observable episodes, transition, or cycles will depend on these goals and tasks. Based on similar considerations, the important influence of task requirements in groups for group process research has been widely acknowledged (Arrow et al. 2000; Cannon-Bowers et al. 1995; Hackman and Morris 1975; McGrath et al. 2000; Tschan and von Cranach 1996).

As group tasks and goals can be very different, they may require different types of episodes or cycles. It is thus necessary to develop a good understanding of the specific goals, sub-goals, and behavioural requirements of the tasks with which groups are confronted. One of the methods used to describe tasks as structures of goals and sub-goals that are hierarchically nested and temporally related is hierarchical task analysis (Annett 2004; Shepherd 1985), which will be presented next.

6.3 Task Analysis of Team Tasks

A task analysis that describes the steps and behavioural requirements necessary to fulfil a task is very useful for a more precise analysis of the relationship between teamwork behaviour and group performance. There are many different procedures for analysing tasks (cf. Konradt et al. 2006). In our research (Gurtner et al. 2007; Tschan 1995, 2002; Tschan et al. 2000) we have used a simplified version of the hierarchical task analysis (HTA) initially developed by Annett and Shepherd (Annett 2004; Shepherd 2001). This method describes tasks in terms of executable goals and is therefore well suited as an approach to the analysis of group behaviour.

HTA describes tasks as a set of steps to be carried out or goals that have to be achieved. Typically, a goal can be divided into sub-goals. HTA therefore describes tasks as a hierarchical structure containing *general goals as well as sub-goals that are related to the general goals*. It also specifies when a goal is attained: For each goal and sub-goal, *the criteria for judging goal attainment* are listed; these normally include qualifiers (e.g. time, correctness, etc.) of good goal attainment. For each goal, HTA specifies in what order (if any) sub-tasks have to be carried out as well as other conditions of goal attainment: For some tasks, a sub-goal can only be started after another has been finished (sequential requirements); for other goals, several sub-goals have to be pursued in a coordinated manner.

The description of the task in terms of a structure of goals and sub-goals, the criteria of goal attainment, and the description of the conditions for sub-goals are part of a classical HTA, which is often used to analyse tasks carried out by individuals. More recently, Annett et al. (2000) extended HTA to the analysis of group tasks. In accordance with recent concepts (Bowers et al. 1997; Marks et al. 2001; Salas et al. 2005) that distinguish between taskwork (what the team does) and teamwork (how the team coordinates its actions), they assess not only the task goals and sub-goals, but also the *teamwork or coordination requirements* related to each goal or sub-goal. In fact, Shepherd (2001) suggested that some teamwork requirements (for example "inform person x") can be seen as goals in themselves and can be included as separate goals in the system.

The main principles of HTA are relatively simple. As tasks become complex, however, conducting full-fledged HTA can become rather difficult. To perform HTA, the analyst has to know the task well. Classical HTA thus uses observation techniques, expert interviews, and document analysis as a basis.

We provide as an example a simplified version of HTA(one that does not use the elaborate notation of the original system and omits some sub-goals) for the resuscitation task. For this task, extended documentation is available, which is based on research that assessed which actions provide the highest chances of patient survival and recovery. The general guidelines for "advanced cardiopulmonary resuscitation" for medical professionals are regularly updated as new research becomes available. We base the task analysis on the European resuscitation guidelines

6 Developing Observational Categories for Group Process Research Based on Task 99

(Nolan et al. 2005; von Planta 2004), adapting it to our specific research context that involves a patient simulator.[2]

As described in the introduction, the "patient" in the simulator suffers from a cardiac arrest in the presence of a medical professional. We programmed the mannequin to display "ventricular fibrillation". At the beginning of a cardiac arrest, the heart often shows rapid electrical activity, which is, however, not synchronized enough to trigger a coordinated contraction of the heart muscle. Such electrical activity is called ventricular fibrillation or ventricular tachycardia. If the patient is connected to a surveillance monitor displaying heart activities, ventricular fibrillation is clearly recognizable by the trained physician. If a patient has this condition, defibrillation (application of electrical countershocks with the use of two panels placed on the chest of the patient) may help to restart synchronized cardiac activity and thus restore the heartbeat.

The main goals of cardiopulmonary resuscitation (see Fig. 6.1) are (1) diagnosing the cardiac arrest, (2) oxygenating the brain, and (3) attempting to reestablish spontaneous circulation. Each of the main goals can be further broken down into sub-goals. Cardiopulmonary resuscitation is best carried out by a group. On the most general level, teamwork and coordination requirements can be specified as the need to establish a shared mental model of the situation and the intervention, and to assign tasks to people.

The first goal, "Diagnose the cardiac arrest", contains three sub-goals: 1-1 confirms the absence of a pulse; 1-2 confirms the absence of breathing; and 1-3 confirms the loss of consciousness. Proper diagnosis has to be established before the group can start on goals 2 and 3. Note that the guidelines for advanced cardiopulmonary resuscitation (Nolan et al. 2005) allow only 10 s for the medical professional to diagnose a cardiac arrest, because it is essential to start cardiopulmonary resuscitation very quickly. Teamwork requirements for the diagnosis are that all team members have to be made aware of the diagnosis "cardiac arrest", because this knowledge should trigger the behavioural script "resuscitation". This script can be seen as an individually stored "shared mental model" that contains the most important aspects of the resuscitation procedure. As all medical professionals have received training in resuscitation, one can assume that the script is available. Given the high time pressure in the diagnostic phase, another important teamwork requirement is to terminate the diagnostic phase rapidly and move into the intervention phase in a coordinated fashion.

The second goal (oxygenating the brain) has three sub-goals. The first sub-goal, 2-1 "open airways", contains the sub-sub-goals "checking the mouth for foreign body" and "removing visible obstructions in the mouth" (not shown in Fig. 6.1). Sub-goal 2-2 is "ventilate" (providing oxygen to the lungs) and 2-3 "cardiac massage" (which substitutes for circulation and transports the oxygenated blood to the brain). Sub-goals

---

[2]In our simulator setting, (1) the patient was branched on a heart surveillance monitor, which facilitates diagnosis; (2) an intravenous line was already established; (3) the patient showed ventricular fibrillation; and (4) differential diagnostics was not required, because the simulator mannequin was programmed to wake up after proper resuscitation.

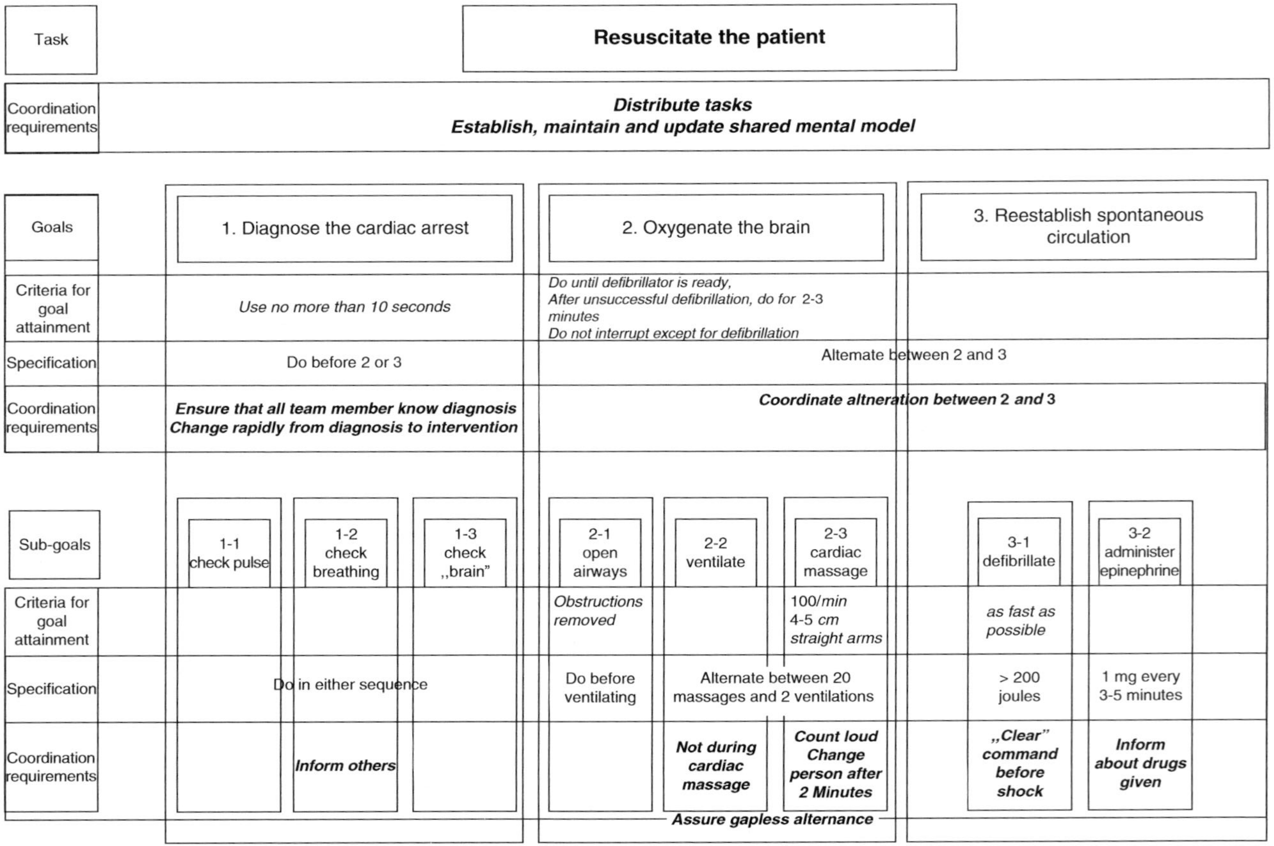

Resuscitate the patient

Distribute tasks
Establish, maintain and update shared mental model

	1. Diagnose the cardiac arrest	2. Oxygenate the brain	3. Reestablish spontaneous circulation
Task			
Coordination requirements			
Goals	1. Diagnose the cardiac arrest	2. Oxygenate the brain	3. Reestablish spontaneous circulation
Criteria for goal attainment	*Use no more than 10 seconds*	*Do until defibrillator is ready. After unsuccessful defibrillation, do for 2-3 minutes. Do not interrupt except for defibrillation*	
Specification	Do before 2 or 3	Alternate between 2 and 3	
Coordination requirements	*Ensure that all team member know diagnosis. Change rapidly from diagnosis to intervention*	*Coordinate altneration between 2 and 3*	

	1-1 check pulse	1-2 check breathing	1-3 check "brain"	2-1 open airways	2-2 ventilate	2-3 cardiac massage	3-1 defibrillate	3-2 administer epinephrine
Sub-goals	1-1 check pulse	1-2 check breathing	1-3 check "brain"	2-1 open airways	2-2 ventilate	2-3 cardiac massage	3-1 defibrillate	3-2 administer epinephrine
Criteria for goal attainment				*Obstructions removed*		*100/min 4-5 cm straight arms*	*as fast as possible*	*1 mg every 3-5 minutes*
Specification		Do in either sequence		Do before ventilating	Alternate between 20 massages and 2 ventilations		> 200 joules	
Coordination requirements		*Inform others*			*Not during cardiac massage*	*Count loud. Change person after 2 Minutes*	*"Clear" command before shock*	*Inform about drugs given*

Assure gapless alternance

Fig. 6.1 Simplified hierarchical task analysis for the cardiopulmonary resuscitation task, including coordination requirements

2 and 3 are closely linked. The guidelines specify an alternate sequence of 30 chest compressions followed by two ventilations. The guidelines also specify criteria for good goal attainment: Cardiac massage has to be done at a rhythm of about 100 beats per minute and at a depth of about 4–5 cm. For patients who are not intubated,[3] cardiac massage and manual ventilation should not be done at the same time, but rather in an alternating but gapless sequence, which results in the coordination requirement to alternate between the person performing chest compressions and the person who ventilates. This coordination can be achieved better if the person performing cardiac massage signals when he or she has finished the 30 compressions; the recommendation is thus to count each compression out loud.[4] Sub-goal 3 (reestablishing spontaneous circulation) has two sub-goals: 3-1 defibrillation (applying electrical countershocks to convert the ventricular fibrillation to a regular heartbeat) and 3-2 administering epinephrine (adrenaline), a drug that constricts the vessels and increases pressure and can thus improve the effectiveness of defibrillation and cardiac massage. Coordination requirements for defibrillation are important, as during defibrillation all helpers should stay away from the patient or the bed, because the electric shock applied could harm bystanders. Thus, before defibrillation, a "clear" command should be given to ensure that none of the helpers touches the patient or bed. Administering epinephrine entails the coordination requirement of informing the group when the drug is given to update the group members' mental model.

The resuscitation guidelines also specify the temporal sequences and conditions to change from goal 2 (oxygenating the brain) to goal 3 (reestablishing spontaneous circulation): first, goal 2 has to be pursued until the defibrillator is ready; after unsuccessful defibrillation, 2 min of ventilation–cardiac massage cycles should be performed and epinephrine should be administered before the next defibrillation. Coordination requirements involve ensuring that the group keeps track of time and changes in a coordinated way between goals 2 and 3. Again, note that we have presented a simplified version here; many more specifications for this task are given in the guidelines.

Task analysis can be performed on different levels of specificity (from only a few general goals to a very elaborate system of goals and sub-goals), and it can be done for very different types of tasks. The resuscitation example describes a team task that usually lasts less than 30 min, but HTA is also suitable for broader and more complex activities. For example, Shepherd (2001, p. 124ff) provides an analysis of the main nursing tasks in a hospital ward, as well as several examples of management tasks (p. 126ff). Thus, HTA is not restricted to "hands-on tasks" but can also be used for the analysis of more cognitive team tasks, such as decision making or problem solving. In the next section, we show how the results of HTA can be helpful in developing process performance measures.

---

[3]Placement of a tube to allow artificial ventilation of a patient.

[4]For advanced life support, intubation (introducing a tube into the trachea of the patient to facilitate ventilation) is suggested as a more efficient way to ventilate the patient. If this is done, cardiac massage and ventilation no longer need to be alternated.

6.4 Assessing Process Performance Measures Based on Task Analysis

The general IPO model of group performance can lead to the assumption that group performance is best measured as the result available once the group has finished its work. Conceptualizations of work performance differentiate, however, between outcome and process (behavioural) aspects of performance (Sonnentag and Frese 2002) and see performance as a multidimensional construct (Campbell et al. 1993). Output aspects of performance are measures of performance effectiveness (how well or to what degree a goal is attained), whereas process aspects of performance can be related to efficiency measures of performance, where the output is related to the resources needed to achieve them (Pritchard and Watson 1992).

For many tasks, outcome performance is not the only, or the most important, aspect of task performance. For example, although the outcome can be that a crew landed a plane safely, crew performance also requires that the pilots descend smoothly on the correct area of the tarmac and perform the correct tasks in the right sequence during the landing approach. Although patient well-being and recovery are crucial outcome measures for judging the quality of surgery, it is also important that during surgery the right procedures are chosen and performed in a timely manner. Process measures of performance are also important because the relationship between process and performance is often less than perfect (Boos 1996), and many aspects apart from process parameters may be important for an outcome. If this imperfect association between process performance and outcome performance is neglected, a positive outcome may erroneously be taken as "proof" that a person, or a team, has acted perfectly.

The necessity to distinguish between process performance measures and outcome measures can be illustrated very well by the cardiopulmonary resuscitation task. One could be tempted to measure resuscitation performance as the outcome "patient survival", as the main goal is to save the patient's life. However, it would be erroneous to assume that if the patient survives, group performance was optimal and, if the patient dies, group performance was bad. Survival of a cardiac arrest depends on many factors other than the performance of the resuscitation team. An analysis of patient survival after in-hospital cardiac arrest has shown that physiological (the underlying problem) and demographic (e.g. age) aspects of patients are much more important predictors of cardiac arrest survival than the timeliness of starting the resuscitation or group skills (Cooper and Cade 1997). Although coordination quality does significantly contribute to the outcome, there is a substantial chance that a patient will die, even if the team does everything "right". In this and many similar cases, relying solely on output performance will not lead to the most valid assessment of group performance.

Group performance, be it process or outcomes measures, is particular to each task, and generalizations across tasks are normally not appropriate (Mathieu et al. 2008). One has thus to develop and define performance measures separately for each task. Task analysis can be very useful here, because it already specifies the

6 Developing Observational Categories for Group Process Research Based on Task 103

Table 6.1 Example of process performance measures for the resuscitation task

Goal	Process performance measure	Possible operationalization
The diagnosis "cardiac arrest" should be done rapidly (goal 1)	Time until diagnosis is called	Coding time between onset of cardiac arrest and a person calling it a cardiac arrest
Resuscitation should be started as soon as possible (transition goal 1 to goals 2 and 3)	Time until the first meaningful intervention is started	Coding of time from cardiac arrest until any one of the following actions: chest compression; ventilation; defibrillation
There should be continuous, uninterrupted oxygenation of the brain (goal 2)	Time until ventilation-cardiac massage starts	Coding time between onset of cardiac arrest and first ventilation or cardiac massage
Cont. goal 2	Percentage of "hands-on" time (uninterrupted ventilation/ cardiac massage), of total time, excluding the episodes during which hands-on is not suitable (during defibrillation)	Second-by-second coding of hands-on time (yes-no); calculating the percentage of hands-on time in whole process
Cont. goal 2	Unnecessary interruptions of ventilation and cardiac massage	Assessing all interruptions of behaviour longer than 5 s, code if necessary or not
Remove potential obstructions in airways (2-1)	Visual control of mount and remove of potential airway obstructions	Dummy-code: yes, if a group member controls mouth visually or with fingers
Appropriate ventilation-cardiac massage cycles (2-2 and 2-3)	No overlapping ventilation – cardiac massage	Behaviour coding – instances cardiac massage and ventilation overlap (recoded)
Cont. goal 2-2/2-3	30:2 cycles	Overall behaviour rating (yes/ partially/no)
Technical aspects of cardiac massage (2-3)	Chest compression rate of 100 p min Depth Arm position	Behaviour coding, rating
Attempt to establish heartbeat (3)		
Defibrillate as soon as defibrillator is available (3-1)	Time elapsed until first defibrillation	Time between cardiac arrest and first defibrillation
Alternate between defibrillations and oxygenating the brain	Number of alternations Time of ventilation-cardiac massage between defibrillations	General rating
Use correct defibrillation strength	More than 200 joules	Note all defibrillations with correct strength; calculate percentage of correct defibrillations (behavioural observation)
Administer epinephrine	Correct dosage	Dummy coding, based on communication

Note: In our scenarios, the defibrillator is always at the same place; otherwise, the time needed to transport the defibrillator would have to be subtracted criteria for good goal and sub-goal attainment, which often allows process performance measures to be derived.

6.4.1 Developing Process Performance Measures for the Cardiopulmonary Resuscitation Task

We will illustrate process performance measures for the resuscitation task (see Table 6.1) used in our research (Hunziker et al. 2009; Lüscher et al. submitted; Marsch et al. 2004a, 2005; Tschan et al. 2006; Vetterli 2006). In the table we refer to the goal or sub-goal of the task analysis in the left column, specify the process performance measure in the middle column, and provide a short description of the operationalization in the right column.

Note that all the process performance measures presented here can be based on the observation of the overt behaviour of group members or communication between them, permitting performance assessment based on video recordings. Our experience has shown that high interrater reliability can be achieved for those codings (kappas between 0.75 and 1), although only after extensive training of the coders. The coders – particularly those who are not medical professionals – first take an online class on cardiopulmonary resuscitation to familiarize themselves with the guidelines and the research behind them to understand the basic resuscitation task. They are then trained on sample tapes based on a coding manual. Typically, a 10- to 15-h investment is necessary until satisfactory reliability of process performance coding is achieved. For time-based codings, good interrater reliability is easier; the other behavioural ratings need more extensive training.

Assessing several measures of process performance raises the question of how to use them. Principally, they could be used separately (and some of them may be omitted depending on the research purpose), or they might be combined into more general indicators or a single performance measure. We mentioned above that theories of performance often regard performance as multidimensional (Pritchard and Watson 1992; Sonnentag and Frese 2002). To the extent that this is true, the different indicators may represent rather different aspects of performance, and a single measure might miss important aspects.

We tested the hypothesis that the process performance markers are not simply different measures of a single overall performance, but represent distinct aspects. We coded four process performance measures in 29 groups of physicians and nurses confronted with a cardiac arrest that developed suddenly during a routine situation (Vetterli et al. 2009). Process performance markers coded in this study were (1) the time elapsed until the first meaningful intervention (transition from goal 1 to goals 2 and 3; see Fig. 6.1); (2) the percentage of hands-on time (goal 2); (3) the time until the first defibrillation (goal 3); and (4) the time until the first resuscitation cycle (goals 1 to 3), including administration of epinephrine, was completed. Table 6.2 shows the intercorrelation between these different process performance measures. Note that we recoded the time measures so that all measures now reflect a higher performance.

6 Developing Observational Categories for Group Process Research Based on Task 105

Table 6.2 Intercorrelations of four process performance measures for the resuscitation task

	1	2	3
1. (reversed) Time until first intervention	–		
2. (reversed) Time until first defibrillation	−0.393*	–	
3. (reversed) Time to complete goals 1–3	−0.274	0.603**	–
4. Percentage of hands-on time	0.275	−0.158	−0.020

Note: $N = 29$ groups
$*p<.05$ $**p<.01$

The results indeed show that the different process performance measures do not represent a global performance construct. Interestingly, a fast "start" in terms of resuscitation does not predict earlier defibrillation; on the contrary, it is associated with later defibrillation. This negative association may be because groups decide to defibrillate early on but neglect the task requirement to engage in cardiac massage and ventilation until the defibrillator is ready. As can be expected, the two performance measures "time to first defibrillation" and "time to complete goals 1–3" are highly intercorrelated. Here, substituting one of these variables with the other is possible. The conclusion that the different performance measures are not simply facets of overall performance is also reinforced by a reliability analysis of the performance measures mentioned in Table 6.2, which yielded a Cronbach's alpha of only 0.21.

This analysis underscores the necessity to assess performance separately for the different goals rather than, or at least in addition to, computing an overall performance measure.

6.4.2 Research Examples Relating Input Factors to Measures of Process Performance

The coding of process performance measures allows the effect of input variables to be tested (e.g. in terms of differences in experience, or in group structure) on group process performance. We will now present two examples of such analyses.

In one study (Hunziker et al. 2009), we manipulated the composition of teams of three physicians at the beginning of the cardiac arrest. The aim of the research project was to investigate whether, and how, the short period of common team experience in preformed teams influences performance in the resuscitation task as compared with the ad hoc teams. Half of the teams were thus "preformed" (condition 1); all three physicians were with the patient 2 min before the cardiac arrest started, and all physicians witnessed the cardiac arrest. For the other half of the teams (condition 2), only one physician witnessed the cardiac arrest, and he or she asked the other two physicians, who waited in the hallway, to join in and help when the emergency situation started. The latter situation, where an ad hoc team is formed, is far more typical of an emergency situation in a hospital, because it is often a first responder who sounds the alarm and summons other people to help. Note that in the ad hoc teams, the incoming team members joined immediately after the emergency occurred (fewer than 7 s after the cardiac arrest). The results show that ad hoc teams had a statistically significant lower percentage of hands-on time (51.7%) than the preformed teams (68.7%). Ad hoc teams took about 40 s longer to defibrillate and more than 50 s longer to administer epinephrine. Ad hoc teams did not need significantly longer to start cardiac massage, nor did they show a worse chest compression rate. These results show that basic aspects of the task (starting cardiac massage) were not influenced by the different group compositions. More complex aspects, however, such as defibrillation or decisions to administer medication, were significantly delayed in ad hoc groups. This indicates that even a short period of previous collaboration was beneficial for later team performance, particularly for the more complex aspects of the task.

In the same study, we also tested whether clinical experience (another input aspect) influenced process performance. Half of the groups studied were composed of general practitioners, and the other half were experienced hospital physicians, who were more likely to have experience with cardiac arrests (Hunziker et al. 2009). Analyses showed no differences between hospital physicians and general practitioners for hands-on time or time elapsed to start cardiac massage. Hospital physicians, however, defibrillated earlier, administered epinephrine more rapidly, and showed better performance in chest compression rates than general practitioners. This indicates that different professional background and experience influence the more complex aspects of the task, but not the more basic aspects. Similar findings emerged in another study where ad hoc groups of inexperienced medical students were compared with ad hoc groups of experienced general practitioners (Lüscher et al. 2010). Although medical students were as fast as general practitioners at diagnosing the situation, and were as good as or better than general practitioners in technical aspects of the task (e.g. chest compression rate), the experienced physicians outperformed the medical students in hands-on time, started cardiac massage about a minute earlier, and defibrillated significantly faster. Thus, although medical students were familiar with the resuscitation procedure (as indicated by their overall technical performance), they had much more difficulty with the more complex aspects of the task, which they performed with significant delay.

The results of these studies underscore again the usefulness of multiple process performance measures, at least for this task. As already mentioned, assessing process performance is especially important if the task outcome does not fully depend on the team process, but on other variables as well – in our case, for instance, patient condition (Cooper and Cade 1997). Similar arguments may actually hold for most tasks. For example, a team of air traffic controllers may have successfully guided all landing and starting planes so that no accidents occurred (outcome performance), but may have created dangerous situations or long delays that – for all kinds of reasons – did not result in an accident. A cockpit crew may land their plane on time (outcome performance), for example, but in order to do so successfully may have to make many flight path corrections during their descent.

6.5 Developing and Testing Hypotheses for the Relationship of Team Behaviour with Process Performance Variables

As explained above, concepts of team processes distinguish between taskwork and teamwork (Bowers et al. 1997; Marks et al. 2001; Salas et al. 2005). A similar distinction is made between the technical and non-technical skills needed for a task (Flin and Maran 2004; Flin et al. 2008). Taskwork, or technical skills, refers to behaviour related to the content of the task; and teamwork, or non-technical skills, refers to the coordinative aspects of the group's process. Usually, research on team performance is interested in the relationship between aspects of teamwork or non-technical skills and performance.

What coordination behaviour should be observed if one is interested in predicting group performance? There is a multitude of such behaviours, and several authors have developed lists. For example, communication, coordination, cooperation, and leadership have been named as major categories for medical teams (Flin and Maran 2004); others have also included gathering and exchanging information, supporting others (Fletcher et al. 2004), or monitoring behaviour (Healey et al. 2004). Some authors have suggested further division of some of these broad categories. For example, Xiao et al. (2004) suggested distinguishing different aspects of leadership, such as strategic planning, reporting and critiquing plans, coaching, maintaining awareness, and requesting information. The examples provided here are by no means exhaustive, and many more coordination behaviours are mentioned in the literature.

Some general coordination behaviours may be important for the performance of most, if not all, teams (Salas et al. 2005). Nevertheless, most authors explicitly or implicitly agree that different situations and tasks require different coordination behaviours. Thus, most researchers adapt their observational categories to a given task or task type. For example, Weingart and colleagues (Weingart et al. 2004) developed behavioural observation categories specifically for negotiation tasks. For hidden profile tasks, information exchange is the teamwork variable observed most frequently (e.g. Larson et al. 1998; Stasser et al. 1995). Boos and colleagues developed specific categories for complex problem solving in groups (Boos et al. 1990); Kolbe et al. (2009) developed their coding system for anaesthesia teams based on an analysis of coordination requirements in anaesthesia, and for complex decision making based on a task analysis of team decision making (see Chap. 9); and even the well-known interaction process analysis coding system (Bales 1950) was initially meant to code only decision-making tasks. There is indeed a general, albeit more implicit than explicit, agreement that the teamwork behaviours observed should be relevant for the task of the group.

We indeed think that the extensions of the general IPO model to include more task-related aspects, especially in combination with the coordination requirements specified in the task analysis, can be very useful for developing hypotheses about what coordination behaviour is important and should be observed, and which behaviours might be most important at what time or phase of the group process (Tschan et al. 2009). For instance, one can assume that directive leadership should be most useful for performance when a group begins a new task, especially when the task requires high initial coordination between team members and coordinated transitions between sub-tasks (Fernandez et al. 2008; Marks et al. 2001; Wageman et al. 2009). On the other land, leadership may not be so crucial for group performance in later, more routine phases when the group is already well coordinated. In routine phases, more direct leadership may even be associated with lower performance, as it may indicate that leadership is necessary in order to correct members' behaviour (Burtscher et al. 2010; Künzle et al. 2010). More educational leadership, however, may be important for later phases (Wageman et al. 2009).

To illustrate how coordination requirements derived from HTA may inform hypotheses on what coordination behaviour may be important and when, we again refer to the resuscitation task. Resuscitation teams need to change rapidly from the diagnostic phase to the hands-on action phase (see Fig. 6.1), and thus they may well profit from strong leadership at the beginning of this task. The concept of multiple IPO episodes or cycles suggests (Futoran et al. 1989; Marks et al. 2001) that leadership may regain importance immediately after a change in input variables, for example, if a new group member joins or leaves (Ballard et al. 2007; Wageman et al. 2009; Waller et al. 2004). Thus, we do not assume the same positive impact of leadership at all moments of the group process.

For the simulated resuscitation task, we investigated the link between leadership and performance for groups of nurses and physicians treating a cardiac arrest. At the beginning of the cardiac arrest, three nurses were alone with the patient, but after a few minutes into the cardiac arrest, a resident physician joined the group; thus, an important input variable changed (Tschan et al. 2006). Based on the assumptions that directive leadership is crucial at the beginning of a task and after input changes, we hypothesized that leadership should enhance performance (1) in the very early stages of the cardiac arrest when nurses respond to this emergency situation, and (2) immediately after the resident enters the group, because the group composition changes: The resident has to assume responsibility, and the group needs to adapt to this situation. Both hypotheses were supported by the data. There was a significant correlation of $r = 0.45$ between leadership utterances and performance when the nurses were alone. After the resident joined the group, more directive leadership by him or her significantly predicted performance ($r = 0.52$) – but only leadership during the first 30 s after he or she joined the group. As predicted, more leadership after this short adaptation phase was no longer a predictor of performance ($r = 0.27$, ns).

This example shows how important it can be to develop specific hypotheses about how and when specific teamwork behaviours may be most important for performance, and it also shows that knowledge about teamwork requirements can be helpful. Indeed, if we had only tested the general hypothesis that "leadership is important for resuscitation performance", we would have rejected this hypothesis based on our data. Note that these results are quite compatible with more general leadership theories, which advocate that leadership is only fruitful if it is adapted to the requirements of the situation and the followers (e.g. House 1971; Vroom and Yetton 1973; Wageman et al. 2009), as well as leadership conceptions for groups

6 Developing Observational Categories for Group Process Research Based on Task 109

(von Cranach, 1985; McGrath 1962). In order to test such general assumptions, however, we need rather fine-grained analyses of specific task requirements.

The specificity of task requirements across time and sub-tasks has far-reaching implications. Among other things, it may imply that a group may be very good with regard to specific task requirements but not with regard to others. For instance, we compared the performance of teams in a routine situation vs. an emergency situation. It turned out that performance in one situation was only weakly related to performance in the other. As an example, some groups took quite some time to gather information. This behaviour was very useful in the routine situation; by contrast, it diminished performance in the emergency situation, where speed of reaction was crucial (Vetterli et al. 2009).

The importance of specific coordination requirements across tasks and time can hardly be overemphasized. We therefore want to illustrate two other examples. In a large study about pilot crew performance, behavioural observations were made for more than 300 flights (Thomas 2004). The researchers analysed predictors of successful error management across different flight phases (pre-departure preparation, takeoff, approach landing). It turned out that error management for the different tasks was predicted by different coordination behaviours. In the pre-departure phase, collaboration between pilot and co-pilot was especially important: During takeoff, contingency planning increased the probability of error management; and during landing, vigilance and problem identification as well as assertiveness of the co-pilot were predictors for good error management.

In another study in a hospital setting, Bogenstaetter et al. (2009) investigated errors made when nurses and physicians transmitted information to a physician who joined an ongoing cardiopulmonary resuscitation. They found that 18% of the information given to the incoming physician was inaccurate and contained errors. Most of these errors were related to information about two tasks, namely, (1) defibrillation and (2) medication. To remember how many defibrillations have occurred, and with how many joules, the best way to store this information is in the same way it needs to be retrieved – for instance, by saying loudly: "This is the third defibrillation". This way of storing information is referred to as *transfer-appropriate processing* (Morris et al. 1977). Usually, however, people do not store information in this way; our study shows, however, that those who do commit fewer errors in transmitting information. Such results are directly transferable to training. Again, they refer to specific task requirements that are not easily detected without fine-grained task analysis.

6.6 Conclusions

In this chapter we showed that knowledge about specific taskwork and coordination requirements of a given task can be useful for small group research. We showed how task analysis can help to define process performance measures that allow us to tap into important aspects of group performance beyond "output" (Pritchard and

Watson 1992). Furthermore, we showed that task analysis can help to develop and test hypotheses about specific teamwork behaviour and performance.

We are not the first to stress the importance of tasks for small group research. It has long been acknowledged that the type of task is an important predictor of group processes and group performance. Accordingly, there is a long standing tradition in distinguishing among different task types (McGrath 1984; Shaw 1976; Steiner 1972) or different types of groups by type of task (Arrow et al. 2000; Hackman and Wageman 2005). Indeed, in early studies, Hackman (1968) and Kent and McGrath (1969) compared influences on group performance across different tasks. They found that task characteristics explain more than ten times the amount of variance than is explained by teamwork aspects.

The well-known task classifications, however, are often not fine-grained enough to uncover such specific requirements as the time when something has to be done, the period during which something is useful or not, etc. Furthermore, these classifications are often not suitable for complex tasks because these often contain different or mixed task types as sub-tasks. For example, in terms of McGrath's circumplex model (1984), the resuscitation task used as an example in this chapter contains the task types "choose", "decide", and "performance of psycho-motor tasks". Furthermore, the task is at the same time "conceptual" as well as "actional".

HTA does not describe task types and is thus more flexible. In HTA, we see the chance to define the task requirements in a flexible and adaptive way for many different tasks in group research. HTA is also flexible with regard to the level of specificity involved, as the level of specificity may be chosen depending on the knowledge and experience of the people involved. For instance, for medical personnel, it may be fine-grained enough to specify the task "defibrillate". By contrast, it may be necessary to present a hierarchical task analysis of the defibrillation sub-tasks for laypeople who are not familiar with that task.

Thus, there are advantages of including task aspects more explicitly in group research. First, more specific analyses about the relationship of teamwork and different task requirements can help to develop more precise knowledge, and can inform theories about the relationship between teamwork behaviours and aspects of group performance. Second, more precise knowledge about task and coordination requirements can be useful for team training. If groups have to carry out complex tasks, training probably has to be task-specific. For example, pilot crews learn takeoff and landing as different tasks; and in the operation room, an appendectomy surgery is a task that is very different from a hip implant surgery. As these different tasks require such different taskwork, it is also likely that coordination requirements vary as well. Gaining more precise knowledge about when which behaviour is needed for which task can help to train teams in a more specific way.

Of course, analysing group processes in terms of task requirements has disadvantages as well. First, generalization from one task to another is not easy, and thus such research may not immediately contribute to a "unified theory of group performance". Actually, a next step may have to consist of developing typologies of task requirements, possibly based on typologies of tasks. Some time ago, Wood (1986) called for a theory of tasks, conceptualizing and defining task complexity.

6 Developing Observational Categories for Group Process Research Based on Task 111

It may be fruitful to pursue such efforts. A second disadvantage is that performing task analysis is an additional step to the already high burden of group process research. Especially for complex tasks for expert actors, hierarchical task analysis can be very time-consuming and may require extended training. On the other hand, having more specific ideas about what behaviour should be observed and when can considerably decrease the time researchers spend on coding. Looking at the pros and cons, we feel that the advantages certainly outweigh the disadvantages.

In their treatises of group process analysis methods, Weingart (1997), McGrath and Altermatt (2001), as well as Zaccaro et al. (2005) all stated that it may not be possible to develop a single, overall behavioural coding system for all group processes. They thus suggested that the research goal should guide which behaviour(s) should be observed. We add to this recommendation that knowledge of the task may be helpful for choosing appropriate observational categories and process performance measures as well as for developing hypotheses about predictors of successful performance.

References

Abella BS, Alvarado JP, Myklebust H, Edelson DP, Barry A, O'Hearn N et al (2005) Quality of cardiopulmonary resuscitation during in-hospital cardiac arrest. J Am Med Assoc 293(3): 305–310

Annett J (2004) Hierarchical task analysis. In: Diaper D, Stanton NA (eds) The handbook of task analysis for human-computer interaction. Lawrence Erlbaum, Mahwah, NJ, pp 67–82

Annett J, Cunningham D, Mathias-Jones P (2000) A method for measuring team skills. Ergonomics 43(8):1076–1094

Antoni C, Hertel G (2009) Team processes, their antecedents and consequences: implications for different types of teamwork. Eur J Work Organ Psych 18:253–266

Arrow H, McGrath JE, Berdahl JL (2000) Small groups as complex systems: formation, coordination, development, and adaptation. Sage Publications, Newbury Park, CA

Bales RF (1950) Interaction process analysis. A method for the study of small groups. Addison-Wesley, Cambridge, MA

Ballard DI, Tschan F, Waller M (2007) All in the timing: considering time at multiple stages of group research. Small Group Res 39:328–351

Bogenstaetter Y, Tschan F, Semmer NK, Spychiger M, Breuer M, Marsch SU (2009) How accurate is information transmitted to medical professionals joining a medical emergency? A simulator study. Hum Factors 51:115–125

Boos M (1996) Entscheidungsfindung in Gruppen: Eine Prozessanalyse [in German]. Huber, Bern

Boos M, Morguet M, Meier M, Fisch R (1990) Zeitreihenanalysen von Interaktionsprozessen bei der Bearbeitung komplexer Probleme in Expertengruppen [in German]. Z Sozialpsychol 1990:53–64

Bowers CA, Brown CC, Morgan BB (1997) Team workload: its meaning and measurement. In: Brannick MT, Salas E, Prince C (eds) Team performance assessment and measurement. Lawrence Erlbaum, London

Brett J, Weingart L, Olekalns M (2004) Bubbles, bangles, and beads: Modeling the evolution of negotiating groups over time. Rese Manag Groups Teams 6:36–64

Burtscher MJ, Wacker J, Grote G, Manser T (2010) Managing nonroutine events in anesthesia: the role of adaptive coordination. Hum Factors. doi:10.1177/0018720809359178

Campbell JP, McCloy RA, Oppler SH, Sager CE (1993) A theory of performance. In: Schmitt N, Borman WC et al (eds) Personnel selection in organizations. Jossey-Bass, San Francisco, pp 35–70

Cannon-Bowers JA, Tannenbaum SI, Salas E, Volpe CE (1995) Defining competencies and establishing team training requirements. In: Guzzo RA, Salas E (Eds) Team effectiveness and decision making in organizations. Jossey-Bass. San Francisco, pp 333–380

Cooper S, Cade J (1997) Predicting survival, in-hospital cardiac arrests: Resuscitation survival variables and training effectiveness. Resuscitation 35:17–22

Fernandez R, Kozlowski SWJ, Shapiro MJ, Salas E (2008) Toward a definition of teamwork in emergency medicine. Acad Emerg Med 15:1104–1112

Fletcher G, Flin R, McGeorge P, Glavin R, Maran N, Patey R (2004) Rating non-technical skills: developing a behavioural marker system for use in anaesthesia. Cogn Technol Work 6: 165–171

Flin R, Maran N (2004) Identifying and training non-technical skills for teams in acute medicine. Qual Saf Health Care 13(Suppl 1):i80–i84

Flin R, O'Connor P, Chrichton M (2008) Safety at the sharp end. A guide to non-technical skills. Ashgate, Aldershot, UK

Futoran GC, Kelly JR, McGrath JE (1989) TEMPO: A time-based system for analysis of group interaction processes. Basic Appl Soc Psych 10(3):211–232

Gurtner A, Tschan F, Semmer NK, Nägele C (2007) Getting groups to develop good strategies: effects of reflexivity interventions on team process, team performance, and shared mental models. Organ Behav Hum Dec 102:127–142

Hackman JR (1968) Effects of task characteristics on group products. J Exp Soc Psychol 4:162–187

Hackman JR, Morris CG (1975) Group tasks, group interaction process, and group performance effectiveness: a review and proposed integration. In: Berkowitz L (ed) Advances in experimental social psychology, vol 8. Academic, New York, pp 45–99

Hackman JR, Wageman R (2005) A theory of team coaching. Acad Manage Rev 30(2):269–287

Healey AN, Undre S, Vincent CA (2004) Developing observational measures of performance in surgical teams. Qual Saf Health Care 13(Suppl1):i33–i40

House RJ (1971) A path-goal theory of leader effectiveness. Admin Sci Quart 12:321–338

Hunziker S, Tschan F, Semmer N, Zobrist R, Spychiger M, Breuer M et al (2009) Hands-on time during cardiopulmonary resuscitation is affected by the process of teambuilding: a prospective randomised simulator-based trial. BMC Emerg Med 9:3

Hunziker S, Bühlmann C, Tschan F, Balestra G, Legret C, Schumacher C et al (2010) Brief leadership instructions improve cardiopulmonary resuscitation in a high fidelity simulation: a randomized controlled trial. Crit Care Med 38:1–4

Ilgen DR, Hollenbeck JF, Johnson MD, Jundt D (2005) Teams in organizations: from input–process–output models to IMOI models. Annu Rev Psychol 56:517–543

Kauffeld S, Lehmann-Willenbrock N, Henschel A, Neininger A (2009) Empirical discussion types: the good, the bad, and the indifferent. Presentation at the Fourth Annual INGRoup Conference, Colorado Springs, July 16–18, 2009

Kent RN, McGrath JE (1969) Task and group characteristics as factors influencing group performance. J Exp Soc Psychol 5:429–440

Kolbe M, Künzle B, Zala-Mezö E, Wacker J, Grote G (2009) Measuring coordination behaviour in anesthesia teams during induction of general anesthetics. In: Flin R, Mitchell L (eds) Safer surgery. Analysing behaviour in the operating theatre. Ashgate, Burlington, UK, pp 203–223

Konradt U, Semmer NK, Tschan F (2006) Aufgabenanalyse [in German]. In: Zimolong B, Konradt U (eds) Enzykolopädie der Psychologie, vol 2, Themenbereich D, Serie III: Wirtschafts-, Organisations- und Arbeitspsychologie (Ingenieurpsychologie). Hogrefe, Göttingen, pp 359–392

Künzle B, Kolbe M, Grote G (2010) Ensuring patient safety through effective leadership behaviour: a literature review. Saf Sci 48:1–17

Larson JR, Christensen C, Franz TM, Abott AS (1998) Diagnosing groups: the pooling, management, and impact of shared and unshared case information in team-based medical decision making. J Pers Soc Psychol 75(1):93–108

Lüscher F, Hunziker S, Gaillard V, Tschan F, Semmer NK, Hunziker P et al (2010) Proficiency in cardiopulmonary resuscitation of medical students at graduation: a simulator-based comparison with general practitioners. Swiss Med Weekly 104:57–61

Lüscher F, Hunziker S, Gaillard V, Tschan F, Semmer NK, Hunziker P, Marsch SU (2010). Proficiency in cardiopulmonary resuscitation of medical students at graduation: A simulator-based comparison with general practitioners. Swiss Medical Weekly, 140, 57-61

Marks MA, Mathieu JE, Zaccaro SJ (2001) A temporally based framework and taxonomy of team processes. Acad Manage Rev 26:356–376

Marsch SU, Hunziker P, Spychiger M, Breuer N, Semmer N, Tschan F (2004a) Teambuilding delays crucial measures in simulated cardiac arrests. In: Paper presented at the Schweizerische Gesellschaft für Intensiv Medizin May 12–15, 2004, Interlaken

Marsch SU, Müller C, Marquardt K, Conrad G, Tschan F, Hunziker PR (2004b) Human factors affect quality of cardiopulmonary resuscitation in simulated cardiac arrests. Resuscitation 60(1):51–56

Marsch SU, Tschan F, Semmer N, Spychiger M, Breuer M, Hunziker PR (2005) Unnecessary interruptions of cardiac massage during simulated cardiac arrests. Eur J Anaesthesiol 22:831–833

Mathieu JE, Maynard M, Rapp T, Gilson L (2008) Team effectiveness 1997–2007: a review of recent advancements and a glimpse into the future. J Manage 34:410–476

McGrath JE (1962) Leadership behaviour: some requirements for leadership training. US Civil Service Commission, Washington, DC

McGrath JE (1964) Social psychology: a brief introduction. Holt, New York

McGrath JE (1984) Groups, interaction and performance. Prentice Hall, Englewood Cliffs, NJ

McGrath JE (1991) Time, interaction, and performance (TIP): a theory of groups. Small Group Res 22:147–174

McGrath JE, Altermatt WT (2001) Observation and analysis of group interaction over time: some methodological and strategic consequences. In: Hogg MA, Tindale RS (eds) Blackwell handbook of social psychology: group processes. Blackwell, Oxford, pp 525–556

McGrath JE, Tschan F (2004) Dynamics in groups and teams: groups as complex action systems. In: Poole MS, van de Ven AH (eds) Handbook of organizational change and development. Oxford University Press, Oxford, pp 50–73

McGrath JE, Arrow H, Berdahl JL (2000) The study of groups: past, present, and future. Pers Soc Psychol Rev 4:95–105

Moreland RL, Fetterman JD, Flagg JJ, Swanenburg KL (2010) Behavioral assessment practices among social psychologists who study small groups. In: Agnew CR, Carlson DS, Graziano WG, Kelly JR (eds) Then a miracle occurs: focusing on behaviour in social psychological theory and research. Oxford University Press, New York, pp 25–56

Morris CD, Bransford JD, Franks JJ (1977) Levels of processing versus transfer appropriate processing. J Verbal Learn Verbal Behav 16:519–533

Nolan JP, Deakin CD, Soar J, Bottiger BW, Smith G (2005) European Resuscitation Council Guidelines for Resuscitation 2005: Section 4. Adult advanced life support. Resuscitation 67 (1):39–86

Pritchard RD, Watson MD (1992) Understanding and measuring group productivity. In: Worchel S, Wood W, Simpson JA (eds) Group process and productivity. Sage, Newbury Park, CA, pp 251–275

Ravakhah K, Khalafi K, Bathory T, Wang HC (1998) Advanced cardiac life support events in a community hospital and their outcome: evaluation of actual arrests. Resuscitation 36:95–99

Salas E, Sims DE, Burke CS (2005) Is there a "Big Five" in teamwork? Small Group Res 36(5):555–599

Shaw ME (1976) Group dynamics: the psychology of small group behaviour. McGraw-Hill, New York

Shepherd A (1985) Hierarchical task analysis and training decisions. Program Learning Educ Tech 22:162–176

Shepherd A (1998) HTA as a framework for task analysis. Ergonomics 41(11):1537–1552

Shepherd A (2001) Hierarchichal task analysis. Taylor and Francis, London

Sonnentag S, Frese M (2002) Performance concepts and performance theory. In: Sonnentag S (ed) Psychological management of individual performance. Chichester, Wiley

Stasser G, Stewart DD, Wittenbaum GM (1995) Expert roles and information exchange during discussion: the importance of knowing who knows what. J Exp Soc Psychol 31:244–265

Steiner ID (1972) Group processes and productivity. Academic, New York

Thomas MJW (2004) Predictors of threat and error management: Identification of core nontechnical skills and implications for training systems design. Int J Aviat Psychol 14(2):207–231

Tschan F (1995) Communication enhances small group performance if it conforms to task requirements: the concept of ideal communication cycles. Basic Appl Soc Psych 17(3): 371–393

Tschan F (2002) Ideal cycles of communication (or cognition) in triads, dyads, and individuals. Small Group Res 33(6):615–643

Tschan F, von Cranach M (1996) Group task structure, processes and outcome. In: West M (ed) Handbook of work group psychology. Wiley, Chichester, UK, pp 95–121

Tschan F, Semmer NK, Nägele C, Gurtner A (2000) Task adaptive behaviour and performance in groups. Group Process Intergroup Relat 3(4):367–386

Tschan F, Semmer NK, Gautschi D, Hunziker P, Spychiger M, Marsch SU (2006) Leading to recovery: group performance and coordinative activities in medical emergency driven groups. Hum Perform 19(3):277–304

Tschan F, McGrath JE, Semmer NK, Arametti M, Bogenstaetter Y, Marsch SU (2009) Temporal aspects of processes in ad hoc groups: a conceptual scheme and some research examples. In: Roe R, Waller MJ, Clegg C (eds) Doing time: advancing temporal research in organizations. Routledge, London

Vetterli M (2006) Les "temps morts" du processus de réanimation cardiaque [unpublished master's thesis] [in French]. University of Neuchâtel

Vetterli M, Tschan F, Semmer NK, Marsch SU (2009) Process performance markers for complex routine and emergency medical tasks: basis for comparing team performance across situations. Paper presented at the 4th Annual INGRoup Conference

von Cranach M (1985) Leadership as a function of group action. In: Graumann CR, Moscovici S (eds) Changing concepts of leadership. Springer, New York

von Planta M (2004) Wissenschaftliche Grundlagen der kardiopulmonalen Reanimation (CPR) [in German]. Schweiz Med Forum 4:470–477

Vroom VH, Yetton PW (1973) Leadership and decision-making. University of Pittsburgh Press, Pittsburgh

Wageman R, Fisher CM, Hackman JR (2009) Leading teams when the time is right: finding the best moments to act. Organ Dyn 38:192–209

Waller MJ (1999) The timing of adaptive group responses to nonroutine events. Acad Manage J 42(2):127–137

Waller MJ, Gupta N, Giambatista RC (2004) Effects of adaptive behaviours and shared mental model creation on control crew performance. Manage Sci 50(11):1534–1544

Weingart LR (1992) Impact of group goals, task component complexity, effort, and planning on group performance. J Appl Psychol 7(5):682–693

Weingart LR (1997) How did they do that? The ways and means of studying group processes. Res Organ Behav 19:189–239

Weingart LR, Olekalns M, Smith PL (2004) Quantitative coding of negotiation behaviour. Int Negot 9:441–455

Wittenbaum GM, Vaughan SI, Stasser G (1998) Coordination in task-performing groups. In: Tindale RS, Heath L, Edwards J, Posavac EJ, Bryant FB, Suarez-Balcazar Y, Henderson-King E, Myers J (eds) Theory and research on small groups. Plenum, New York, pp 177–204

Wood RE (1986) Task complexity: definition of the construct. Organ Behav Hum Dec 37:60–82

Xiao Y, Seagull FJ, Mackenzie CF, Klein C (2004) Adaptive leadership in trauma resuscitation teams: a grounded theory approach to video analysis. Cognit Tech Work 6(3):158–164

Zaccaro SJ, Caracraft M, Marks M (2005) Collecting data in groups. In: Leong FTL, Austin JT (eds) The psychology research handbook: a guide for graduate students and research assistants. Sage, Thousand Oaks, CA, pp 227–237

Part II
Assessing Coordination in Human Groups – Concepts and Methods

Chapter 7
Assessing Coordination in Human Groups: Concepts and Methods

Thomas Ellwart

Abstract This integrating chapter summarises different coordination constructs and methods for assessment in human group research. Because of the oversized number of coordination constructs, they are clustered along first-order variables of coordination, such as impersonal coordination instruments, personal coordination, tacit behaviours, team knowledge, team attitudes, and coordination as outcome. This overview is grounded in both a functional and temporal perspective of coordination and offers a pattern of orientation in the variety of coordination variables. The second part of this chapter introduces methodological streams to be found in the research for assessing group coordination in the laboratory and the field and will refer to authors of Part II in this book to give an outlook for the following chapters.

7.1 Introduction

The aim of Part II of this book is to give insights into the concepts and methods of human group coordination in the social sciences. The reader will find chapters that focus on variables such as coordination potential (Chap. 8), shared mental models and team knowledge (Chaps. 9 and 10), and methods such as micro-analyses for measuring different coordination mechanisms (Chap. 11). The following overview structures different conceptual and methodological approaches to coordination into a single comprehensive classification. The taxonomy offered in this integrating chapter does not represent a theoretical or functional model of coordination processes (cf. Chaps. 2 and 4) but does offer a pattern of orientation in the variety of coordination variables. Since specific concepts of coordination and methods of assessment are covered in detail in this volume, this chapter will merely summarise and highlight the essentials.

T. Ellwart
University of Trier, Department of Economic Psychology, D-54286 Trier, Germany
e-mail: ellwart@uni-trier.de

M. Boos et al. (eds.), *Coordination in Human and Primate Groups*,
DOI 10.1007/978-3-642-15355-6_7, © Springer-Verlag Berlin Heidelberg 2011

This integrating chapter is divided into two parts. The first part (Sect. 2, "Perspectives on Coordination in Human Group Research") defines the framework of the taxonomy, which is based on influential models of human group coordination (e.g. Arrow et al. 2000; Espinosa et al. 2004; Tschan 2000 and Chap. 6; Wittenbaum et al. 1998). This perspective classification is grounded in both a functional and temporal view of coordination (cf. Arrow et al. 2000; McGrath and Tschan 2004; Wittenbaum et al. 1998) and distinguishes first-order variables of coordination according to three dimensions: (1) coordination as a process or as an outcome variable; (2) a temporal focus on coordination as pre-, in-, and post-processes; and (3) coordination as explicit or implicit coordination processes. Throughout the text, examples from the chapters in Part II are included in order to integrate their presented concepts into the coordination taxonomy. The second part of this chapter (Sect. 3, "Methods for Assessment: How Is Coordination Measured?") introduces methodological streams to be found in the research for assessing group coordination in the laboratory and the field. Again, this section will refer to authors of Part II in this book to give an outlook for the following chapters.

7.2 Perspectives on Coordination in Human Group Research

7.2.1 Coordination as Process or Outcome Variable

The major problem in drawing a systematic overview of coordination concepts and its measures is the lack of a consistent definition due to the differences between the existing characterisations (Malone and Crowston 1994). The fuzziness in the terminology results in a wide spectrum of coordination variables and related concepts, depending on the perspective of the researchers. An important first-order distinction in classifying coordination is whether the focus is on coordination processes or coordination outcomes (see Fig. 7.1). Looking at the definition of Malone and Crowston (1994), coordination can be understood as the management of dependencies among tasks, tools (resources), and people. From a measurement perspective, researchers can consider the process of coordination as the way in which group members synchronise their actions in order to complete the group task successfully (Wittenbaum et al. 1998). On the other hand, researchers may choose to focus on the outcome (i.e. state of coordination), which represents the extent to which team members have managed interdependencies to varying degrees of success (Espinosa et al. 2004).

Both perspectives – processes and outcomes of coordination – build the framework of the taxonomy of coordination (see Fig. 7.1). Most of the empirical and theoretical research has focussed on processes of coordination. The key interest is on the activities carried out by team members when managing dependencies towards a super-ordinate goal (e.g. planning to develop a software product, communication to reach a team decision). These activities focus on synchronising

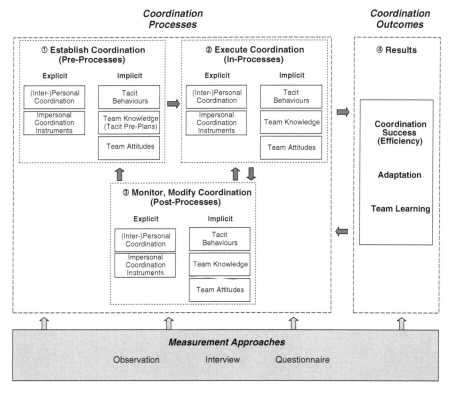

Fig. 7.1 Integrative taxonomy of coordination mechanisms and outcomes (cf. Arrow et al. 2000; Espinosa et al. 2004; Tschan 2000; Wittenbaum et al. 1998)

(1) the group goals, (2) the sub-tasks to a collective set of tasks, (3) the interaction processes between team members, and (4) resources within the group (Arrow et al. 2000; Tschan 1995; Zalesny et al. 1995). In psychological research, coordination processes are classified (1) with regard to their temporal occurrence during group functioning (pre-, in-, or post-process; see Sect. 2.2) and (2) the degree of explicitness or implicitness (explicit or implicit/tacit coordination; see Sect. 2.3).

Coordination as an outcome (i.e. state of coordination) represents a more distal view on coordination as a result of team coordinating processes (cf. Fig. 7.1 and Table 7.1). Measurable outcome indicators of coordination are, for example, coordination success/effectiveness (e.g. Hoegl and Gemuenden 2001; Chap. 11), adaptation (e.g. Burke et al. 2006), or team learning (e.g. Sarin and McDermott 2003). Coordination success describes the result, for instance, that all sub-tasks are integrated well and that delivery is on schedule and within budget (Hoegl and Gemuenden 2001). In this sense, coordination success can be understood as a conceptualisation of team efficiency (Hoegl and Gemuenden 2001). Efficiency and effectiveness are two often-used variables to describe the extent of team performance (Denison et al. 1996; Hackman 1987). Whereas effectiveness refers

Table 7.1 Overview of key variables in group coordination theory

	Variables in research	Explicitness	phase in framework	References for measures
Impersonal Coordination Instruments	Administrative coordination (programming) Technical tools	Explicit	Pre-, in-, post-process	Faraj and Sproull (2000), Kraut and Streeter (1995), Van der Aalst and van Hee (2004), Stout et al. (1999); Massey et al. 2003, Georgakopoulos et al. (1995)
(Inter-) Personal Coordination	Planning behaviour	Explicit	Pre-, in-process	Hoegl and Gemuenden (2001), Kolbe (2007), Kraut and Streeter (1995)
	Information exchange (temporal phase, flow, format, frequency, openness, formalisation, quality)	Explicit	Pre-, in-, post-process	Hoegl and Gemuenden (2001), Fiore et al. (2003), Clampitt and Downs (2004), Hess et al. (2000); O'Reilly and Roberts (1976)
	Feedback (seeking, giving, receiving)	Explicit	Post-process	Steelman et al. (2004), Dickinson and McIntyre (1997), Kluger and DeNisi (1996), Hackman and Oldham (1975)
	Leadership (locomotion, cohesion, affect and conflict management, shared leadership)	Explicit	Pre-, in-, post-process	Arnold et al. (2000), Bass and Avolio (2004), Kacmar, et al. (2003), Pearce and Sims (2002), Weinkauf and Hoegl (2002)
Tacit Behaviours	Anticipated backup behaviour (information, knowledge, helping behaviour)	Implicit	In-, post-process	Dickinson and McIntyre (1997), McIntyre and Salas (1995), Rico et al. (2008), Wittenbaum et al. (1998), MacMillan et al. (2004), Marks and Panzer (2004), Burke et al. (2006)
	Monitoring (team-mates activities, progress, system)			
	Adaptation (dynamic adjustment)			
Team Knowledge	Team mental models Team situation models Transactive memory	Implicit	Pre, in-, post process	Ellwart (Chapter 9 in this book); Cooke et al. (2000), Wegner (1995), Mohammed and Dumville (2001)
Team Attitudes	Trust Collective efficacy Cohesion Collective orientation	Implicit	Pre, in-, post process	Fiore et al. (2003), Rico et al. (2008), Kozlowski and Ilgen (2006), Beal et al. (2003)
Coordination as Outcome	Coordination success Team adaptation Team learning		Output	Edmondson (1999), Burke et al. (2006), Hoegl and Gemuenden (2001), Lewis (2003)

7 Assessing Coordination in Human Groups: Concepts and Methods 123

to the degree to which the team meets expectations regarding the quality of the outcome (Hoegl and Gemuenden 2001), efficiency is closely related to coordination success. The team's efficiency is assessed in terms of adherence to schedules and budgets (Hoegl and Gemuenden 2001, p. 438). In recent literature, team adaptation is classified as another concept of coordination output (Burke et al. 2006). Regarding Campbell (1990), performance is not solely the consequence of action; it is the action itself (p. 704). Extending this view to group coordination, team adaptation can be conceptualised as a dependent variable in a rather longitudinal enactment of processes (Burke et al. 2006; Kozlowski and Bell 2003). Adaptation can be defined as change in team performance in response to a salient cue that leads to a functional outcome for the entire team (Burke et al. 2006). Thus, adaptive team performance is manifested in the innovation of new or the modification of existing structures and/or group actions. Besides coordination success and adaptation, team learning describes another output variable of coordination. In most definitions it represents a process in which a team takes action, obtains and reflects upon feedback, and makes changes to adapt or to improve performance (Edmondson 1999; Argote et al. 2001). Hence, team learning will result in knowledge being embedded within the team, with the intent that it ultimately promotes performance improvement (Argote et al. 2001). In this taxonomy of coordination, activities such as feedback, information exchange, and trust are mechanisms that lead to team learning. Therefore, it is difficult to differentiate the processes of team learning (Argote et al. 2001; Edmondson 1999) from team learning as an output (Ellis et al. 2003; Sarin and McDermott 2003). As an output variable, team learning is defined as a "permanent change in the team's collective level of knowledge and skills" (Ellis et al. 2003, p. 821).

In sum, two major distinctions in coordination research are the perspectives of coordination processes and the coordination outcomes. While coordination processes are explained in detail ahead, coordination outcomes represent important variables from a distal view. In Part II of this book, Kristina Lauche reflects on a coordination outcome: namely, team innovation. She points out important coordination processes that are necessary to achieve high-quality innovation outcomes in problem-solving tasks. While she predicts team outcomes by rather external conditions for the team (e.g. autonomy, local control, involvement in problem setting, organisational support), approaches by Ellwart et al. or Kolbe et al. (Chaps. 9 and 11, respectively) focus on within-team processes during task completion.

In the following section, the focus is on coordination processes and its classification into explicit and implicit processes and also from a temporal perspective.

7.2.2 Temporal Perspective on Coordination: Pre-, In-, and Post-processes

Another first-order classification of coordination processes is the distinction among processes that occur before, during, and after the actual team task. This temporal perspective on group coordination overcomes the shortcomings of the traditional single-loop view on group performance (input–process–output models; e.g. Hackman 1987) and can be found in different theoretical conceptualisations of group coordination (cf. Arrow et al. 2000; Burke et al. 2006; Marks et al. 2001; Fiore et al. 2003; Wittenbaum et al. 1998; see also Chap. 2). Different than the classical input–process–output models, group work is characterised by several input–process–output cycles of goal-directed activities. For example, Tschan (2000) draws on theories of action regulation (cf. Hacker 2003) and describes three recurrent sequential regulatory processes (pre-, in-, post-processes) in order to reach a team goal. Using slightly different terminology, Arrow et al. (2000, 2004) distinguish three similar phases in their theoretical framework: elaboration, enactment, and modification, as do Fiore et al. (2003), who use the descriptors pre-, in-, and post-process coordination. In the first phase (prior interaction), "elaboration" takes place, where the focus is on orientation, goal choice, and planning of the group task. In the "enactment" phase, the task is executed and in-process performance takes place. In the third and final "modification" phase, results are evaluated for both the maintenance of actions and their modification. Despite the different terminology (cf. also Marks et al. 2001: action and transition phases, or Wittenbaum et al. 1998: pre- and in-process synchronisation described in Chap. 4), the key essence is that coordination research should not only focus on what is termed in-process coordination (e.g. interactions during the task), but also include both antecedents and consequences of coordination efforts, such as preparatory behaviour and post-interaction (e.g. reflection or rumination on performance).

In sum, the temporal classification of group coordination distinguishes among coordination processes that take place before, during, and after task actions. This integrating overview chapter on coordination concepts subsumes coordination processes into three temporal phases (see Fig. 7.1): pre-processes, in-processes, and post-processes. For example, because Lauche as well as Badke-Schaub and colleagues have grounded their views of coordination on action regulation theory (Chaps. 8 and 10, respectively), this integrating taxonomy categorises their coordination processes of planning as "pre-process", team mental models as "in-process", and feedback as "post-process". In Chap. 5, the anaesthesia team processes laid out by Kolbe and colleagues focus on what this integrating taxonomy subsumes as "in-process" explicit coordination, where they analyse the occurrences of group interactions such as collective actions (synchronising interdependent role-related activities) or exchanging information (lay out personal knowledge).

7.2.3 Explicit and Implicit Coordination Processes

The other first-order classification of group coordination processes covered in this perspectives section is the distinction between explicit and implicit coordination processes (e.g. Espinosa et al. 2004; Fiore et al. 2003; Wittenbaum et al. 1998 described in Chap. 4). Although this distinction is discussed in the literature regarding some theoretical and conceptual problems (Chap. 4; Grote et al. 2003), it represents a useful approach for structuring the taxonomy of group coordination, as the terminology is applied in many theoretical and empirical studies (e.g. Rico et al. 2008).

Explicit coordination mechanisms or processes are known from many studies in classical organisational research (March and Simon 1958; Van de Ven et al. 1976). Groups coordinate explicitly by using (1) impersonal coordination instruments (i.e. tools, schedules, plans, manuals) and (2) verbal and written communication (personal coordination). These mechanisms are applied explicitly with the intention to manage task dependencies (Espinosa et al. 2004; Wittenbaum et al. 1998). Table 7.1 displays some key variables of explicit coordination to be found in the literature as well as some references for further readings. The impersonal mode of explicit coordination requires minimal verbal communication and takes the form of administrative coordination or programming (e.g. formal policies and procedures, project milestones and delivery schedules, project documents and memos, regularly scheduled team meetings, requirement review meetings, or design inspections in software development teams; cf. March and Simon 1958; Faraj and Sproull 2000; Kraut and Streeter 1995). Another type of impersonal coordination is technical tools, for example, in workflow management systems (Georgakopoulos et al. 1995; van der Aalst and van Hee 2004; Massey et al. 2003).

The (inter-)personal mode of coordination depends on communication between team members and is useful when things are not anticipated and scheduled (Van de Ven et al. 1976). On a measurement level, one can identify several constructs, such as planning, information exchange, feedback, and leadership (see Table 7.1). The concept of planning involves the transmission of new information through communication between members or within the group (Van de Ven et al. 1976; Wittenbaum et al. 1998). Conceptually, planning poses a difficulty with regard to the classic impersonal/interpersonal differentiation because planning can be established through written (impersonal) memoranda as well as through verbal (personal) instruction. The second concept of interpersonal coordination is information exchange, which represents a dimension of communication. "Communication" provides a means for the exchange of information among team members (Pinto and Pinto 1990) and is often used synonymously with the term "information exchange" in research. Dickinson and McIntyre (1997) describe communication as the "glue" of teamwork, a component that links the other components. Very close to the concept of information exchange is feedback. Feedback involves giving, seeking, and receiving information between team members (Dickinson and McIntyre 1997; Kluger and DeNisi 1996) and is often hardly differentiated from the concept of "information exchange" during task accomplishment. However, the key difference is based on the temporal perspective because feedback represents information exchange as an explicit post-process concept (see Sect. 2.2). The fourth important construct in personal and group coordination to be covered is direct supervision or leadership (Mintzberg 1980). Team leaders are of great value in group coordination because they serve as coordinators of tasks or members, as connection to external teams, and as guides for setting the team's goals (Zaccaro and Marks 1999).

During all stages of the regulatory cycle (pre-, in-, post-processes) they play a key role in facilitating a team's coordination through influencing the attitude and behaviour of individuals and the interaction within and between groups (Bass 1990; Day et al. 2004; Hackman and Wageman 2005). Two main functions of leadership were defined by Cartwright and Zahnder (1968): goal achievement (locomotion) and group maintenance (cohesion). These functions address specific behaviours or activities of team leaders to coordinate a team's actions that are also measurable in group research.

Implicit coordination processes, on the other hand, suggest that control and synchronisation of members' actions can be based on unspoken assumptions about what other team members are likely to do (Wittenbaum et al. 1996, p. 23). In this view, implicit coordination includes non-verbal aspects of communication or actions such as the anticipation of members' needs without explicit communication (tacit behaviours; Wittenbaum et al. 1998). Tacit coordination occurs when members tacitly adjust their own behaviour to fit with the observed behaviour of the other members during an interaction (Wittenbaum et al. 1998). This manner of implicit coordination is described as team members anticipating actions, needs, and task demands and dynamically adjusting their own behaviour accordingly without having to communicate directly with each other or to plan the activity (Cannon-Bowers et al. 1993; Espinosa et al. 2004; Rico et al. 2008; Wittenbaum et al. 1996). In the literature, one can differentiate several behaviours that characterise tacit coordination, such as performance monitoring (e.g. McIntyre and Salas 1995; Dickinson and McIntyre 1997; Jentsch et al. 1999; Salas et al. 1995), anticipated backup behaviour (e.g. Dickinson and McIntyre 1997; Marks and Panzer 2004), and mutual adaptation (e.g. Cannon-Bowers et al. 1995) (for an overview, see Rico et al. 2008). Performance monitoring has been defined as team members' ability to "keep track of fellow team members' work while carrying out their own... to ensure that everything is running as expected and... to ensure that they are following procedures correctly" (McIntyre and Salas 1995, p 23). Anticipated backup behaviour involves assisting team members to perform their tasks (Dickinson and McIntyre 1997). This assistance may occur by (1) providing relevant information, knowledge, or feedback to other team members without a previous request, or (2) by helping a teammate behaviourally, for example, through proactive sharing of workload or correcting mistakes. Mutual adaptation describes the dynamic adjustment of behaviour to expected or unexpected actions and situations (Cannon-Bowers et al. 1995). From the temporal perspective, tacit pre-process coordination takes place before interaction (Wittenbaum et al. 1998). That is the case when team members make assumptions about the task demands and contributions and adjust their own actions accordingly. For example, resource allocation depends on the anticipated group task completion (Wittenbaum et al. 1996). During in-process coordination, members show tacit coordination with anticipated backup behaviour, monitoring, or adaptation during interaction. Tacit post-process coordination takes place when team members give feedback or share information without being explicitly instructed to do so because they anticipate the need of this behaviour.

7 Assessing Coordination in Human Groups: Concepts and Methods 127

To facilitate tacit behaviours in groups, team members need to have a shared understanding about the task and the group (team knowledge), as well as the attitudes that motivate for tacit behaviour. These pre-conditions of tacit behaviour are sometimes also labelled as implicit coordination mechanisms in the literature (Rico et al. 2008) and therefore are included in this overview. Team knowledge (sometimes termed "team cognition") represents the shared or common knowledge that team members have about the task and about each other (e.g. Cannon-Bowers et al. 1993; Kraiger and Wenzel 1997). Although team knowledge is sometimes labelled "implicit coordination process" (e.g. Espinosa et al. 2004), implicit (tacit) coordination and team knowledge are not the same. Implicit coordination is a process of coordinating team interaction behaviour in the absence of overt communication, whereas team knowledge refers to the team-level knowledge structures facilitating those behaviours (Rico et al. 2008). Looking at specific concepts of team cognition, there is a large variety of different terms and conceptualisations. In Chap. 9, Ellwart and colleagues will introduce three conceptualisations: team mental models (Klimoski and Mohammed 1994; Langan-Fox et al. 2000; Langan-Fox 2003), team situation models (Cooke et al. 2000, 2003; Rico et al. 2008), and transactive memory systems (Moreland and Myaskovsky 2000; Wegner 1987).

Team attitudes can be understood as another implicit coordination process and include both affective and motivational aspects. They represent the group's belief in the capability and goodwill within the team. Team attitudes influence what a group chooses to do, how much effort will be exerted, what persistence will be applied during difficulties or failures, and how much divergence will occur between the team members (see Kozlowski and Bell 2003, p. 352; Fiore et al. 2003). Most prominent in the literature are team attitudes such as trust, cohesion, group efficacy, and collective orientation. These attitudes represent affective-motivational mechanisms (Kozlowski and Bell 2003; Kozlowski and Ilgen 2006; Martins et al. 2004; Fiore et al. 2001; Kraiger and Wenzel 1997). When team members show high interpersonal trust, they will consider the interaction with others as safe and feasible. In turn, this will increase their motivation to participate in the team processes and helps to manage task interdependencies (Edmondson 2003; Jones and George 1998). Team cohesion represents a complex, multidimensional team attitude that describes the degree to which team members want to remain on the team (Cartwright 1968; Beal et al. 2003; Mullen and Copper 1994). Team coordination can hardly be achieved without a sufficient level of cohesion. As several studies have shown, including many cited here, cohesion has a strong influence on behavioural performance indicators (e.g. explicit and implicit coordination variables) as well as team effectiveness (coordination output) (Beal et al. 2003). Collective efficacy is defined as "a group's shared belief in its conjoint capabilities to organise and execute the course of action required to produce given levels of attainment" (Bandura 1997, p. 477). These perceptions regarding the joint ability to coordinate and communicate within the team have a positive effect on team performance, especially when interdependence is high (Gully et al. 2002; Jung and Sosik 2003). Finally, collective orientation is defined as the degree to which group members value teamwork and are willing to engage in teamwork behaviours

(Eby and Dobbins 1997). As research suggests, high degrees of collective orientation result in improved performance and higher cooperative team behaviours (Eby and Dobbins 1997; Driskell and Salas 1992). Explicit coordination behaviours such as information exchange are especially influenced by collective orientation (Eby and Dobbins 1997). Hence, this attitude towards teamwork of collective orientation influences coordination during all stages of the coordination process and represents an influential construct within the coordination taxonomy.

Taken together, implicit coordination is the process of team interaction behaviour "that is coordinated in the absence of overt communication" (Rico et al. 2008, p. 167), whereas team knowledge and attitudes refer to the team-level knowledge structures and affect the facilitation of those behaviours. Differentiated in this way, team knowledge and attitudes represent antecedents that are not included in some definitions of coordination processes (e.g. Wittenbaum et al. 1998). However, they are integrated in the coordination taxonomy presented in this chapter because of the close functional and normative relationship between implicit coordinating activities and team knowledge as well as attitudes (Cannon-Bowers and Salas 2001; Espinosa et al. 2004; Rico et al. 2008; Mathieu et al. 2000).

Chapters in Part II of this book will discuss various explicit and implicit processes of group coordination. For example, Chap. 11 introduces a micro-analytical tool to capture and measure the effectiveness effect of explicit coordination in teams through observation of communication patterns (e.g. addressing, instructions, and questions). Chapter 8 introduces personal coordination processes such as planning and feedback as essential mechanisms to achieve high-quality innovations in teams. Contributions in Chaps. 9 and 10 focus on implicit coordination processes by introducing techniques to measure different types of team knowledge such as transactive memory systems and shared mental models, respectively.

7.3 Methods for Assessment: How Is Coordination Measured?

Similar to the numerous concepts within the coordination taxonomy, there is great diversity with respect to methods of measurement. It is outside the scope of this chapter to illustrate each method in detail. However, three common approaches to assessing human coordination are touched upon: observation, interview, and questionnaire.

7.3.1 Observation

The first and widespread approach to assessing coordination in human groups (and non-human primate groups as well) is observation of behaviour such as communication or actions. These approaches differ with regard to the resolution in which the behaviour of interest is observed and analysed on a micro- vs. macro- level. On a

7 Assessing Coordination in Human Groups: Concepts and Methods   129

micro-analytic level, observable and defined behaviours of group interaction are captured, coded, and quantified during a defined time interval (e.g. Kolbe 2007; Chap. 11; Zalesny et al. 1995). In micro-analyses, Zalesny et al. (1995) differentiate between different approaches to specifying the quantification and classification of coordination patterns. For example, Bales' interaction process analysis (Bales 1950) represents a classification of a behavioural stream into predetermined categories with or without an explicit reference to the temporal ordering of the sequence. The onset and offset of any coded unit can be determined by a change in the observed category or by specific time intervals.

On the macro-analytical level, the observed group actions are rated in a more global way, using behavioural observation scales (e.g. Dickinson and McIntyre 1997). For macro-analyses, the focus is not on coding and quantifying each single behavioural act or interaction, but on giving an overall rating about a specific category. For example, Dickinson and McIntyre (1997) developed different behaviourally anchored scales to measure coordination processes such as backup behaviour, team orientation, feedback, and communication. These scales were developed in a four-stage process, starting with workshops collecting critical incidents of coordination, with a clarification and matching of different statements until the final measures were constructed. For example, behavioural observation scales can be used to rate coordination behaviours (e.g. backup behaviour, feedback, team orientation, leadership, communication) on a five-point Likert scale ("almost always" to "almost never"), identifying whether and how often specific behaviours occur within a particular team. Another example in the Dickinson and McIntyre (1997) work is behavioural summary scales, which can be used to rate the quality or skill of coordination behaviour in a team. Different from behavioural observation scales, this measure uses a behaviourally anchored rating scale to evaluate high (5), medium (3), or low (1) skill showed in the specific action. For example, high skills in backup behaviour are anchored by statements such as, "When team members have difficulties, this member steps in to assist" and low skills are anchored by statements such as, "This member is unwilling to ask for help even when it is available" (Dickinson and McIntyre 1997, p. 29).

7.3.2 Interviews and Questionnaires

The second most widely used approach to assess coordination in human groups is the interview of team members regarding coordination processes during teamwork. This way of assessment draws on subjective data of the interviewed partners and often represents an initial qualitative access to coordination in human groups. For example, Edmondson (1999) interviewed team members with the objective to verify that the constructs of team learning behaviour (e.g. feedback seeking, modification, trust) could be operationalised, developing survey items to assess these constructs in a language that is meaningful to the team members.

A third and very economic widely used approach to assess coordination is the use of questionnaire data based on self-ratings of team members or evaluations of external observers (e.g. team leader). There are numerous scales to be found in the literature that capture specific coordination processes and outcomes. Constructs include, for example, administrative coordination (Faraj and Sproull 2000), planning behaviour (Hoegl and Gemuenden 2001), feedback (Steelman et al. 2004), leadership (Weinkauf and Hoegl 2002), team mental models (Eby et al. 1999), team situational models (Biemann et al. 2009; Webber et al. 2000), collective efficacy (Edmondson 1999), and coordination success (Lewis 2003; Hoegl and Gemuenden 2001).

7.3.3 Multiplex Approaches to Coordination

Because coordination in groups reflects individual variables as well as group variables, researchers face the question of where to measure the construct of interest (group level vs. individual level). Throughout the literature there are multiplicities of approaches to be found. There are global measures of the constructs assessing the group construct directly (e.g. observing and rating group interaction, such as Dickinson and McIntyre 1997). In other cases, researchers combine responses of individuals to an aggregated group-level construct (e.g. group sharedness indices in team mental models; see Chap. 9). Both methods are appropriate to capture coordination and its processes. A third methodological approach is to model both individual- and group-level variables simultaneously. In this respect, multilevel theory and methodology [e.g. hierarchical linear modelling (HLM)] offer promising ways to analyse coordination processes on the individual as well as group level (e.g. DeShon et al. 2004; Klein and Kozlowski 2000; Klein et al. 1999; see also Chap. 6). For example, Ellwart and colleagues introduce in Chap. 9 a shared mental model index to measure team knowledge. This index reflects sharedness of expertise knowledge in teams and was modelled in HLM as a group variable to explain individual-level variance in trust and self-efficacy.

Other chapters in Part II will introduce specific methods to assess coordination processes and outcomes in a far more elaborated way. For example, Kolbe and colleagues in Chap. 11 introduce an observational approach to explicit coordination that allows the detection and coding of specific coordination behaviours in a micro-analytical way. In Chap. 8, Lauche describes a body of research that measures various explicit and implicit coordination processes by interviews; and in Chap. 9, Ellwart and colleagues apply questionnaire-based methods to analyse team knowledge. From an integrative perspective, it seems important to underline that the variety of methodological approaches is a major benefit for coordination research. This variety facilitates validation of coordination measures through application of different methods (cf. Chaps. 9 and 10) and offers the opportunity to view group coordination from very different perspectives. The micro-analytical approach by observation (cf. Chap. 11) allows the tracking of coordination behaviours to single episodes in a specific task, which is informative regarding both the functioning of coordination and deductive-induced ways in which teams can adjust and optimise their behaviour. The analysis of more unspecific views captured by questionnaires and scales of coordination can be applied independently from specific tasks and teams and should help, given the right constructs, to screen what coordination processes work insufficiently. One of the more important messages of Part II in this book is that psychological research on coordination processes is made more robust by this pluralism with regard to its methods as well as to its constructs. As long as different approaches find a common language or taxonomy on coordination, they will benefit through innovative combinations of methods and new methodological developments.

References

Argote L, Gruenfeld D, Naquin C (2001) Group learning in organizations. In: Turner M (ed) Groups at work: theory and research. Lawrence Erlbaum, Mahwah, NJ, pp 369–411

Arnold JA, Arad S, Rhoades JA, Drasgow F (2000) The empowering leadership questionnaire: the construction and validation of a new scale for measuring leadership behaviours. J Organ Behav 21:249–269

Arrow H, McGrath JE, Berdahl JL (2000) Small groups as complex systems: formation, coordination, development, and adaptation. Sage, Thousand Oaks, CA

Arrow H, Poole MS, Henry KB, Wheelan SA, Moreland R (2004) Time, change, and development: a temporal perspective on groups. Small Group Res 35:73–105

Bales RF (1950) Interaction process analysis – A method for the study of small groups. Addison-Wesley, Cambridge, MA

Bandura A (1997) Collective efficacy. In: Bandura A (ed) Self-efficacy: the exercise of control. Freeman, New York, pp 477–525

Bass BM (1990) Bass and Stogdill's handbook of leadership. Free Press, New York

Bass BM, Avolio BJ (2004) Multifactor leadership questionnaire. Mindgarden, Menlo Park, CA

Beal DJ, Cohen RR, Burke MJ, McLendon CL (2003) Cohesion and performance in groups: a meta-analytic clarification of construct relations. J Appl Psychol 88:989–1004

Biemann T, Rack O, Ellwart T (2009) Measuring shared mental models: a random group resampling approach. Manuscript submitted for publication.

Burke CS, Stagl KC, Salas E, Pierce L, Kendall D (2006) Understanding team adaptation: a conceptual analysis and model. J Appl Psychol 91:1189–1207

Campbell JP (1990) Modeling the performance prediction problem in industrial and organizational psychology. In: Dunnette MD, Hough LM (eds) Handbook of industrial and organizational psychology, vol 1, Consulting Psychologists Press. Palo Alto, CA, pp 687–732

Cannon-Bowers JA, Salas E (2001) Reflections on shared cognition. J Organ Behav 22:195–202

Cannon-Bowers JA, Salas E, Converse S (1993) Shared mental models in expert team decision making. In: Castellan NJ Jr (ed) Individual and group decision making: current issues. Lawrence Erlbaum, Hillsdale, NJ, pp 221–246

Cannon-Bowers JA, Tannenbaum SI, Salas E, Volpe CE (1995) Defining team competencies: implications for training requirements and strategies. In: Guzzo R, Salas E et al (eds) Team effectiveness and decision making in organizations. Jossey-Bass, San Francisco, pp 333–380

Cartwright D (1968) The nature of group cohesiveness. In: Cartwright D, Zander A (eds) Group dynamics: research and theory. Tavistock, London, pp 91–109

Cartwright D, Zahnder A (1968) Group dynamics – research and theory. Harper and Row, New York

Clampitt PG, Downs CW (2004) Down-Hazen communication satisfaction questionnaire. In: Downs CW, Adrian AD (eds) Assessing organizational communication. Guilford, New York, pp 139–157

Cooke NJ, Salas E, Cannon-Bowers JA, Stout RJ (2000) Measuring team knowledge. Hum Factors 42:151–173

Cooke NJ, Kickel PA, Salas E, Stout R, Bowers C, Cannon-Bowers J (2003) Measuring team knowledge: a window to the cognitive underpinnings of team performance. Group Dyn-Theor Res 7:179–199

Day D, Zaccaro SJ, Halpin SM (eds) (2004) Leader development for transforming organizations: growing leaders for tomorrow. Lawrence Erlbaum, Mahwah, NJ

Denison DR, Hart SL, Kahn JA (1996) From chimneys to crossfunctional teams: developing and validating a diagnostic model. Acad Manage J 39:1005–1023

DeShon RP, Kozlowski WJ, Schmidt AM, Milner KR, Wiechmann D (2004) Multiple-goal, multilevel model of feedback effects on the regulation of individual and team performance. J Appl Psychol 89:1035–1056

Dickinson TL, McIntyre RM (1997) A conceptual framework for team measurement. In: Brannick MT, Salas E, Prince C (eds) Team performance and measurement: theory, methods, and applications. Lawrence Erlbaum, Mahwah, NJ, pp 19–43

Driskell JE, Salas E (1992) Collective behavior and team performance. Hum Factors 34:277–288

Eby LT, Dobbins GH (1997) Collectivistic orientation in teams: an individual and group-level analysis. J Organ Behav 18:275–295

Eby LT, Meade AW, Parisi AG, Douthitt S (1999) The development of an individual-level teamwork expectations measure and the application of a within-group agreement statistic to assess shared expectations for teamwork. Organ Res Methods 2:366–394

Edmondson A (1999) Psychological safety and learning behavior in work teams. Admin Sci Quart 44:350–383

Edmondson A (2003) Speaking up in the operating room: how team leaders promote learning in interdisciplinary action teams. J Manage Stud 40:1419–1452

Ellis APJ, Hollenbeck JR, Ilgen DR, Porter COLH, West BJ, Moon HK (2003) Team learning, collectively connecting the dots. J Appl Psychol 88:821–835

Espinosa JA, Lerch J, Kraut R (2004) Explicit vs. implicit coordination mechanisms and task dependencies: One size does not fit all. In: Salas E, Fiore SM (eds) Team cognition: understanding the factors that drive process and performance. APA Books, Washington, DC, pp 107–129

Faraj S, Sproull L (2000) Coordinating expertise in software development teams. Manage Sci 46:1554–1568

Fiore SM, Salas E, Cannon-Bowers JA (2001) Group dynamics and shared mental models development. In: London M (ed) How people evaluate others in organizations. Lawrence Erlbaum, Mahwah, NJ, pp 309–336

Fiore SM, Salas E, Cuevas HM, Bowers CA (2003) Distributed coordination space: toward a theory of distributed team process and performance. Theor Issues Ergon Sci 4:340–364

Georgakopoulos D, Hornick M, Sheth A (1995) An overview of workflow management: from process modeling to workflow automation infrastructure. Distrib Parallel Dat 3:119–153

Grote G, Zala-Mezö E, Grommes P (2003) Effects of standardization on coordination and communication in high workload situations. In: Dietrich R (ed) Communication in high risk environments. Helmut Buske, Hamburg, pp 127–154

Gully SM, Incalcaterra KA, Joshi A, Beaubien JM (2002) A meta-analysis of team-efficacy, potency, and performance: Interdependence and level of analysis as moderators of observed relationships. J Appl Psychol 87:819–832

Hacker W (2003) Action regulation theory: a practical tool for the design of modern work processes? Eur J Work Organ Psychol 12:105–130

Hackman JR (1987) The design of work teams. In: Lorsch JW (ed) Handbook of organizational behavior. Prentice Hall, Englewood Cliffs, NJ, pp 67–102

Hackman J, Oldham G (1975) Development of the job diagnostic survey. J Appl Psychol 60:159–170

Hackman JR, Wageman R (2005) A theory of team coaching. Acad Manage Rev 30:269–287

Hess KP, Entin EE, Hess SM, Hutchins SG, Kemple WG, Kleinman DL, Hocevar SP, Serfaty D (2000) Building adaptive organizations: a bridge from basic research to operational exercises. In: Proceedings of the 2000 Command and Control Research and Technology Symposium, Monterey

Hoegl M, Gemuenden HG (2001) Teamwork quality and the success of innovative projects: a theoretical concept and empirical evidence. Organ Sci 12:435–449

Jentsch F, Barnett J, Bowers CA, Salas E (1999) Who is flying this plane anyway? What mishaps tell us about crew member role assignment and air crew situational awareness. Hum Factors 41:1–14

Jones GR, George JM (1998) The experience and evolution of trust: implications for cooperation and teamwork. Acad Manage Rev 23:531–546

Jung DI, Sosik J (2003) Group potency and collective efficacy: examining their predictive validity, level of analysis, and effects of performance feedback on future group performance. Group Organ Manage 28:366–391

Kacmar KM, Witt LA, Zivnuska S, Gully SM (2003) The interactive effect of leader-member exchange and communication frequency on performance ratings. J Appl Psychol 88:764–772

Klein KJ, Kozlowski SWJ (eds) (2000) Multilevel theory, research, and methods in organizations. Jossey-Bass, San Francisco

Klein KJ, Tosi H, Cannella AA Jr (1999) Multilevel theory building: benefits, barriers, and new developments. Acad Manage Rev 24:243–248

Klimoski RJ, Mohammed S (1994) Team mental model: construct or metaphor? J Manage 20:403–437

Kluger AN, DeNisi A (1996) The effects of feedback interventions on performance: a historical review, a meta-analysis, and a preliminary feedback intervention theory. Psychol Bull 119:254–284

Kolbe M (2007) Koordination von Entscheidungsprozessen in Gruppen. Die Bedeutung expliziter Koordinationsmechanismen [in German]. VDM Verlag Dr. Müller, Saarbrücken

Kozlowski SWJ, Bell BS (2003) Work groups and teams in organizations. In: Borman WC, Ilgen DR, Klimoski RJ (eds) Handbook of psychology: industrial and organizational psychology, vol 12. Wiley, London, pp 333–375

Kozlowski SWJ, Ilgen DR (2006) Enhancing the effectiveness of work groups and teams. Psychol Sci Public Inter 7:77–124

Kraiger K, Wenzel LC (1997) Conceptual development and empirical evaluation of measures of shared mental models as indicators of team effectiveness. In: Brannick MT, Salas E, Prince C (eds) Team performance assessment and measurement: theory, methods, and applications. Lawrence Erlbaum, Mahwah, NJ, pp 45–84

Kraut R, Streeter L (1995) Coordination in large scale software development. Commun ACM 38:69–81

Langan-Fox J (2003) Team mental models and group processes. In: West MA, Tjosvold D, Smith KG (eds) International handbook of organizational teamwork and cooperative working. Wiley, London, pp 321–360

Langan-Fox J, Code S, Langfield-Smith K (2000) Team mental models: techniques, methods, and analytic approaches. Hum Factors 42:242–271

Lewis K (2003) Measuring transactive memory systems in the field: scale development and validation. J Appl Psychol 88:587–604

MacMillan J, Paley M, Entin EB, Entin EE (2004) Questionnaires for distributed assessment of team mutual awareness. In: Stanton NA, Hedge A, Brookhius K, Salas E, Hendrick H (eds) Handbook of human factors and ergonomics methods. Taylor and Francis, London, pp 511–519

Malone TW, Crowston K (1994) The interdisciplinary study of coordination. ACM Comput Surv 26:87–119

March J, Simon HA (1958) Organizations. Wiley, New York

Marks MA, Panzer FJ (2004) The influence of team monitoring on team processes and performance. Hum Perform 17:25–42

Marks MA, Mathieu JE, Zaccaro SJ (2001) A temporally based framework and taxonomy of team processes. Acad Manage Rev 26:356–376

Martins LL, Gilson LL, Maynard TM (2004) Virtual teams: what do we know and where do we go from here? J Manage 30:805–835

Massey AP, Montoya-Weiss MM, Hung YT (2003) Because time matters: temporal coordination in global virtual project teams. J Manage Inform Syst 19:129–155

Mathieu JE, Heffner TS, Goodwin GF, Salas E, Cannon-Bowers J (2000) The influence of shared mental models on team process and performance. J Appl Psychol 85:273–283

McGrath JE, Tschan F (2004) Temporal matters in social psychology: examining the role of time in lives of groups and individuals. American Psychological Association, Washington, DC

McIntyre RM, Salas E (1995) Measuring and managing for team performance: Emerging principles from complex environments. In: Guzzo R, Salas E (eds) Team effectiveness and decision making in organizations. Jossey-Bass, San Francisco, pp 9–45

Mintzberg H (1980) The nature of managerial work. Prentice Hall, Englewood Cliffs, NJ

Mohammed S, Dumville BC (2001) Team mental models in a team knowledge framework: Expanding theory and measurement across disciplinary boundaries. J Organ Behav 22:89–106

Moreland RL, Myaskovsky L (2000) Exploring the performance benefits of group training: Transactive memory or improved communication? Organ Behav Hum Dec 82:117–133

Mullen B, Copper C (1994) The relation between group cohesiveness and performance: an integration. Psychol Bull 115:210–227

O'Reilly CH, Roberts KA (1976) Relationships among components of credibility and communication behaviors in work units. J Appl Psychol 61:99–102

Pearce CL, Sims HP (2002) Vertical versus shared leadership as predictors of the effectiveness of change management teams: an examination of aversive, directive, transactional, transformational and empowering leader behaviors. Group Dyn-Theor Res 6:172–197

Pinto MB, Pinto JK (1990) Project team communication and crossfunctional cooperation in new program development. J Prod Innovat Manag 7:200–212

Rico R, Sánchez-Manzanares M, Gil F, Gibson C (2008) Team implicit coordination process: a team knowledge-based approach. Acad Manage Rev 33:163–184

Salas E, Prince C, Baker DP, Shrestha L (1995) Situation awareness in team performance: implications for measurement and training. Hum Factors 37:123–136

Sarin S, McDermott C (2003) The effect of team leader characteristics on learning, knowledge application, and performance of cross functional new product development teams. Decision Sci 34:707–739

Steelman LA, Levy PE, Snell AF (2004) The feedback environment scales (FES): construct definition, measurement and validation. Educ Psychol Meas 64:165–184

Stout RJ, Cannon-Bowers JA, Salas E, Milanovich DM (1999) Planning, shared mental models, and coordinated performance: an empirical link is established. Hum Factors 41:61–71

Tschan F (1995) Communication enhances small group performance if it conforms to task requirements: the concept of ideal communication cycles. Basic Appl Soc Psych 17:371–393

Tschan F (2000) Produktivität in Kleingruppen. Was machen produktive Gruppen anders und besser? [in German]. Huber, Bern

Van de Ven AH, Delbecq LA, Koening RJ (1976) Determinants of coordination modes within organizations. Am Sociol Rev 41:322–338

van der Aalst W, van Hee K (2004) Workflow management: models, methods, and systems. MIT, Cambridge, MA

Webber SS, Chen G, Payne SC, Marsh SM, Zaccaro SJ (2000) Enhancing team mental model measurement with performance appraisal practices. Organ Res Methods 3:307–322

Wegner DM (1987) Transactive memory: a contemporary analysis of the group mind. In: Mullen G, Goethals G (eds) Theories of group behavior. Springer-Verlag, New York, pp 185–208

Wegner DM (1995) A computer network model of human transactive memory. Soc Cognition 13:319–339

Weinkauf K, Hoegl M (2002) Team leadership activities in different project phases. Team Perform Manage 8:171–182

Wittenbaum GM, Stasser G, Merry CJ (1996) Tacit coordination in anticipation of small group task completion. J Exp Soc Psychol 32:129–152

Wittenbaum GM, Vaughan SI, Stasser G (1998) Coordination in task-performing groups. In: Tindale RS, Heath L, Edwards J, Posavac EJ, Bryant FB, Suarez-Balcazar Y, Henderson-King E, Myers J (eds) Theory and research on small groups. Plenum, New York, pp 177–204

Zaccaro S, Marks M (1999) The roles of leaders in high-performance teams. In: Sundstrom E et al (eds) Supporting work team effectiveness: best management practices for fostering high performance. Jossey-Bass, San Francisco, pp 95–125

Zalesny MD, Salas E, Prince C (1995) Conceptual and measurement issues in coordination: Implications for team behavior and performance. In: Ferris GR (ed) Research in personnel and human resources management, vol 13, JAI Press. Greenwich, CT, pp 81–115

Chapter 8
Assessing Team Coordination Potential

Kristina Lauche

Abstract Unlike previous chapters that offer measurement approaches for the process or outcome of coordination, this chapter addresses the context and organisational setting in which team coordination occurs. The organisational design presents both opportunities and constraints for teams to manage themselves. This chapter describes a methodological approach for analysing team coordination potential within the organisational context. Focussing on teams in product innovation, a set of five criteria has been developed: (1) autonomy and local control; (2) involvement in problem setting; (3) feedback; (4) team self-regulation; and (5) organisational support for innovation. The actual working conditions are assessed using semi-structured interviews and observations of teamwork.

8.1 Introduction

This chapter proposes a framework for analysing team coordination in terms of constraints and opportunities of the organisational setting. This means that the emphasis is on assessing the external conditions for team coordination rather than team cognition or behaviour. In this, the chapter follows the methodological approach of job analysis as it is used in work psychology for the assessment, evaluation, and design of jobs (Hacker 2003): An external specialist assesses to what extent a given job offers the potential for self-regulation and learning. While most job analysis methods focus on the individual, this approach specifically addresses the potential for team coordination. If the organisational context shows a high potential for team coordination, members are enabled and empowered to coordinate who does what

K. Lauche (✉)
Nijmegen School of Management, Radboud University Nijmegen, Thomas van Aquinostraat 3, 6500 HK Nijmegen, The Netherlands
e-mail: k.lauche@fm.ru.nl

M. Boos et al. (eds.), *Coordination in Human and Primate Groups*,
DOI 10.1007/978-3-642-15355-6_8, © Springer-Verlag Berlin Heidelberg 2011

among themselves, making it more likely that teams will engage in explicit and implicit coordination behaviour. If the organisational context for team coordination potential is low, team members may still try to liaise with others, but they are likely to encounter obstacles in making contact or will be told to mind their own business or follow instructions. In the following example, the organisational context offered high coordination potential for certain aspects: The team members worked co-located in one open-plan office, which enabled frequent interactions and easy access to the same information about their project. Roles and responsibilities were decided within the team with little interference from management. The team proposed solutions for the project and campaigned for a more appropriate software system. When the company initially purchased only one license, team members decided to implement shift work to make best use of it until eventually the company agreed to buy more licenses. Internally, this team managed to address the challenges quite well. However, for the strategic directions of the project, the team was not authorised to make decisions nor did it have access to relevant information. Instead, their task and information input was dependent upon other departments that placed contradictory requirements on their work. This made it virtually impossible for the team to coordinate the more strategic aspects of their work.

The method described in this chapter was developed for the domain of product development, in other words, a context in which 'innovation' is part of the primary task. This domain was chosen because today's complex products are rarely the outcome of individual efforts: Although great inventions are often associated with the names of individuals, innovating is typically a collective process (Hargadon and Bechky 2006). Therefore, innovating typically requires a substantial amount of team coordination and cross-functional collaboration (Edmondson and Nembhard 2009). Product development is also an open-ended task that involves uncertain processes and experimentation (Eisenhardt and Tabrizi 1995), which means that teams can shape the scope of their activity to a larger extent than in other domains such as aviation or medicine. This also means that if innovation teams are granted a high degree of decision latitude, the form of team coordination can largely depend upon the team itself. Figure 8.1 illustrates typical outcome variables for this domain: A high degree of team coordination in innovation projects should lead to better integration of different knowledge sources, improve feasibility and viability of the product, and reduce time to market.

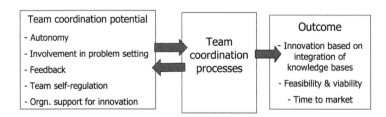

Fig. 8.1 Five criteria for team coordination potential

The chapter is structured as follows: Sect. 8.2 introduces five criteria that will be used to assess team coordination potential as it relates to organisational context and explains how these were derived from the innovation management literature and theoretical concepts in work psychology. Then the procedure for collecting and analysing data via interviews and observations is described in Sect. 8.3. The use of the method is illustrated in Sect. 8.4 with case descriptions from a field study. The chapter concludes with a discussion in Sect. 8.5 of the contribution and limitation of the method and possible transfer to other domains.

8.2 Criteria for Team Coordination Potential in Innovation

The criteria for organisational context factors that support team coordination in innovation were developed on the basis of general models of human-centred job design, on a model of collective self-regulation in teams, and the organisational literature on structures that enhance innovation. While these are different streams of literature, they provide converging arguments for conditions under which team coordination should be both more likely and more effective.

The criteria for human-centred job design specify how tasks should be designed to promote both efficiency and human well-being at work (Hacker 1995; Warr 1994), originally on the individual level but more recently also on the team level (Weber 1997). The criteria are based on empirical studies (Patterson et al. 2004; Wall et al. 1990) and on action regulation theory (Frese and Zapf 1994; Hacker 1994, 2003). The underlying model is dialectical: It proposes that one's work environment provides affordances and constraints that influence what people perceive they can and should do, but people also shape and re-define constraints and opportunities (see Fig. 8.1). For an overview of methods, see Dunckel (1999).

In order to adapt the method to the domain of product development, the innovation management literature was consulted regarding appropriate organisational structures to enhance innovation (Brown and Eisenhardt 1995; Fagerberg et al. 2005). There are conflicting theories that depict innovation either as a chaotic process that requires creative freedom, or as an extraordinary phenomenon in otherwise inert organisations that requires managerial vision and intervention. This chapter draws on Dougherty (2008), who argued that from the perspective of structuration theory (Giddens 1982), these contradicting streams of literature are two incarnations of fundamental principles on how organisations function: social action and social constraint. Arguments from Dougherty's (2008) review and findings from interviews with product developers (Lauche 2001) were used to specify the generic criteria for team coordination potential for the domain of product innovation. This resulted in a set of five criteria for team coordination potential (see Fig. 8.1), which will be described in more detail in the following sections.

8.2.1 Autonomy

The first criterion and prerequisite for team coordination potential is autonomy or local control: being able to monitor and influence one's work process effectively. Autonomy has been conceptualised as an antecedent for accepting responsibility and ownership of the task (Frese and Zapf 1994; Hackman and Oldham 1974). More local control makes work systems more effective because problems can be remedied faster (Wall et al. 1990). In the innovation literature, the same idea is expressed as 'energising work by directly resourcing innovation' in terms of access to others' time and attention, control over application of one's own expertise, and access to multiple options for problems (Dougherty 2008). This means that designers can make or influence decisions on the innovation strategy and product portfolio, are involved in shaping the way innovation is managed, and are included in the selection of strategic partners.

8.2.2 Involvement in Problem Setting

This second criterion refers to involvement in identifying and defining which problems should be solved and which aims are to be achieved. This criterion appears as goal setting and planning in the action regulation literature (Frese and Zapf 1994; Hacker 2003) and is frequently addressed in job design assessments. In the product innovation domain, the notion of planning and goal setting has been further specified as involvement in problem setting for the context of product use and strategic intentions (Lauche 2005a, b). From a psychological perspective, this means that innovation teams can access or obtain information about the innovation task in order to generate an appropriate conceptual representation regarding the intended outcome (Oschanin 1976). This conceptual representation serves as a motivator (a 'vision' of the new product) and as a criterion against which the outcome is checked. The concept is similar to that of mental models, which has also been used to explain how design teams achieve coordination (Badke-Schaub et al. 2007). For innovation work, information about the object of innovation (e.g. market demands and the context of use, emerging technological options, the strategic direction of the company) is vital to avoid solving the wrong problem. Designers need to understand the context of use to make valid assumptions regarding how this context may change with the introduction of the product they are designing (Lauche 2005a, b; McCarthy 2000; Schmidt 2000). If the information that designers have access to is inappropriate or incomplete, this can result in product failures or product uses different than anticipated, as assumptions about the market and/or product were based on inappropriate or incomplete information. Very often, product developers only receive a design brief with second-hand information from marketing, which makes it difficult to empathise with the prospective user, as the findings from user studies are typically not communicated in a form that conveys the richness of data (Postma et al. 2009). Preferably, product developers should be in a position to actively obtain information by engaging in user observation, desk research, and negotiation about the innovation strategy.

8.2.3 Feedback

Feedback as a criterion for team coordination potential is considered important in virtually all models of job design (Warr 2002). It is defined as knowledge about outcomes, which includes test results, performance indicators, and peer review. Feedback enables monitoring of achievements and adaptation of strategies and mental models (Hacker 2003; Hackman and Oldham 1974). For product innovation, early feedback can be used to mitigate the risk inherent to any innovation by testing the feasibility and viability of concepts. Feedback can help to define the scope of an innovation project and act as an opportunity for corrective action and learning. Limited testing facilities, late or insufficient critical review from colleagues and management, and slow feedback from production and sales can delay and hamper innovation projects. However, if product developers engage in field testing and are in a position to obtain feedback from marketing, production, and management during the innovation process, this should shorten the time to market and enable organisational learning within and beyond specific projects.

8.2.4 Potential for Team Self-Regulation

The fourth criterion explicitly refers to the potential to act and self-regulate as a team above and beyond individual autonomy. In action regulation theory, self-regulation in teams has been conceptualised as collective action regulation (Weber 1997), specifically referring to the need to connect individual autonomous agents. Zölch's (2001) concept of interlacing of actions addresses the coordination needs that arise between semi-autonomous teams, which should also be the responsibility of the teams if the company does not want to revert back to a hierarchical structure. Similar concepts for white-collar work can be found in the organisational literature under the concept of 'communities of practice' (Lave and Wenger 1991). Organisations that treat their members as a community of self-managed, competent practitioners can strengthen formal institutions of professional practice and create a basis on which teams can innovate (Dougherty 2008).

As a criterion for the assessment of the organisational context relative to how it either supports or hinders team coordination potential, this means that with a high degree of team self-regulation, members have the possibility and authority to collectively decide about the way they manage their work. This includes the planning and allocation of work, the selection of team members and decisions about roles, internal coordination and leadership functions, as well as boundary negotiations with other teams or management. Innovation is not treated as a sequential process with tasks carried out in isolation, as the high degree of self-regulation enables the team to address innovation from a cross-disciplinary perspective and to resolve technical problems or delays as they arise.

8.2.5 Organisational Support for Innovation

As technology and science studies have shown, the myth of the genius inventor is typically an insufficient explanation for innovation (Miettinen 1996). Product innovation cannot be achieved on an 'individual' basis if the organisational context is not supportive and there is insufficient interdisciplinary communication and understanding for design (Edmondson and Nembhard 2009; Lauche et al. 1999). The criterion of organisational support for innovation has been conceptualised as the commitment to and attributed importance of innovation that senior management show both in their interaction with product developers and in their allocation of resources (West 2002). A recent meta-analysis confirmed support for innovation as one of the strongest team-level predictors of innovation (Hülsheger et al. 2009). While most studies measure support for innovation as individual perceptions in a survey, here it is treated as an assessment of the objective context conditions in the organisation. Under high-support conditions, innovation projects are appropriately resourced and their strategic alignment and performance receive managerial attention. Under low-support conditions, developers will be fighting for resources, which means that they have to find time for innovation work amid their daily workload and may encounter interruptions, time pressure, and multi-tasking (Hacker 1995), which can be detrimental to good design practice.

8.2.6 Effect of Criteria on Coordination Processes

Ideally, teams should find themselves under supportive conditions for all five criteria in order to coordinate their innovation efforts successfully. For unfavourable conditions, there are specific detrimental effects; however, to a certain extent these can be compensated for by other criteria. For instance, a lack of organisational support can be compensated for by a high degree of autonomy, high involvement in problem setting, and sufficient feedback. In this case, teams should show a high degree of coordination within the team and limited coordination with the rest of the organisation, akin to the situation at most universities. A lack of team self-regulation tends to lead to less frequent team coordination, as this is either not possible or not considered necessary in the organisation. This condition of less frequent team coordination could still lead to acceptable outcomes in cases of low task interdependency but it will create problems and delays at the interfaces of more complex products. To some extent, involvement in problem setting and feedback can compensate for each other: If team members have access to a broad range of information sources in defining the scope of their task, they are less dependent on feedback; and if the sources of feedback are plentiful, the team can pursue a more exploratory approach and compensate for the lack of involvement in goal setting. A lack of involvement in problem setting has the tendency to reduce the scope and amount of coordination, as there are fewer issues that require coordination. A lack of feedback should not affect coordination as such but will make the team less

8 Assessing Team Coordination Potential

adaptive. A lack of autonomy should reduce team coordination quite drastically, as there is nothing to coordinate: Members simply await their respective orders.

8.3 Description of the Methodological Approach

The following section explains the methodological approach for assessing the potential for team coordination as affected by the organisational setting. Similar to other instruments for assessing organisational conditions (Dunckel 1999), the approach consists of expert ratings based on a combination of observations and interviews with job holders. Observable behaviour and verbal statements from interviewees are treated as indicators for underlying conditions, and it remains the role of the researcher to make a judgement. The theoretical model is normative in the sense that it presupposes certain organisational conditions to be more favourable for team coordination and innovative outcomes. The assessment can therefore be expressed as an ordinal scale, but one should be aware that the data are essentially qualitative and refer to types of conditions.

What kind of data should be collected, and how should it be analysed?

8.3.1 Data Collection

As is the case for most job analysis methods, the assessment of team coordination potential requires the presence of the researcher during periods of normal work as a non-participating observer. The most suitable time samples are team meetings during which the team works on an innovation project, such as defining the scope, generating ideas, consolidating options, reviewing the progress of a project, and making decisions. The researcher should take notes on the five criteria (see Fig. 8.1) without interfering with what the team does.

If permission can be obtained, the ideal way is to capture the team interaction on video or audio to allow for a more detailed analysis and consolidation of coordination potential ratings within the research team. However, in the domain of new product development, negotiating access can often be a challenge in itself, as the content of the work is commercially sensitive. Field observations in this area therefore require rapport building and professional conduct on behalf of the researcher to establish trust. It is also advisable that the observers are sufficiently familiar with the domain to be able to understand and make sense of the technical discussions. We typically worked with interdisciplinary teams of one domain expert (designer or engineer) and one psychologist.

Since transcribing and analysing video or audio recordings is a very time-consuming process, it is useful to also take field notes to generate an overview of longer meetings and then decide which passages to analyse in more detail. It has proven useful to record notes on a standard form that tabled time, actors, and what they said or did, and also contained a separate free-form space for observer comments (Lauche et al. 2001) (see Fig. 8.2).

inno **pro**	**Online Observation**	Subtopic:_____	page:___ /___
Time Who	Content, verbal interaction		Memos, meta-Observations

Fig. 8.2 Standard form for recording verbal team interaction and notes for further analysis

Observing overt behaviour typically provides some evidence for assessment, but other aspects may be difficult to observe or may not occur in a given situation. It is therefore common practice in job analysis to complement the observation with a semi-structured interview. Semi-structured interview guidelines with possible response categories can be found in Lauche (2001). The data can be further complemented by analysing documents and meeting minutes and by using questionnaire scales such as the scale for collective action regulation in innovation teams (Weber and Lauche 2010).

8.3.2 Data Analysis

As is typical for qualitative research, the analysis for team coordination potential is an iterative process of forming assumptions based on part of the material and consolidating these assumptions through a systematic analysis of the remaining material. The researcher's notes will form the starting point to determine which parts of a meeting should be analysed in more detail. The most efficient form is to first map out the overall process, decide which sections contain the most relevant information on team coordination (e.g. initiation for new phases, critical situations, allocation of tasks at the end), and then proceed with detailed analysis. The selected parts are then reviewed from the video recording, again taking notes for evidence of the five coordination potential criteria. If the coordination potential assessment forms part of a larger study that also addresses the content of team interaction, it is advisable to transcribe all relevant video and audio recordings in order to make them more accessible for further analysis and discussions within the research team.

The transcripts are then coded using the five criteria for team coordination potential in innovation work (see Fig. 8.1). Each criterion is rated on a scale of 1–6 based on a description of three levels: impoverished, conventional, and expansive (see Table 8.1).

Impoverished conditions refer to organisational conditions that make it difficult for teams to coordinate their innovative activities due to lack of managerial support, unavailability of necessary information about threats and opportunities, or shortage of resources. Innovation is not valued, and product development is not an established organisational function. Anyone attempting to innovate under impoverished conditions will not be in a position to obtain an adequate picture of the requirements for a new product and its viability, and the impoverished nature of the organisational culture will also not be supportive of anyone spending time to interact with others to scope a project or generate ideas.

Conventional conditions refer to a traditional functional structure with a departmentalised, Tayloristic work organisation. The main focus of the organisation is on the production and reproduction of existing strengths. New products are developed as part of a strategy that is handed down to the designers without any involvement of their expertise. Marketing and production only become involved as part of codified procedures and a sequential execution of projects. This type of organisation can produce new products but is less likely to engage in radical innovation or expand its capability to innovate.

Expansive conditions are based on the ideas of integrated product development (Ehrlenspiel 1995) and support and empowerment for innovation (West and Farr 1990). They enable designers to expand and shape the innovation activity, resembling the idea of expansive learning (Engeström 1987). Designers and managers share a vision of expanding the scope of their business, and the product development process is managed as the trajectory for the future of the business. If required, new technology is developed in cooperation with a network of suppliers. Product developers closely interact with marketing and production specialists and are actively involved in strategic discussions to identify threats and opportunities for innovation. Expansive conditions have more potential for radical innovation, as the scope of innovation itself can be transformed.

8.4 Illustration of Method with Case Descriptions from a Field Study

The application of the proposed assessment methodology will be illustrated using three examples from a field study (Lauche and Erez 2009). The cases broadly represent the 'impoverished', 'conventional', and 'expansive' levels of team coordination potential explained in the category system in Sect. 8.3.

Table 8.1 Five criteria for coordination potential plus definitions of ratings

Autonomy and local control	1–2 Impoverished: lack of autonomy and control and undue dependence on other actors that result in slow progress and feelings of helplessness. 3–4 Conventional: control and decision latitude over one's own work but little influence on the overall project or strategy. 5–6 Expansive: control to make or influence decisions on innovation strategy and product portfolio, involved in selection of strategic partners, and required to shape the way innovation is managed.
Involvement in problem setting	1–2 Impoverished: no involvement in problem setting, information about requirements inappropriate or incomplete. 3–4 Conventional: designers receive a brief and work out detailed specifications from the information they are given. Information on the context of the product is handed down to them from other actors such as marketing representatives. 5–6 Expansive: designers involved in negotiation of innovation strategy; have access and time to gather or actively obtain information on user needs, market developments, strategic intentions, technological developments.
Feedback	1–2 Impoverished: limited testing facilities, late or insufficient critical review from colleagues and management, and slow feedback from production and sales. 3–4 Conventional: some information on the fulfilment of specifications as part of the design task (testing, simulation) but limited feedback after handover (production problems, sales figures, customer feedback). 5–6 Expansive: testing facilities and active involvement in field-testing, access to feedback from users, marketing, production and management during the innovation process.
Team self-regulation	1–2 Impoverished: teams do not have the opportunity to communicate or authority to decide on how they coordinate their work; problems such as role duplication and unclear responsibilities remain unresolved. 3–4 Conventional: teams work according to a waterfall model where they hand over the work sequentially. The organisational division of labour partly replaces the need for direct communication, yet it leaves little opportunity for an integrated product innovation approach and any delays cannot be compensated for. 5–6 Expansive: innovation seen as a task that necessitates team interaction for content of product and detailed planning of project; empowerment of teams to self-regulate and maintain a professional community.
Organisational support for innovation	1–2 Impoverished: New Product Development not established as a function, importance of product innovation not recognised. Designers have to find time for innovation work amidst their daily workload and constantly fight for resources. 3–4 Conventional: established product development function but no specific support for innovation and no involvement in company strategy on innovation. 5–6 Expansive: innovation seen as essential for company strategy, commitment from senior management, understanding for the needs of R&D; proactive investment of resources during the early stages of product innovation.

8.4.1 Example 1: Impoverished Innovation Coordination Opportunities

The company produced customised investment goods for the manufacturing industry, and a new generation of machines was developed about every 5 years. While the engineering work was carried out in-house, the sales organisation was located off-site in two different locations and the R&D of new machines was outsourced. As part of a strategy change, the company aimed to establish product innovation as an internal competence and head-hunted an R&D team from another company. As a result, the team members already knew each other but were new to the organisational task. Figure 8.3 shows the rating for all five coordination criteria: 1 is the lowest rating, and 6 is the highest.

Autonomy was rated as 2: The team had complete control regarding its internal organisation but could only attempt to influence decisions about the scope and timing of its project, requirements, budget, and support. It prepared 20 different concepts, but the choice was made by a review team on the management level. Team coordination was focussed on the technical aspects of the product.

Involvement in problem setting was rated as 1. The strategic aims were defined by the market organisation and the team was not in a position to question or amend the requirements because of a lack of information and power. The physical distance to the market organisations hindered coordination. As the lack of information was not something that the team itself could compensate for, it concentrated its coordination efforts on explicating open questions. However, some major issues were still left undefined by management at the end of the concept stage.

Feedback was rated as 3. The team received feedback from the management review and from semi-annual technical sales meetings. It also made obtaining feedback part of its own coordination efforts: Members were tasked to perform

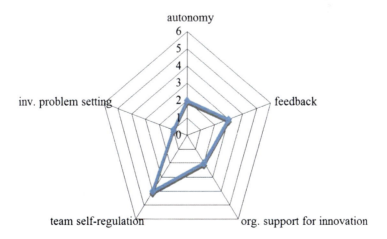

Fig. 8.3 Rating for team coordination potential for impoverished example

numerical simulations and tests of crucial components, which the company supported by allocating the required funds for a prototype.

Team self-regulation was rated as 4, as it was high within the newly recruited team but clearly restricted to internal task allocation. The team worked co-located in one room and resorted to the tried and tested roles it had acquired beforehand in order to meet the deadlines. Team members' interactions with other parts of the company were limited, and only the team leader took part in reviews with management. The company held infrequent meetings with the sales staff and engineers from all three locations, and when they did, the meetings mainly served coordination purposes.

Organisational support for innovation was rated as 2. This was the first attempt to establish new product development as a function. While managers were keen to see the project succeed, the team felt that its requests were not met with much understanding. As the lack of support was something the team could not compensate for with its own coordination efforts, it caused not only delays but also frustration.

8.4.2 *Example 2: Conventional Innovation Coordination Opportunities*

The company in the conventional example formed part of a larger holding. The task was to develop the next generation of automated special-purpose machines for high-volume manufacturing. Compared to the existing portfolio, the new machine would employ more technological sophistication and provide more flexibility. Also, the sales and engineering process of the new machine type was to be innovated by new tools for customer-driven design. After mainly relying on externals for the concept development, an R&D team was gradually established with newly recruited staff and delegates from other departments for cross-functional input (Fig. 8.4).

Autonomy was rated as 4: The project manager was given a lot of freedom to manage the project and managed any technical issues in liaison with an external design consultancy. The team members contributed to coordination by developing their own guidelines for project management and programming styles. However, for issues other than technology, the team could not coordinate resources and was dependent upon other functions in the company (such as, HR, Sales).

The involvement in problem setting was rated as 3, as it was high for the initial team of externals but restricted to requirement for subsystems for those who joined later. A market analysis and customer surveys were planned but were dropped due to time constraints. The core team was well informed about new technical developments and regularly attended training events, but the cross-functional members of the wider team were less clear about their task and back-delegated some of their tasks to management.

Feedback was rated as 5, as it was available from extensive simulation and testing as well as the sales team and customers. The new process involved much iteration, which allowed for a learning process to happen, as well as provided built-in opportunities for course-correcting. This led to high coordination in the

8 Assessing Team Coordination Potential

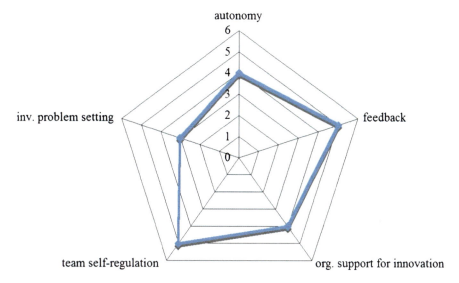

Fig. 8.4 Rating for team coordination potential for conventional example

cross-functional task force, as service staff returned customers' and operators' feedback, and assembly staff also readily returned any issues.

Team self-regulation was rated as 5. The project management involved a lot of coordination with other players inside and outside the company. A cross-functional task force was drawn in to help coordinate the project before the R&D team was fully established.

Organisational support for innovation was rated as 4. The project was clearly recognised as important, but the team was also confronted with doubts about its technical success and fears about the implications for the rest of the company. The lack of support was something that the team members could not compensate for with their own coordination efforts: They scheduled review meetings, but some failed to attend; as a result, important decisions could not be made. Team members supported each other but could not replace the practical resources such as of journal or Internet access.

8.4.3 Example 3: Expansive Innovation Coordination Conditions

The expansive example comes from a company with an established R&D function. Innovation was a managed process with a continuous stream of new products. The project addressed here was special in the sense that it was aimed at radical innovation with new customer benefits and new markets. This had implications for the underlying business model, which required additional persuasion to convince senior management (Fig. 8.5).

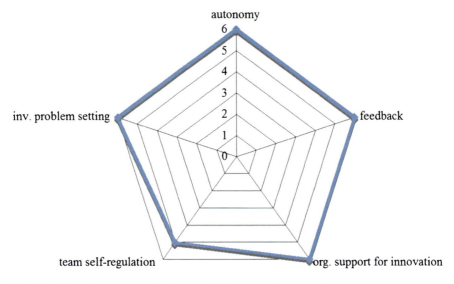

Fig. 8.5 Rating for team coordination potential for expansive example

The team was given a high degree of autonomy (rated as 6) as long as it managed to convince budget holders that the plans would be successful. The project was team members' own initiative – they were not given a brief. The team members coordinated the definition of the scope themselves and enlisted external facilitators as they saw fit.

Involvement in problem setting was rated as 6, as the definition of the task was part of the project activity. The team scanned technological developments and conducted its own observations of end users before defining the scope of the project. Team members coordinated the project planning by spelling out implications of backward compatibility and software integration with other products.

Feedback was rated as 6, as the designers actively sought and had good means to obtain feedback through feasibility tests, prototyping, and presentations to sales staff. Team members gave feedback during meetings, and senior management regularly reviewed the project.

Team self-regulation was rated as 5. It was an explicit issue and the team always discussed allocation of tasks during its meetings. However, coordination was complicated by the fact that most team members were also assigned to other projects, which were given priority as they approached product launch.

Organisational support for innovation was rated as 6. There was good understanding of innovation management, professional practice was valued, and financial resources were available once a project had been approved. It was part of the organisational culture to embrace the uncertainty of innovation with the flow of idea generation as well as the drawbacks and long nights of detailed design. This enabled the team to focus its coordination efforts on the innovation itself rather than the necessary infrastructure.

8.5 Discussion

Unlike previous chapters that offer measurement approaches for the process or outcome of coordination, this chapter addresses the organisational setting in which team coordination occurs. While most research analyses the coordination processes as such in some sort of context, this approach zooms in on the context as a relevant independent variable that can affect the likelihood and form of the coordination processes. While the organisational context does not fully determine what people do, it provides parameters of opportunities or constraints for potential actions to either occur or be assuaged. This approach to team coordination assessment therefore broadens the scope of the coordination model introduced in this book in terms of organisational context as an input factor, and stipulates effects of better coordination on outcome variables such as successful implementation of innovation. The approach can also be complemented with other measures that more specifically address the *form* of coordination.

The model is not conceptualised as a deterministic relationship; instead, it assumes that constraints and opportunities afford certain behaviours and thus make them more *likely* to occur, but they do not directly determine occurrences of initiative, creativity, perseverance, or performance. All three examples, in fact, illustrate that innovators struggle with the circumstances they encounter and try to pursue an innovation that their company does not fully support. However, under low- or conventional innovation conditions, it is more difficult, and at times impossible, to achieve radical innovation.

The specific contribution of this assessment methodology lies in its focus on complex tasks in the workplace and that the assessment is contextualised in the domain of product innovation. Based on action regulation theory and innovation management literature, five criteria for coordination potential as it relates to the organisational context were formulated: autonomy, involvement in problem setting, feedback, team self-regulation, and support for innovation. Ideally, the organisational context should support coordination on all five aspects. However, a high degree of autonomy, involvement in problem setting, and feedback can compensate for a lack of support for innovation, and team self-regulation is less critical in projects with less task interdependency. The model also assumes that there cannot be too much coordination potential: More opportunities to engage in coordination when it is required should help the team to achieve its task. The effort involved in coordinating more complex projects may add to the workload, but should also enable the team to address the challenges that arise from these complex tasks (Edmondson and Nembhard 2009).

On the basis of field observation and interviews, each criterion is rated on a six-point scale by the researchers. The measure thus relies on an expert judgement rather than on self-reported data. This means that observers need to be trained in the methodology and become sufficiently knowledgeable about the content in order to be able to rate the potential for team coordination. The reliability and validity of the assessment should be confirmed in independent double-analysis, using different investigators at different times, focussing on different subtasks of a job and different employees (Oesterreich and Bortz 1994).

The assessment approach presented here does not claim to offer a universal measure for coordination processes of humans and non-human primates. The methodology was developed for studying human activities in the workplace as an analytical tool for diagnosing the strengths and weaknesses of an organisation's ability to foster team coordination potential. If applied in a human sample, it can be combined with other measures that address team attitudes, cognition, and behaviour during the actual coordination processes. For research on non-human primates, the approach could potentially be applied to conditions of explorative and proactive behaviour, as it relies on an assessment of conditions and observations rather than on cognition. The definition of the five dimensions – autonomy, involvement in problem setting, feedback, team self-regulation, and support for innovation – could be adopted to describe to what extent the social organisation in non-human primate groups enables them to explore new territory, tools, food, or shelter.

References

Badke-Schaub P, Neumann A, Lauche K, Mohammed S (2007) Mental models in design teams: a valid approach to performance in design collaboration? CoDesign 3:4–19

Brown S, Eisenhardt K (1995) Product development: past research, present findings, and future directions. Acad Manage Rev 20:343–378

Dougherty D (2008) Bridging social constraint and social action to design organizations for innovation. Organ Stud 29:415–434

Dunckel H (ed) (1999) Handbuch psychologischer Arbeitsanalyseverfahren [in German]. VDF, Zürich

Edmondson AC, Nembhard IM (2009) Product development and learning in project teams: the challenges are the benefits. J Prod Innov Manage 26:123–138

Ehrlenspiel K (1995) Integrierte Produktentwicklung—Methoden für Prozessorganisation, Produkterstellung und Konstruktion [in German]. Hanser Verlag, München

Eisenhardt K, Tabrizi BN (1995) Accelerating adaptive processes: product innovation in the global computer industry. Admin Sci Quart 40:84–110

Engeström Y (1987) Learning by expanding: an activity-theoretical approach to developmental research. Orienta-Konsulit, Helsinki

Fagerberg J, Mowery DC, Nelson RR (2005) The Oxford handbook of innovation. Oxford University Press, Oxford

Frese M, Zapf D (1994) Action as the core of work psychology: a German approach. In: Triandis HC, Dunnette MD, Hough LM (eds) Handbook of industrial and organizational psychology. Consulting Psychologists Press, Palo Alto, CA, pp 271–340

Giddens A (1982) Profiles and critiques in social theory. University of California Press, Berkeley

Hacker W (1994) Action regulation theory and occupational psychology. Review of German empirical research since 1987. Ger J Psychol 18:91–120

Hacker W (1995) Arbeitstätigkeitsanalyse. Analyse und Bewertung psychischer Arbeitsanforderungen [in German]. Asanger, Heidelberg

Hacker W (2003) Action regulation theory: a practical tool for the design of modern work processes? Eur J Work Organ Psychol 12:105–130

Hackman JR, Oldham GR (1974) The job diagnostic survey: an instrument for the diagnosis of jobs and the evaluation of job redesign projects. Yale University, New Haven, CT

8 Assessing Team Coordination Potential

Hargadon AB, Bechky BA (2006) When collections of creatives become creative collectives: a field study of problem solving at work. Org Sci 17:484–500

Hülsheger UR, Anderson N, Salgado JF (2009) Team-level predictors of innovation at work: a comprehensive meta-analysis spanning three decades of research. J Appl Psychol 94: 1128–1145

Lauche K (2001) Qualitätshandeln in der Produktentwicklung. Theoretisches Modell, Analyseverfahren und Ergebnisse zu Förderungsmöglichkeiten [in German]. VDF Hochschulverlag, Zürich

Lauche K (2005a) Collaboration among designers: analysing an activity for system development. Comp Supp Coop Work 14:253–282

Lauche K (2005b) Job design for good design practice. Design Studies 26:191–213

Lauche K, Erez M (2009) Expansive innovation: what do teams do who redefine their company's strategy? Delft University of Technology, Delft

Lauche K, Verbeck A, Weber WG (1999) Multifunktionale Teams in der Produkt- und Prozessentwicklung [in German]. In: Zentrum für Integrierte Produktionssysteme der ETH Zürich (ed) Optimierung der Produkt- und Prozessentwicklung Beiträge aus dem Zentrum für integrierte Produktionssysteme. VDF, Zürich, pp 99–118

Lauche K, Ehbets Müller R, Mbiti K (2001) Understanding and supporting innovation in teams. In: 13th International Conference on Engineering Design. Design Society, Glasgow, pp 395–402

Lave J, Wenger E (1991) Situated learning. Legitimate peripheral participation. Cambridge University Press, Cambridge

McCarthy J (2000) The paradox of understanding work for design. Int J Hum-Comput Stud 53:197–219

Miettinen R (1996) Theories of invention and an industrial innovation. Sci Stud 2:34–48

Oesterreich R, Bortz J (1994) Zur Ermittlung der testtheoretischen Güte von Arbeitsanalyseverfahren [in German]. ABO Aktuell 3:2–8

Oschanin DA (1976) Dynamisches operatives Abbild und konzeptionelles Modell [in German]. Probleme und Ergebnisse der Psychologie 59:37–48

Patterson MG, West MA, Wall TD (2004) Integrated manufacturing, empowerment, and company performance. J Organ Behav 25:641–665

Postma C, Lauche K, Stappers PJ (2009) Trialogues: a framework for bridging the gap between people research and design. Designing pleasurable products and interfaces. Compiègne, France

Schmidt K (2000) The critical role of workplace studies in CSCW. In: Luff P, Hindmarsh J, Heath C (eds) Workplace studies recovering work practice and information system design. Cambridge University Press, Cambridge, pp 141–149

Wall TD, Corbett MJ, Martin R, Clegg CW, Jackson PR (1990) Advanced manufacturing technology, work design and performance: a change study. J Appl Psychol 75:691–697

Warr P (1994) A conceptual framework for the study of work and mental health. Work Stress 8:84–97

Warr P (2002) The study of well-being, behaviour and attitudes. In: Warr P (ed) Psychology at work. Penguin, London, pp 1–25

Weber WG (1997) Analyse von Gruppenarbeit [in German]. Kollektive Handlungsregulation in soziotechnischen Systemen. Huber, Bern

Weber WG, Lauche K (2010) FEZT – Fragebogen zur erlebten Zusammenarbeit in multifunktionalen Teams [in German]. In: Sarges W, Wottawa H, Roos C (eds) Handbuch wirtschaftspsychologischer Testverfahren Band 2: Organisationspsychologische Instrumente. Pabst, Lengerich, pp 87–95

West MA (2002) Sparkling fountains or stagnant ponds: an integrative model of creativity and innovation implementation in work groups. Appl Psychol Int Rev 51:355–424

West MA, Farr JL (1990) Innovation and creativity at work: psychological and organisational strategies. Wiley, Chichester, UK

Zölch M (2001) Zeitliche Koordination in der Produktion [in German]. Aktivitäten der Handlungsverschränkung. Huber, Bern

Chapter 9
Measurement of Team Knowledge in the Field: Methodological Advantages and Limitations

Thomas Ellwart, Torsten Biemann, and Oliver Rack

Abstract Team knowledge is seen as an important element in the understanding of coordination processes in teams. Congruent with the taxonomy of coordination mechanisms (cf. Chaps. 2 and 7), the construct of team knowledge refers to shared team-level knowledge structures facilitating implicit processes such as tacit behaviours as well as coordination success. This chapter answers three major questions: (1) What are the challenges of measuring team knowledge in organizational settings compared to more controlled laboratory settings? (2) What concepts of team knowledge exist in the psychological literature, and how are they related to coordination processes? (3) What methods can be applied to measure team knowledge in the field? Although there are several approaches to identifying and measuring team knowledge in a laboratory setting, applications in an organizational context are rare. Thus, this chapter discusses three types of team knowledge: team mental models, team situation models, and transactive memory systems. The advantages and limitations of techniques for capturing team knowledge are discussed and current directions are introduced.

9.1 Introduction

As described in Chap. 7, which is the integrating chapter for Part II of this book, successful coordination processes rely on team knowledge, which is defined as commonly shared knowledge that team members have about a task and about each

---

T. Ellwart (✉)
University of Trier, Department of Economic Psychology, D-54286 Trier, Germany
e-mail: ellwart@uni-trier.de

T. Biemann
Economics and Social Sciences, University of Cologne, 50923 Cologne, Germany
e-mail: biemann@wiso.uni-koeln.de

O. Rack
School of Applied Psychology, University of Applied Sciences Northwestern Switzerland, Riggenbachstrasse 16, 4600 Olten, Switzerland
e-mail: oliver.rack@fhnw.ch

M. Boos et al. (eds.), *Coordination in Human and Primate Groups*,
DOI 10.1007/978-3-642-15355-6_9, © Springer-Verlag Berlin Heidelberg 2011

other (e.g. Cannon-Bowers et al. 1993; Kraiger and Wenzel 1997).[1] In this way, team knowledge is thought to help team members anticipate the needs and actions of others in order to "implicitly" coordinate group behaviour and improve team effectiveness. In most of the present coordination research, team knowledge is applied in the context of work teams and to a somewhat lesser degree with regard to sports teams. Team knowledge is not a coordination process per se as is tacit behaviour or feedback. However, in the literature it is often labelled as implicit coordination because it represents a team-level knowledge structure that facilitates implicit coordination behaviours such as monitoring, anticipated backup, or dynamic adjustment (Rico et al. 2008). Thus, the concept is of interest in many different domains of group interaction, such as those occurring in families, organizations, and communities. But how can researchers capture the shared knowledge of a group? What aspects can be identified and measured, and what methods are appropriate?

In this chapter we will discuss the concepts and measurements of team knowledge as follows: In the first section we will highlight the challenges of measuring team knowledge in organizational settings compared to more controlled laboratory settings. In the second section we will give an overview of different theoretical concepts of team knowledge and thus explain what concepts of team knowledge can be measured. In the third section we will introduce specific methods to assess team knowledge in a more detailed way. These common methodological approaches to team knowledge will be explained and evaluated in terms of their usefulness in field settings. Finally, in the general discussion we will outline directions for a valid assessment of team knowledge in organizational settings, which can complement laboratory studies and enrich our understanding of implicit team coordination.

9.1.1 Team Knowledge and Its Current Research Status in the Literature

Although team knowledge is seen as an important prerequisite to a comprehensive understanding of coordination processes in teams, its reflection in psychological research lags behind the importance of the concept. Several empirical studies have shown that team knowledge and indicators of "explicit" team coordination and performance are clearly related (e.g. Edwards et al. 2006; Chap. 11; Lim and Klein 2006; Marks et al. 2000, 2002; Mathieu et al. 2000, 2005; Smith-Jentsch et al. 2005; Stout et al. 1999; for a review, see DeChurch and Mesmer-Magnus 2010), but it seems that when regarding team knowledge, there is much more understanding to be gained from a theoretical perspective than from manifold empirical evidence.

---

[1]The term "team knowledge" is a defined concept in team as well as group research. Although debates exist regarding the differences between teams and groups, we use the term "team knowledge" synonymously for both. Therefore, team knowledge as it is used here represents the shared knowledge of team/group members.

Moreover, at present, there is only slight evidence that team knowledge directly influences implicit team coordination such as anticipation (e.g. Ellwart and Konradt 2007a; see also Chap. 5). There are at least two reasons for this lack of empirical research: First, various competing methods and tools have been developed to capture team knowledge (e.g. Cooke et al. 2000; Langan-Fox et al. 2000; Mohammed et al. 2000), which can potentially yield different facets of team knowledge and thus hinder an integrative picture of team knowledge (Mohammed et al. 2000). Second, small group/team research is mostly limited to controlled laboratory situations, as well as small and distinct groups with identical and specific tasks (Lewis 2003). This makes it difficult for applied psychological research to transfer the concepts of team knowledge into organizational teams and enrich the empirical foundation. Hence, the purpose of this chapter is to give a summary of common measurement techniques to capture team knowledge of organizational teams, with a special focus on the practicability in field settings.

9.1.2 Challenges to Measure Team Knowledge in Field Settings

There are different measurement approaches to capturing team knowledge. Most of them have been successfully applied in highly standardized experimental settings (Cooke et al. 2000; Langan-Fox et al. 2000; Mohammed et al. 2000). However, the majority of these measurement approaches have important limitations for assessing team knowledge in field settings due to the difficulties associated with transferring theoretical methods and tools into field settings. The first problem is that experimental methods depend on tasks being identical across teams in order to apply content-specific tools for group comparison (for detailed information, see Sect. 3). Second, the researcher needs to label (and therefore identify) the shared knowledge of interest precisely prior to the task in order to measure its specific content (Lewis 2003). But organizational teams hardly ever work on tasks that comprise such straightforward characteristics, as tasks vary across projects and teams to a large degree. Applied psychological research investigates heterogeneous teams fulfilling heterogeneous tasks to draw valid and functional conclusions about coordination processes and team knowledge. Thus, as with any other coordination entity, measurement techniques of team knowledge need to take into account different requirements in field applications: (1) methods to identify and quantify team knowledge need to be *less task- and team-specific* in order to allow a comparison between groups; and (2) field research tools need to be efficient with *low material and effort costs* to stakeholders as well as participants in order for researchers to be granted access.

To illustrate the specific needs of field measures, one can think of a scenario where the aim is to evaluate the functional relationship between planning processes and team knowledge. In a laboratory setting, the experimenter can define a task and a group that will align with the constructs of this interest. For example, Stout et al. (1999) designed a surveillance/defence mission task that lasted approximately 1.5 h with a team knowledge measure that involved 190 paired-comparison judgments. In these judgments, participants were asked to rate to what extent specific concepts were related (c.g. "Second in Command tells Mission Command what target looks like and how many miles away it is" and "Mission Command tells Second in Command what weapon to use"). Quantitative analyses lead to a team knowledge indicator of a shared mental model. Stout and colleagues were able to show a relationship between explicit planning and implicit team knowledge.

If researchers want to replicate this study in a field setting, the above-described technique for operationalizing team knowledge would not be applicable in organizational teams. First of all, the content of team knowledge needs to be known before it can be integrated into the paired-comparison measure, an impractical constraint in field research (see Chap. 6). Second, the content of team knowledge needs to be similar across different teams and their tasks in order to apply a comparable measurement approach for all teams. However, in many settings there is a lack of the statistically required number of teams necessary to compare task and team characteristics. Third, many companies (as well as their employees) refuse to participate in investigations where team members work on queries that take longer than 30 min.

An alternative measurement approach to team knowledge in field settings is represented in the team coordination study of Ellwart and Konradt (2007b). Thirty-seven project teams were investigated in a field setting using Likert scales to assess planning, team knowledge, and coordination success where measurements were taken twice during the project. The measurement of team knowledge was neither task- nor team-specific and consisted of a five-item scale that was transferred into a shared mental model index (cf. 9.3.3.1; e.g. "I have a good "idea" of the responsibilities of individual team members"). Both studies addressed a similar question and showed that shared mental models (i.e. team knowledge) mediate between planning and coordination success (Ellwart and Konradt 2007b; Stout et al. 1999).

In sum, the multifaceted nature of team knowledge dictates that different measures will yield different information about team knowledge. Moreover, different methods are more or less applicable, depending on the sample and the task. As shown previously, laboratory-based methods may be difficult to transfer into field settings because of the constraints of a common task and the team characteristics, as well as the efforts and costs of such procedures. However, for the further development of research and theory, it would be of great importance to compare the results found in the lab (mostly experimental) to the results found in the field setting (mostly correlational). This integrative approach would allow a combination of methods to benefit from the strengths and to compensate for the weaknesses of each method.

Before answering the question "How can we measure team knowledge in the field?" the following section will address the question "Which concepts of team knowledge can be measured?"

9.2 Concepts of Team Knowledge

In the literature there exist several definitions of team knowledge. It has been frequently referred to as shared knowledge and – in similar contexts – as shared mental models, shared cognition, and shared understanding (Blickensderfer et al. 1997; Cannon-Bowers et al. 1993; Cooke et al. 2000). Building on these distinctions from the literature, this section will introduce three types of team knowledge: (1) team mental models, which represent the shared team- and task-relevant knowledge of the group; (2) team situation models, which develop dynamically when the group is actually engaged in the task (dynamic understanding) (Cannon-Bowers et al. 1993); and (3) transactive memory systems, which represent the team's knowledge on individual expertises within the team (Wegner 1987).

In research, most conceptualizations of team knowledge refer to the first concept of *team mental models* (e.g. Edwards et al. 2006; Mathieu et al. 2005; Langan-Fox et al. 2000). Team mental models are the organized and shared understanding and mental representation of knowledge about central elements of the team, its tasks, and its environment (Klimoski and Mohammed 1994). Cannon-Bowers et al. (1993) defined four content domains underlying team mental models: (1) knowledge of the equipment and tools the group uses in the task (equipment model); (2) understanding of the task, such as strategies or goals (task model); (3) awareness of the team members themselves, such as roles, skills, and knowledge (team member model); and (4) understanding of effective team processes or interactions (team interaction model). This classification represents one approach to order the various content domains of team mental models and may differ from other classifications.

Team situation models emerge whenever a team is actually engaged in a specific task (Cooke et al. 2000).[2] A team situation model is the team's collective understanding of the specific situation, and should change in alignment with modifications of the situation (dynamic understanding). Whereas the function of team mental models is embedded in a collective knowledge base that leads to common expectations, the function of team situation models is to interpret specific situations in a compatible way (Cooke et al. 2000). A shared team situation model helps to coordinate team actions according to a specific situation and to determine strategies, supporting the anticipation of other members' needs and actions in selecting the appropriate action (e.g. backup behaviour, information exchange, actions). Team situation models are based on knowledge from existing team mental models and also include characteristics of the specific situation, the second aspect

---

[2]One can argue that the labels of these different types of team knowledge are from a classification by Cooke et al. (2000) and can therefore vary among authors. Team situation models are, like all types of team knowledge, a mental representation of the task and the team. However, the focus here is on this very specific situation. As introduced in Chap. 7, the integrating chapter for Part II of this book, this type of team knowledge may be especially relevant during performance (in-process).

indicating the qualitative difference between the two concepts (Cooke et al. 2000; Rico et al. 2008).

The third type of team knowledge, *transactive memory,* is conceptualized as a set of distributed, individual memory systems that combines the knowledge possessed by particular members with shared awareness of who knows what (Wegner 1995). When each team member learns in a general sense what the other team members know, the team can draw on the detailed knowledge distributed across members. Each member keeps track of other members' expertise, directs new information to the matching member, and uses that tracking to access needed information (Mohammed and Dumville 2001; Wegner 1987, 1995). Given the presumed distribution of specialized memories across team members, transactive memory systems reduce the individual's cognitive load and thereby are more efficient for the individual regarding cognitive labour (Brauner and Becker 2004; Hollingshead 1998). From a theoretical perspective, the team knowledge component of transactive memory can be seen as a type of team mental model (Mohammed and Dumville 2001). Because transactive memory systems capture a shared understanding about who knows what within a team, it refers to the awareness of the team members regarding roles, skills, and knowledge – what Cannon-Bowers et al. (1993) termed "team member model" (Mohammed and Dumville 2001). However, transactive memory also underlines team processes of specialization within a team (Lewis 2003). It therefore represents a separate category of team knowledge with a strong link to team mental models.

Overall, team knowledge can be classified as team mental models, team situation models, and transactive memory systems. All three conceptualizations describe different facets of team knowledge; and their measurement approaches vary in terms of how team knowledge is defined, elicited, and analysed. Whereas team mental models describe rather long-lasting aspects of team knowledge that exist prior to the task, team situation models refer to the specific situation and change accordingly. Transactive memory, as the shared awareness of who knows what in the team, describes a kind of specific aspect of team mental models.

9.3 Common Measures of Team Knowledge

Methods for measuring team knowledge reported in the literature vary in terms of how team knowledge is elicited (e.g. observation, interviews) and analysed (scaling techniques, quantification of indicators) (for an overview, see Cooke et al. 2000; Langan-Fox et al. 2000; Mohammed et al. 2000). This section is oriented towards the terminology of previous reviews and focuses on the applicability of the measures in field settings, offering some updates on new developments (cf. DeChurch and Mesmer-Magnus 2010). A central distinction between different measurement techniques of team knowledge is the question of whether they capture the content (elicitation methods) and/or the structure of knowledge (representation methods). Regarding this issue, there is an inconsistent use of the terms "elicitation methods"

9 Measurement of Team Knowledge in the Field

Table 9.1 Methodological approaches to measuring team knowledge (TK)

	Content Elicitation of TK	Concept Analysis of TK	
		Modelling structure and sharedness of TK	Group agreement as indicator of TK
Application	Determine the content of TK (qualitative level, intra-team comparison)	Reveal structure/ relationship between contents of TK (quantitative level)	Quantify the degree of sharedness *without* capturing structure/ relationship
Indicators	*Qualitative Data* Content of TK, comparison of TK to team referent	*Quantitative Data* Sharedness/Similarity Accuracy	*Quantitative Data* Sharedness (e.g. r_{WG}, r_{GR} Biemann et al. 2009)
Instruments	Observation, interviews and surveys, process tracing, card sorting	Multidimensional Scaling Pathfinder QAP (UCINET)	Likert Skales (e.g. knowledge ratings, behaviour ratings)
Advantages for Field Application	Offers information about the domain of interest May structure further quantitative investigations	Useful for highly structured and well defined tasks	less task-/team- specific Little efforts to implement in organizations
Limitations for Field Application	No metrics to quantify structure of TK Often intensive and costly Most methods depend on verbally expressible knowledge	Very task-/team-specific content domains Complex implementation Overestimates common agreement (except r_{GR}) High number of comparisons $\rightarrow$ high efforts	Only indicator of sharedness, but no information about absolute level of knowledge Overestimates common agreement (except r_{GR})

and "representation methods" (cf. Cooke et al. 2000; Langan-Fox et al. 2000). For the purpose of this chapter, we draw on the work of Langan-Fox et al. (2000) and Cooke et al. (2000) to distinguish between *qualitative* content elicitation methods and *quantitative* concept analysis methods. Table 9.1 gives an overview of the methical approaches for measuring team knowledge.

9.3.1 Content Elicitation of Team Knowledge

Content elicitation methods explicate a team's domain-related knowledge in a qualitative way. The aim of these methods is to map out the content of team knowledge at a qualitative level, for example, to reveal exactly what team knowledge is needed for a specific task. Methods for content elicitation of team knowledge are manifold (cf. Cooke et al. 2000; Langan-Fox et al. 2000; Mohammed et al. 2000). In this chapter we briefly introduce observation, interviews and surveys, and process tracing as methods for eliciting team knowledge. Card sorting[3] represents an approach that captures the content of team knowledge but also refers to aspects of structure and representation of the team knowledge domain.

Observation of team knowledge can be applied in the field context and can be based on written, audio, and/or video forms. It provides a large amount of information on both the form and content of communication, coordination, and performance. Through deduction, it facilitates the drawing of inferences about concept domains and the relationship between them. For example, Badke-Schaub et al. (see Chap. 10) applied observation of communication patterns as an indicator of team mental model development. The authors concluded that the less communication that took place regarding specific content domains (planning, roles), the better the team mental model developed. For application in the field, observation is a very extensive method that is excellent for gaining a general understanding of the situation, as well as for generating and verifying hypotheses. However, as in other approaches, it relies on the skills of the researcher to identify important concepts of team knowledge at a qualitative level. Moreover, there might be a problem of validity when researchers deduce from observed performance (e.g. communication) a specific team knowledge, due to the questionable theoretical link between performance in a task and team knowledge structure (Langan-Fox et al. 2000).

Standardized *interviews* (and also written *surveys)* are systematic ways to elicit complete representations of individual and team knowledge. Respondents are asked to explain key elements or causal relations of specific and relevant knowledge domains. In field application, surveys and questionnaires are easier to administer and to conduct than interviews because they are independent and participants can decide when and where to fill out the forms. However, questionnaires require more preparation time than (unstructured) interviews and depend on sufficient context knowledge to adequately formulate the survey or questionnaire. Generally, interviews and surveys are a valuable starting point to clarify the content of team knowledge in the field because they offer a first explication of team knowledge such as extents, distribution, and tracking tendencies among team members. In contrast to laboratory experiments, the researcher cannot define the content domains of team knowledge a priori. In most cases, it is necessary to build a complete and comprehensive map of team knowledge and its associations. Thus, interviews are a valuable way to outline team knowledge, but they are only a starting point and are inadequate for providing detailed, complex knowledge as well as other important information that cannot be explicitly expressed.

---

[3]Following previous reviews, we conceptualize card sorting as an elicitation tool but highlight that it is also applicable in terms of a structural analysis of shared mental models.

9 Measurement of Team Knowledge in the Field 163

Process tracing techniques are field methods for collecting data on team knowledge concurrently with data on task performance and can be based on verbal or non-verbal data. In verbal protocol analyses, respondents are "thinking aloud" to explain their behaviour and the teams' behaviour during task performance (van Someren et al. 1994). These retrospective reports are useful for garnering data on intellectual tasks naturally involving verbalization (Langan-Fox et al. 2000) that do not involve physical task performance (e.g. decision making, general reasoning processes). However, in complex field applications, there is the problem of varying degrees of individual awareness regarding cognitive structures that underlie behaviour, and it is therefore difficult to compare team member protocols systematically. Process tracing based on non-verbal data includes, for example, actions, facial expressions, gestures, and general behavioural events to trace cognitive processes (Cooke et al. 2000).

Visual *card sorting* represents a tool that is helpful when eliciting team knowledge and developing a structure or relative representation of the team's operative concepts. Moreover, this approach can be applied in a group context where group members develop the team knowledge structure together. Participants name all the concepts that they consider relevant to the domain of interest, and then write them on cards. When concepts have been pre-explicated by an alternative technique, the researcher can provide cards containing concepts to the participants beforehand. The participants then sort the concepts individually or as a group and arrange related aspects closer together, and less related concepts farther apart. This tool can be easily applied in a field context and provides good face validity for the team (Langan-Fox et al. 2000). No statistical procedures are needed to elicit or structure the team knowledge concepts, and this card-sorting method can also be used to measure a team mental model through group sessions. However, the application is limited to concepts that can be compared on the basis of feature matching or spatial distance. For example, to visualize a transactive memory system of a team, cards with expertise domains can be assigned to cards of team members. The expertise of team members is then indicated by a low distance between the expertise domain and the member's name. This approach becomes difficult, for example, when concepts of interest represent complex processes or strategies that cannot be plotted visually.

9.3.2 Concept Analysis of Team Knowledge

Whereas content elicitation methods reveal team knowledge at a qualitative content level, concept analysis approaches probe the quantitative structural relationships of team knowledge within a team. Thus, structure elicitation methods aim at revealing how different knowledge aspects are related to each other. There are two approaches in concept analyses: The first approach models the structure and relationship of team knowledge concepts and reveals whether individual mental representations are similar (shared) between the group members. The second method ignores the relationship and structure of team knowledge concepts on the individual as well as team levels, focusing on group agreement regarding more specific characteristics of team knowledge that are interpreted as shared understanding. We will explain both approaches in the following sections.

9.3.2.1 Modelling Structure and "Sharedness" of Team Knowledge

The following methods are valuable for quantifying the representations of concepts and their relationships. The researcher collects similarity ratings on each possible pair of team knowledge concepts from each team member. These ratings indicate whether the concepts are related positively or negatively and to what extent they are related positively or negatively. In the next step, these relationships between the concepts are compared at a team level. The procedures are based on *proximity matrices* designed to capture components and organizational structures of cognitive models by applying techniques such as Pathfinder networks (Stout et al. 1999), the quadratic assignment procedure (QAP; Mathieu et al. 2005), and multidimensional scaling (see Mohammed et al. 2000; Cooke et al. 2000). For example, in a study by Lim and Klein (2006), participants were asked to rate the relatedness of various statements describing their team's taskwork. The resulting proximity matrices of each team member were then compared to those of the other team members to assess team mental model similarity by employing Pathfinder and QAP correlations.

Multidimensional scaling (MDS) gives a pictorial representation of how items are clustered. The inputs are pairwise-similarity ratings of all concepts. The MDS analyses then search for the best placement in the space relative to their similarity or contrariness, resulting in a set of geometric models. The idea is that geometric distance represents psychological distance. In a team knowledge context, MDS can be used to illustrate relative comparisons between mental models that exist among the different team members. However, there are some methodological limitations and restrictions (Langan-Fox et al. 2000).

Pathfinder represents a computerized networking technology that displays team knowledge as an associative network based on the relationship between specific concepts of team knowledge. It results in a network structure of nodes and links, the nodes representing the concepts and the links representing the pairwise relationship between the concepts. This method offers a graphic representation of the team's knowledge structure, along with quantitative indices (e.g. spatial coordinates, dimension weights, pairwise distances between concepts). An important advantage of Pathfinder is that the complexity of the data is reduced via simplified, illustrative techniques, thus making the data more comprehensible than by any of the other techniques (e.g. multidimensional scaling). This simplification is achieved because the link between concepts in a network is eliminated if it does not represent the shortest pathway between the two concepts (Cooke et al. 2000; Schvaneveldt et al. 1989). Thus, the focus is on the closest and strongest relationship between concepts.

An alternative approach to measuring the perceived importance and similarity of team knowledge structures are *quadratic assignment procedures (QAP)*, comparing correlations integrated by the UCINET software (Borgatti et al. 2002). QAP calculates the simple matching coefficient between corresponding cells of data matrices from two team members (this method is limited to dyads: If no match is made, no calculation is made). Several quantitative indicators give information about the individual and team mental model. For example, the centrality index for each concept is a measure of the importance of a concept to the overall network of concepts. Similar to the results of Pathfinder or MDS, this method analyses individuals' pattern of ratings throughout the matrix (Mathieu et al. 2005) and indicates to which extent team members' models show similar patterns of relationship.

Despite the value of these quantitative methods based on proximity matrices, there are some disadvantages regarding their application in the field. First, it is necessary that each concept can be rated relative to all other concepts of team knowledge in order to ready the matrix for calculation. For example, Mathieu et al. (2005) operationalized task mental models of teams in a flight simulator task with eight attributes: (1) diving/climbing, (2) banking/turning, (3) airspeed, (4) selecting/shooting weapons, (5) reading/interpreting radar, (6) intercepting enemy, (7) escaping enemy, and (8) dispensing chaff and flares. Team members then rated each relationship between all attributes using a nine-point scale from -4 (negatively related, a high degree of one requires a low degree of the other), 0 (unrelated), to $+4$ (positively related, a high degree of one requires a high degree of the other). Shared team knowledge was indicated once all team members achieved similar ratings, for example, once all agreed that airspeed is positively related to escaping from enemies. This approach is limited in a practical sense because even though the reduction to eight single attributes such as climbing or airspeed can be applied in that specific laboratory task, in complex field environments it is often difficult to extract a definitive number of concepts that represent the key elements of team knowledge that can therefore be applied across different teams and tasks.

But many organizations are interested in a more elaborate picture of the team knowledge that analyses aspects of sharedness of knowledge concepts rather than their relativity to each other or to task. For example, do team members have both a shared and an *accurate* understanding of how to behave in certain situations, and if so, to what degree? Therefore, proximity matrices do not seem to be adequate methods, because it is of less interest how several elements are related to each other or conceptually mapped in mental representations of team members. The approach via proximity ratings strictly focuses on the structural relationship between single tasks or team concepts. In many applied cases, teams are less interested in the deep-level analysis of the structural relationship of team knowledge concepts than they are in knowing whether team members share the same ideas about the team or the task, such as agreement on a specific goal. In this case, for example, instruments are needed that can extract the overall agreement of the team regarding the team goal. In this instance, the focus has shifted from the *structure* of the representation to the *content* of sharedness of specific concepts, regardless of their structural representation. Group agreement indices based on Likert scales may be a more suitable approach for obtaining such measurements and will be described in the next section.

9.3.2.2 Group Agreement as Indicator of Team Knowledge

This methodological approach measures aspects of team knowledge "sharedness" using agreement indices derived from Likert-type questionnaires. (Note: In this section the terms "shared mental model" and "sharedness" are used as synonyms for "agreement.") Although the term "shared" is not always defined precisely and distinctively (Mohammed et al. 2000), agreement in team knowledge reflects the degree to which team members share a similar view, and can be evaluated using different team- or task-related statements in a questionnaire. For example, team members are asked to rate statements regarding the contents of a mental representation (e.g. "How useful is this strategy XY to reach the goal?"). The ratings of all group members are compared and an agreement index is computed that indicates the degree of similarity between team member ratings. In most cases, indices are based on the concept of *within-group agreement* (e.g. r_{wg} by James et al. 1984) and are used to quantify team mental model agreement or similarity (e.g. Eby et al. 1999; Levesque et al. 2001; Webber et al. 2000). Eby et al. (1999), for example, developed a questionnaire to measure shared expectations regarding teamwork. Each individual team member rated 28 items on teamwork (e.g. "The team develops a task strategy") on a five-point Likert-type scale. Webber et al. (2000) used a similar approach to measure consensus on strategic team mental models of basketball players and asked team members about the effectiveness of actions in specified situations. In a recent approach, Johnson et al. (2007) developed a rating scale instrument of 42 items that are linked to the five emergent factors of shared mental models, including general task and team knowledge ("My team knows the relationship between various task components"), general task and communication skills ("My team communicates with other teammates while performing team tasks"), attitude toward teammates and task ("My teammates take pride in their work"), team dynamics and interactions ("My team undertakes interdependent tasks"), and team resources and working environment ("There is an atmosphere of trust in my team"). Additionally, Eby et al. (1999) and Webber et al. (2000) applied within-group agreement indices to determine the similarity of the team members' mental models using the $r_{wg(j)}^{*}$ index (Lindell et al. 1999) for each team based on member responses. This index, as well as the widely used r_{WG} index (James et al. 1984), compares the average *obtained* variance in a team to the *expected* variance under a specified distribution of random responses. High levels of agreement between *obtained* and *expected* indicate high agreement within the team. Besides the focus only on agreement, Johnson et al. (2007) discussed (1) the calculation of average ratings of team knowledge (mean scores) as well as (2) indices of agreement. To calculate shared team knowledge, also known as team agreement, the average evaluation for each item was computed for each team in order to first calculate the degree of knowledge among the team members (absolute knowledge). The standard deviation of the average score among team members represents how closely aligned each team member is on any particular item (team agreement). However, they do not discuss how average ratings of absolute knowledge and agreement indices could be combined into one single index, which would represent interesting information about team knowledge for both laboratory and field research.

Thus, in the following section, we will discuss an integrating approach to how absolute evaluation of team knowledge concepts and consequent agreement scores can be integrated into a valid similarity coefficient for field application (Ellwart and Konradt 2007a). Moreover, we discuss a statistical procedure to improve the team-specific validity of the team knowledge measurement (Biemann et al. 2009).

9.3.3 Further Perspectives on Field Applications

The following sections will introduce two new perspectives on the use of agreement measures in team knowledge research. First, the shared mental model index is introduced, which combines absolute knowledge in a team with the agreement between team members; second, the distinction between general and team-specific agreement is discussed.

9.3.3.1 Combining Absolute Team Knowledge and Agreement: The Shared Mental Model Index (SMM Index) of Expertise Location

As discussed in the previous section, indices of agreement from Likert scales are one valuable research approach to measuring the similarity of team knowledge representations in applied field settings. However, it is important to recognize that a singular focus on agreement indices gives an insufficient picture of the team knowledge model, because agreement just tells if the team shares the knowledge but not to what degree (quantitative measure) or whether they simply agree on having no shared knowledge at all (qualitative measure). Thus, we will discuss two specific conditions of team knowledge that are indicators of team or situational mental models, as well as of transactive memory systems: (1) the absolute knowledge in the team (do they or don't they know it?) and (2) agreement (to what extent do they agree in knowing or not knowing it?). In this section we want to give an empirical example of a Likert-based measurement approach that combines absolute knowledge and agreement. The example is from research on a team mental representation regarding the level of expertise within a team (Faraj and Sproull 2000; Hollingshead 1998; Lewis 2003, 2004). This type of team knowledge regarding the expertise status and specific know-how of the team relates to transactive memory systems (who knows what in the team) discussed earlier in this chapter.

Regarding team knowledge and expertise location in groups, absolute knowledge (the extent of team member knowledge re experts in different domains, who they can ask for help, etc.) and agreement (team members hold similar views on who is the expert on what) are two important indicators. At the individual level, team members need to have an accurate mental representation about the expertise domains of the other team members (Hollingshead 1998) in order to coordinate expertise efficiently. At the group level, there needs to be agreement (i.e. *team agreement*) with regard to individual expert representations in order for the group to be successful (Mohammed and Dumville 2001). With classical Likert scale approaches, such as the ones applied by Eby et al. (1999) or Webber et al. (2000), questions would be posed regarding team member knowledge about the expertise domains of their teammates (e.g. "We know the specific knowledge team members possess"), and then only the agreement (r_{WG}) of these ratings between the team members would be analysed. This score, however, only gives information as to whether there is *variance* between team members in their ratings – not whether they do or do not know about the expertise domains within the team, or who holds these domains within the team. To illustrate this important difference, Fig. 9.1 displays "agreement" and "knowledge" exemplarily. Think of a hypothetical group with three members who are asked to rate the item "I know the expertise domains of my colleagues" between 1 and 5 (1 = *I do not know*; 5 = *I do know*). When researchers just compare agreement scores, it remains unclear whether "agreement" was in knowing the experts (case 4, all members give high ratings = high mean score) or in not knowing the experts (case 2, all members give low ratings = low mean score). The same holds true when comparing cases 1 and 3. Both cases would yield the same low agreement score because of different evaluations. However, in case 3 there are two members with high knowledge, whereas in case 1 there is only one member with knowledge. This indicates that both "agreement" (variance scores) and "knowledge" (mean scores) are necessary indicators of team mental representations to gain a comprehensive evaluation of team agreement.

In sum, to our knowledge there are no field measures in team mental model research and related areas that assess and integrate agreement on knowledge about expertise location, combined with absolute team knowledge in one single index. Therefore, Ellwart and Konradt (2007a) developed the *shared mental model index (SMM index)*, which integrates these two indicators of team agreement, adopting a scale from Faraj and Sproull (2000). The original scale asks, from a team-level

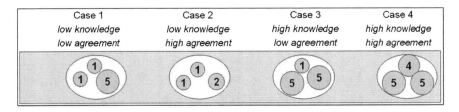

Fig. 9.1 Combinations of agreement and knowledge within a team

perspective, whether the team has knowledge about the experts within the group (e.g. "The team has a good map of each others' talents and skills"). Because the group-level perspective of the items does not reflect the individual representation of each member's knowledge (cf. Klein et al. 2001), the SMM index changes the original group focus to an individual perspective (e.g. "I have a good map of other team members' talents and skills"). The scale (four items) captures the individual meta-knowledge of expertise location: who knows what in the team. Using the four Cannon-Bowers et al. (1993) content domain labels outlined in Sect. 2 of this chapter, the SMM index addresses the concept of team mental models, specifically the awareness of the team members themselves regarding roles, skills, and knowledge (team member model). It is therefore related to the transactive memory concept (Wegner 1987, 1995), reflecting meta-knowledge within a team.

To calculate agreement between the individual scores at the team level, Ellwart and Konradt's (2007a) SMM index uses the average deviation score (*AD;* Burke et al. 1999; Burke and Dunlap 2002). In comparison to other indices for estimating interrater agreement (for an overview, see Brown and Hauenstein 2005), the average absolute deviation has two major advantages: First, the AD indices do not require the determination of a null random response distribution such as r_{wg}. Second, AD is computed relative to the mean of an item and therefore provides more direct conceptualizations in the same metric of the original measurement scale (Burke and Dunlap 2002). The same metric allows team agreement scores on expertise location (average deviation) to be related to the group members' absolute knowledge (meta-knowledge on expertise location) in one single coefficient. The aim of the SMM index of expertise location is to integrate knowledge and agreement in a single score. Therefore, the average deviation score is subtracted from the mean score (low average deviation scores = high agreement; high average deviation scores = low agreement). This means that the team's SMM index on expertise location reflects its absolute knowledge about its expertise minus the degree of agreement within the team. Teams that reveal high absolute knowledge but high disagreement show a lower shared mental model than teams with high agreement. To provide validity of this approach, the SMM index of expertise location should be sensitive to both (1) group differences regarding different levels of meta-knowledge, and (2) high and low team consensus.

Research results showed that teams differ regarding the relationship between agreement and knowledge, underlining the validity of the SMM index (Ellwart and Konradt 2007a). Moreover, experimental and field testing of the SMM index yielded construct validity as well as proof that it predicts team coordination success in both experimental ($N = 120$ students in 40 teams) and field settings ($N = 130$ participants in 37 project teams) (Ellwart and Konradt 2007a, b). The SMM index relates (1) to accuracy and consensus scores from objective expertise ratings, (2) to subjective coordination success (self-perceptions that processes were executed in a coordinated way) and team performance at the group level, and (3) to knowledge credibility and task-related self-efficacy at the individual level. These results provide evidence from experimental and field data that the SMM index is a useful and valid measure of expertise location in teams. The empirical data support the assumption that the SMM index is a conceptually and statistically valid measure of knowledge and agreement regarding the location of expertise within teams. Its independence from task performance – and therefore its appropriateness for field settings and comparisons between teams and tasks – are its main advantages. Convergent and criterion-related validity was shown through its relationship to established and objective measures of transactive memory accuracy and consensus introduced by Austin (2003). Moreover, these experimental and field tests of the SMM index demonstrate that it is significantly related to constructs of team coordination such as coordination success and knowledge credibility (Lewis 2003).

In sum, the SMM index of expertise location offers an economic but valid measurement approach that can be used across various types of teams in field settings. Although the SMM index cannot capture the specific underlying organizational structure of the specific knowledge domains, it can be used as a screening tool prior to more extensive investigations in specific teams. Moreover, this approach is applicable to other team knowledge concepts, for example, task-related knowledge (Ellwart and Konradt 2007b). In this particular study, team members rated statements concerning their knowledge about tasks, for example, the goals (e.g. "I know how much progress has been made towards achieving team goals"), responsibilities (e.g. "I have a good 'idea' of the responsibilities of individual team members"), or interdependencies (e.g. "I know how the tasks of my team members are related to each other"). Similar to the SMM index regarding expertise location, absolute knowledge and agreement were combined in a single score. Ellwart and Konradt (2007b) were able to show that the task-related mental model can predict task and team conflicts as well as coordination success over time. Moreover, task-related shared mental models mediated the relationship between explicit planning and coordination success. Applied in the field, the SMM index can help to explain, diagnose, and circumvent problems in teams, particularly in organizational teams whose performance depends on optimizing knowledge assets.

9.3.3.2 Identifying Team Specific Agreement: Improving the Validity of Team Knowledge Quantifications

In this section we point out an important statistical limitation of most team knowledge measurements introduced so far. In the beginning of this chapter, two sets of methodological approaches to analysing the structure and sharedness of team knowledge were introduced: The first set focuses on the analysis of the relationship between various team knowledge elements using *proximity matrices* (Stout et al. 1999; Mathieu et al. 2005). The second set uses agreement indices derived from Likert-type questionnaires (*variance-based approaches*). However, both approaches suffer from conceptual problems: Neither differentiates between team-specific and general agreement, a shortcoming that biases correct estimations of the existence and significance of shared team knowledge (Biemann et al. 2009). We argue that these are two different sources of agreement that are erroneously treated equally in most of the methods applied and should instead be separated when sharedness of team-specific knowledge is the focus of the analysis.

On the one hand, agreement can derive from team processes (e.g. planning, reflexion, trust) that are very specific to the team. On the other hand, there can be statistical agreement that is *not* team-specific but also results in a high within-group agreement. Consider an example from a team knowledge measurement that focuses on strategies in a team situation model. One item may be "Insulting the opponents to make them nervous", which represents a strategy used in a competitive team-based computer game. The average r_{WG} as well as the r_{WG}^* would show relatively high values (0.69, 0.86; cf. Biemann et al. 2009), representing high agreement within the teams. However, Biemann and colleagues showed by means of statistics that regardless of the team membership, *all* participants agreed that insulting the opponents is not a good strategy to avert losing the game. It is not surprising that the statistical agreement within each team is high, since the agreement among all participants across all teams is also high. The argument against this representing team-specific agreement is that the agreement scores are an expression of a common understanding that does not depend on team boundaries. Only if there is a consensus within some teams that an action is useful, while other teams disregard the same action as useful, is there a *team-specific* agreement. Unfortunately, this discrepancy between team-specific and general consensus is not reflected in the existing measures of group agreement.

Thus, Biemann et al. (2009) introduced *random group resampling (RGR)* as an easy-to-apply method to differentiate between *team-specific* agreement and *common* agreement. In prior research, the RGR was used as a post hoc significance test to estimate whether indices of sharedness were the result of group-specific variance or of a general phenomenon (Bliese and Halverson 2002; Ellwart and Konradt 2007a, b). Basically, RGR uses a random group resampling to compare within-group agreement data garnered from the actual observed groups with within-group agreement data from hypothetical simulated groups (Bliese et al. 1994; Bliese and Halverson 1996, 2002; Castro 2002). As an addition to other post hoc testing, this procedure introduced by Biemann et al. (2009) offers a direct indicator of group-specific shared variance that can be applied to variance-based approaches (e.g. Likert scales) as well as to proximity matrices (e.g. Pathfinder, MDS). The idea behind RGR is simple: Sharedness of team knowledge is only considered team-specific when it differs from the shared variance of unspecific random teams of the entire population. This random-based variance provides an unbiased statistical estimator of the population variance (Biemann et al. 2009). Thus, the actual population variance of participants can be validated before being integrated into calculations estimating team-specific within-group agreement indices, as well as calculations estimating proximity matrices.

Moreover, the indices based on RGR have an interpretable value useful for the measurement of influences related to team knowledge. RGR ratings over zero (zero = no team-specific agreement) indicate the existence of team-specific agreement and therefore an RGR rating-related potential for positive influence on team coordination and performance (Salas and Fiore 2004).

9.4 General Discussion: Measures of Team Knowledge in Field Research

Team knowledge represents the shared or common knowledge that team members have about a task and about each other (Cannon-Bowers et al. 1993; Kraiger and Wenzel 1997). We have introduced three types of team knowledge found in psychological research: team mental models, team situation models, and transactive memory systems. For these three types of team knowledge, we have discussed two methodological approaches that focus on either the *qualitative* elicitation of team knowledge or on the *quantitative* analysis of structure and/or agreement (cf. Table 9.1).

Methods such as observation, interviews, surveys/questionnaires, process tracing, or card sorting aim at determining the content of team knowledge on a qualitative level. They help create a starting point for understanding the knowledge that a team possesses and shares in order to perform its task, enabling researchers to begin identifying and comparing knowledge domains between or within the team(s). For field application, these methods of content elicitation are a valuable starting point for becoming aware of the specific knowledge within the team. However, most methods are time-consuming and costly because each member must be interviewed, surveyed, observed, or otherwise analysed to illicit the necessary data. Nonetheless, once assembled, the resulting qualitative data yield essential information to the researcher for further quantitative analyses such as concept analyses.

Concept analysis methods such as multidimensional scaling, Pathfinder, or UCINET reveal the structure (relationship) and sharedness of team knowledge at a quantitative level (see Sect. 3.2.1 for a description of their operabilities). Because these methods require objective and valid descriptions of team knowledge that fit to all participants in the study, they are very team- and task-specific and therefore make them difficult to apply in a field setting. Comparative investigations between organizational teams make it especially unreasonable to reduce the relevant knowledge to 10 or 15 dimensions that are meaningful to all teams and their members. Moreover, the often high numbers of pairwise ratings make these approaches very intensive and costly in the field. This might be the reason why methods such as Pathfinder or UCINET are mostly applied in laboratory studies of highly structured, well-defined team tasks involving small numbers of team members.

In field research, the quantitative analyses of team knowledge can be done reasonably effectively using Likert scale ratings. One benefit of these scale-rating approaches is that ratings of single items are less intensive and costly. Moreover, many content domains of team knowledge can be addressed by items at a task- and team-independent level (Lewis 2003). The drawback is that these qualitative indicators are solely focused on sharedness of the ratings within a team, failing to capture the underlying structure and relationship of the shared concepts. Nevertheless, the advantages for field application are significant, as these Likert-based ratings of team knowledge offer valuable indicators of commonly shared mental representations of different types of team knowledge at relatively low costs in terms of time and complicity. From our perspective, these indicators are especially useful when large numbers of teams are assessed in terms of the same knowledge concept. For example, if multi-team companies want to decide which of their teams may participate in trainings to improve knowledge exchange, a short indicator scale allows them to screen many heterogeneous teams with the same method and to then pick out noticeable groups for more specific diagnostics and treatments. More elaborated and specific techniques at this early stage in a change process would surely cause organizational and implementation-related difficulties compared to this rather economic Likert-based approach. Close attention is needed to validate indicators of sharedness with other approaches in order to apply them as valuable screening tools in organizational field research. From a methodological perspective, classical indicators of sharedness from Likert items (r_{WG}) are limited to giving information about the extent to which team members agree or disagree with regard to specific team knowledge domains vs. the absolute level and accuracy of team knowledge. The shared mental model index (SMM index) introduced by Ellwart and Konradt (2007a) offers a practicable way to combine both types of information into a single index. But there are also limitations to this approach. First of all, the ratings given by team members represent their subjective perception regarding their knowledge of the task or the team. The instrument provides no way to prove whether they really know it or just think they know it. Another potential problem is the aspect of social desirability. Especially in field context surveys, it is problematic and therefore highly unlikely (whether true or not) for team members to disagree with statements such as, "I know how the tasks of my team members are related to each other", Nonetheless, applications in field and laboratory settings should treat the SMM index as a screening tool and combine it with other approaches to team knowledge analysis.

In sum, there are a variety of methods for researchers to assess different types of team knowledge in laboratory and field settings. These methods differ with regard to their specific focus (knowledge elicitation, concept analyses) and their practicability, both of which depend on the specific team and task setting. This plurality of methods allows the researcher to cross-validate the instrument of choice, applying efficient but valid approaches in a field context. However, using numerous approaches means capturing different facets of team knowledge that apply very heterogeneous statistic strategies. Thus, a macro-prospective research strategy may be in order to avoid difficulties in comparing results across different studies and teams. A rather contemporary study conducted by DeChurch and Mesmer-Magnus (2010) showed in a current meta-analysis that team knowledge operationalization impacts the observed relationship between the mental models and team process. Perhaps their most important finding was that methods that model the structure or organization of knowledge are the most predictive. Even though the magnitude of the relationship differed across measurement method, indicators of team knowledge were positively related to team performance, regardless of the manner in which operationalization was performed (DeChurch and Mesmer-Magnus 2010).

As discussed in the integrating chapter by Ellwart (Chap. 7), team knowledge is an important condition for implicit coordination in groups. Team knowledge allows teams to anticipate actions of team members, to provide help and guidance without explicit communication, and represents the common understanding of the group about their task and team. In human team research, team knowledge is mostly captured by language-based approaches, limiting their application in non-human investigations. However, it is conceivable that also in non-human groups there are mental representations that coordinate the behaviour of the group (cf. Chap. 14). Although these non-human representations are not coded in language and are outside the range of self-reflection, they are comparable to the team knowledge concepts in human small group research. Both concepts represent the mental map that helps the group to behave in a coordinated fashion. If this map is not similar between the members of the group, there will be a lack of synchronicity in group behaviour, regardless of whether they are human or non-human groups.

References

Austin JR (2003) Transactive memory in organizational groups: the effects of content, consensus, specialization, and accuracy on group performance. J Appl Psychol 88:866–878

Biemann T, Ellwart T, Rack O (2009) Quantifying the similarity of team mental models: shortcomings and advancement. In: Academy of Management Best Paper Proceedings 2009, AOM Annual Meeting, August 7–11, Chicago

Blickensderfer E, Cannon-Bowers JA, Salas E (1997) Does overlap of team member knowledge predict team performance? In: Paper presented at the Human Factors and Ergonomics Society Annual Meeting, Albuquerque (September 1997)

Bliese PD, Halverson RR (1996) Individual and nomothetic models of job stress: an examination of work hours, cohesion, and well-being. J Appl Soc Psychol 26:1171–1189

Bliese PD, Halverson RR (2002) Using random group resampling in multilevel research: an example of the buffering effects of leadership. Leadership Quart 13:53–68

Bliese PD, Halverson RR, Rothberg JM (1994) Within-group agreement scores: using resampling techniques to estimate expected variance. Acad Manage J 13:303–307

Borgatti SP, Everett MG, Freeman LC (2002) UCINET 6 for Windows: Software for social network analysis. Analytic Technologies

Brauner E, Becker A (2004) Wissensmanagement und Organisationales Lernen: Personalentwicklung und Lernen durch transaktive Wissenssysteme [in German]. In: Hertel G, Konradt U (eds) Human resource management im Inter-und Intranet. Hogrefe, Göttingen, pp 235–252

Brown RD, Hauenstein NMA (2005) Interrater agreement reconsidered: an alternative to the r_{wg} indices. Organ Res Methods 8:165–184

Burke MJ, Dunlap WP (2002) Estimating interrater agreement with the average deviation index: a user's guide. Organ Res Methods 5:159–172

Burke MJ, Finkelstein LM, Dusig MS (1999) On average deviation indices for estimating interrater agreement. Organ Res Methods 2:49–68

Cannon-Bowers JA, Salas E, Converse S (1993) Shared mental models in expert team decision making. In: Castellan JN Jr (ed) Individual and group decision making: current issues. Lawrence Erlbaum, Hillsdale, NJ, pp 221–246

Castro SL (2002) Data analytic methods for the analysis of multilevel questions: A comparison of intraclass correlation coefficients, $r_{wg(j)}$, hierarchical linear modelling, within- and between-analysis, and random group resampling. Leadership Quart 13:69–93

Cooke NJ, Salas E, Cannon-Bowers JA, Stout RJ (2000) Measuring team knowledge. Hum Factors 42:151–173

DeChurch LA, Mesmer-Magnus JR (2010) Measuring shared team mental models: a meta-analysis. Group Dyn-Theor Res 14:1–14

Eby LT, Meade AW, Parisi AG, Douthitt S (1999) The development of an individual-level teamwork expectations measure and the application of a within-group agreement statistic to assess shared expectations for teamwork. Organ Res Methods 2:366–394

Edwards BD, Day EA, Arthur W, Bell ST (2006) Relationships among team ability composition, team mental models, and team performance. J Appl Psychol 91:727–736

Ellwart T, Konradt U (2007a) Measuring shared mental models of expertise location in teams: two validation studies. In: Paper presented at the 2nd Annual INGRoup Conference (interdisciplinary network for group research), July 2007, Lansing, MI

Ellwart T, Konradt U (2007b) Explicit and implicit team coordination: Influences of planning and shared mental models on team conflicts and coordination success. In: Paper presented at the 13th Congress of Work and Organizational Psychology, May 2007, Stockholm, Sweden

Faraj S, Sproull L (2000) Coordinating expertise in software development teams. Manage Sci 46:1554–1568

Hollingshead AB (1998) Retrieval processes in transactive memory systems. J Pers Soc Psychol 74:659–671

James DL, Demaree RG, Wolf G (1984) Estimating within-group interrater reliability with and without response bias. J Appl Psychol 69:85–98

Johnson TE, Lee Y, Lee M, O'Connor D, Khalil M, Huang X (2007) Measuring sharedness of team-related knowledge: design and validation of shared mental model instrument. Hum Resource Develop Intl 10:437–454

Klein KJ, Conn AB, Smith DB, Sorra JS (2001) Is everyone in agreement? An exploration of within-group agreement in employee perceptions of the work environment. J Appl Psychol 86:3–16

Klimoski RJ, Mohammed S (1994) Team mental model: construct or metaphor? J Manage 20:403–437

Kraiger K, Wenzel L (1997) Conceptual development and empirical evaluation of measures of shared mental models as indicators of team effectiveness. In: Brannick MT, Salas E, Prince C (eds) Team performance assessment and measurement. Lawrence Erlbaum, Mahwah, NJ, pp 63–84

Langan-Fox J, Code S, Langfield-Smith K (2000) Team mental models: techniques, methods and analytic approaches. Hum Factors 42:242–271

Levesque LL, Wilson JM, Wholey DR (2001) Cognitive divergence and shared mental models in software development project teams. J Organ Behav 22:135–144

Lewis K (2003) Measuring transactive memory systems in the field: scale development and validation. J Appl Psychol 88:587–604

Lewis K (2004) Knowledge and performance in knowledge-worker teams: a longitudinal study of transactive memory systems. Manage Sci 50:1519–1533

Lim BC, Klein KJ (2006) Team mental models and team performance: a field study of the effects of team mental model similarity and accuracy. J Organ Behav 27:403–418

Lindell MK, Brandt CJ, Whitney DJ (1999) A revised index of interrater agreement for multi-item ratings of a single target. Appl Psych Meas 23:127–135

Marks MA, Zaccaro SJ, Mathieu JE (2000) Performance implications of leader briefings and team-interaction training for team adaptation to novel environments. J Appl Psychol 85:971–986

Marks MA, Sabella MJ, Burke CS, Zaccaro SJ (2002) The impact of cross-training on team effectiveness. J Appl Psychol 87:3–13

Mathieu JE, Heffner TS, Goodwin GF, Salas E, Cannon-Bowers JA (2000) The influence of shared mental models on team process and performance. J Appl Psychol 85:273–283

Mathieu JE, Heffner TS, Goodwin GF, Cannon-Bowers JA, Salas E (2005) Scaling the quality of teammates' mental models: Equifinality and normative comparisons. J Organ Behav 26:37–56

Mohammed S, Dumville BC (2001) Team mental models in a team knowledge framework: expanding theory and measurement across disciplinary boundaries. J Organ Behav 22:89–106

Mohammed S, Klimoski R, Rentsch JR (2000) The measurement of team mental models: we have no shared schema. Organ Res Methods 3:123–165

Rico R, Sánchez-Manzanares M, Gil F, Gibson C (2008) Team implicit coordination process: a team knowledge-based approach. Acad Manage Rev 33:163–184

Salas E, Fiore S (2004) Team cognition. American Psychological Association, Washington, DC

Schvaneveldt RW, Durso FT, Dearholt DW (1989) Network structures in proximity data. In: Bower GH (ed) The psychology of learning and motivation: advances in research and theory, vol 24. Academic, New York, pp 249–284

Smith-Jentsch KA, Mathieu JE, Kraiger K (2005) Investigating linear and interactive effects of shared mental models an safety and efficiency in a field setting. J Appl Psychol 90:532–535

Stout RJ, Cannon-Bowers JA, Salas E, Milanovich DM (1999) Planning, shared mental models, and coordinated performance: an empirical link is established. Hum Factors 41:61–71

van Someren MW, Barnard YF, Sandberg JA (1994) The think aloud method: a practical guide to modelling cognitive processes. Academic, London

Webber SS, Chen G, Payne SC, Marsh SM, Zaccaro SJ (2000) Enhancing team mental model measurement with performance appraisal practices. Organ Res Methods 3:307–322

Wegner DM (1987) Transactive memory: a contemporary analysis of the group mind. In: Mullen B, Goethals GR (eds) Theories of group behavior. Springer, New York, pp 185–208

Wegner DM (1995) A computer network model of human transactive memory. Soc Cognition 13:1–21

Chapter 10
An Observation-Based Method for Measuring the Sharedness of Mental Models in Teams

Petra Badke-Schaub, Andre Neumann, and Kristina Lauche

Abstract This chapter explores the role and development of mental models in coordination. We introduce a theoretical framework on the development of shared mental models and a measurement approach based on observational data. The basic assumption is that individual mental models are shared through verbal communication. At the beginning of a task, this is likely to be explicit and thus observable. Once the team members assume that they hold a shared mental model, less verbal communication will be required and team members will continue their coordination in an implicit fashion. The methodology is illustrated using data from observations of two meetings of a design team. The analysis largely confirms our hypotheses. Implications for using the model and method in other contexts are discussed.

10.1 Introduction

This chapter aims to contribute to the conceptualisation of the measurement of sharedness, widely considered as one of the basic characteristics of team mental models that influence coordination (Cannon-Bowers et al. 1993; see also Chaps. 1, 2, 5, 7, and 11). Our framework addresses coordination as an activity in its social context, and team mental models as the corresponding cognitive representation. Coordination refers to a wide range of activities in different phases of aligning task

P. Badke-Schaub (✉) and A. Neumann
Faculty of Industrial Design Engineering, Delft University of Technology, Landbergstraat 15, 2628 CE Delft, The Netherlands
e-mail: p.g.badke-schaub@tudelft.nl; a.neumann@tudelft.nl

K. Lauche
Nijmegen School of Management, Radboud University Nijmegen, Thomas van Aquinostraat 3, 6500 HK Nijmegen, The Netherlands
e-mail: k.lauche@fm.ru.nl

M. Boos et al. (eds.), *Coordination in Human and Primate Groups,*
DOI 10.1007/978-3-642-15355-6_10, © Springer-Verlag Berlin Heidelberg 2011

or problem-solving processes, aiming to arrive at a desired goal (see Chap. 2 for an overview of the theory of coordination and Chap. 7 for conceptual and methodological approaches). Activities subsumed under coordination include impersonal coordination such as standardisation (van de Ven et al. 1976; Kieser and Kubicek 1992), standardised operating procedures, organisational rules (Grote et al. 2004), and personal coordination such as team interaction, negotiation, and mutual arrangements. Coordination can take explicit and implicit forms. The relevance and degree of explicit coordination depend on the complexity of the task, the distribution of knowledge and competencies of the persons involved, the history (both content and extent) of the team working together, the use of supporting tools, and the specific environment. Implicit coordination typically develops over time, and we argue that this process is enabled through the development of shared representations – also known as *shared mental models.*

We define coordination as planning and monitoring the process of problem solving in terms of task and team: who is going to work on what; which sub-task and when; delivering which output; and using what kind of equipment, tools, etc. The assumption is that explicit coordination activities lead to better performance due to the development of a shared understanding in the team (Stout et al. 1999; see also Chap. 11). Thus, effective coordination is based on the explication of mental models between team members regarding what needs to be done and how, sharing individual and discipline-oriented knowledge and procedures. Our approach to measuring coordination draws on these explications of mental models in verbal communication: Talking about the content of their mental models helps team members to achieve coordination; once shared mental models have been established, less discussion is required.

Numerous studies have analysed the influence of different coordination processes in teams on performance (e.g. Eccles and Tenenbaum 2004; Entin and Serfaty 1999; Espinosa et al. 2007; Hoegl and Gemuenden 2001; Orasanu 1994). Most of these studies found that effective coordination enhances team performance (Stewart 2006). Moreover, it has been shown in high-risk environments such as medical teams (Chap. 5), aviation (National Transportation Safety 1994), and the nuclear industry that deviations from procedural standards are the most frequent category in crew- or operator-related accidents. The results mentioned above indicate a causal relationship between coordination and shared mental models, as well as between shared mental models and group performance. In these studies, the range of behaviours in emergency situations was typically characterised as standard operating procedures with pre-defined steps that form shared rules to be followed by the team. Our question is, how can shared mental models in teams be measured in situations with less defined routines of behaviour and fewer standard operating procedures – what kind of coordination is relevant in these situations?

We postulate that explicit and implicit coordination are two forms of activities that occur in different phases of the problem-solving process. According to Wittenbaum et al. (1998), 'in-process planning may occur rarely without inducement because of the natural tendency to coordinate tacitly and task demands that make communication difficult while performing' (p. 199). Teams who have worked together for a certain period of time develop shared mental models. On the basis of this common understanding, implicit coordination can evolve – which is advantageous in high-risk, time-pressure situations that require avoiding wasting time and/or concentration resources (Espinosa et al. 2004). The time and extent required for a team to develop shared mental models depends on the complexity of the task, the motivation of the team members to reach the common goal, the expectations of the team members derived from the history of the team, and the distribution of knowledge in the team (Arrow et al. 2000). Given the dynamic nature of team processes, it is of further interest to understand how sharedness develops over time and how exactly this process affects coordination and team performance. Little is known about how sharedness develops and how it affects the coordination process. This limited theoretical understanding and the small number of empirical studies to date could be due to the lack of convincing methods to measure these processes (Badke-Schaub et al. 2007). Further research is needed to arrive at more specific knowledge, but this requires better ways of measuring sharedness of team mental models in different situations and of assessing the development of sharedness all along the process.

In this chapter we want to address the following question: How do mental models in teams develop over time? How can we measure sharedness? How do changes in mental models affect the type of coordination? Do these phenomena concern the entire group, or are they mainly found in certain parts of the team?

10.2 Theoretical Framework of Sharedness, Mental Models, and Coordination

Any measurement must be based on an idea of what should be measured and how the measured variables are linked to each other. The measurement of sharedness also requires such a conceptual framework, which we present in the following subsection.

10.2.1 Concept of Mental Models

The concept of mental models was first proposed as an individual cognitive concept, defined as internal representations that humans build in order to cope with the world around them (Craik 1943; Gentner and Stevens 1983; Johnson-Laird 1980). While interacting with the environment, with other people, and with artefacts such as products, sketches, or prototypes, humans develop and adapt mental models (simplified representations) in order to understand, predict, and act in a world of continuously incoming and sometimes contradicting new information. Mental models include concepts, propositions, scripts, frames, and mental images.

These simplified representations enable a person to quickly assimilate fresh information and act even in new and unknown situations. Mental models are specific for a given task at a particular time, and change dynamically as they are updated with new information. As mental models are interpretations of reality, they can be more or less accurate and appropriate, and be more or less similar to other people's mental models. When individuals meet in a team context, these different mental models come together, ideally developing into shared mental models, which we will discuss in further detail ahead.

Shared mental models are defined as the degree of convergence among team members with regard to the content of known elements, as well as the structure between elements (Mohammed et al. 2000). They form an organised understanding of relevant knowledge that is shared by team members (Cannon-Bowers et al. 1993; Klimoski and Mohammed 1994). There are different assumptions about the various kinds of representations, yet most researchers agree on at least two basic types of representations (Cooke et al. 2000; Klimoski and Mohammed 1994; Rentsch and Hall 1994): Their definition of 'task mental model' encompasses all aspects related to the execution of the task, while the 'team mental model' is defined as all representations related to the team and the team members that are essential for working together. It has been shown that teams perform better when they have a shared mental model of the task (e.g. Lim and Klein 2006; Mathieu et al. 2005, 2008; Mesmer-Magnus and DeChurch 2009; Salas et al. 2008).

10.2.2 Types of Mental Models

In this section we introduce our theoretical framework of the development of shared mental models (see Fig. 10.1). The model builds on and further differentiates the distinction between taskwork and teamwork (Rousseau et al. 2006). This distinction explicates that taskwork includes working on the content and also managing the process, while teamwork refers to the coordination of roles and responsibilities in the team, as well as the creation and maintenance of a cohesive team climate. Thus, we differentiate among four types of mental models relating to the task, the process, the team, and the climate. For each type of mental model, we will spell out how explicit coordination can help to arrive at a shared mental model, and how these shared mental models in turn affect coordination processes.

The *task model construct* of Fig. 10.1 refers to representations of the problem at hand. Communicating about task-related content such as defining the problem, generating ideas, evaluating solutions, and making decisions should enhance the team's sharedness of the task mental model. The communication in the team gives further information about discrepancies in task-related knowledge between the team members. For example, explanations and questions are of particular interest, as they serve as indicators for missing knowledge: Team members provide each other with explanations or they elicit information often based on specialised expert knowledge about a subject held by team members. By asking questions and

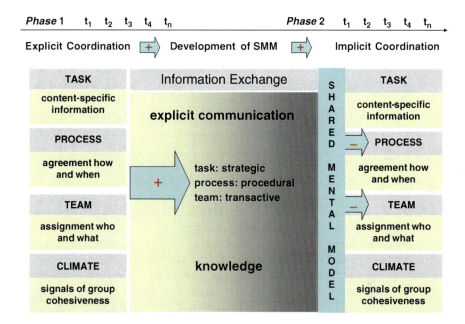

Fig. 10.1 A model of the development of shared mental models in teams. +: increases, −: decreases

providing information, the team creates a common knowledge base. The literature cited above on shared mental models contends that the more information that is exchanged, the richer the sharedness between the mental models of all team members about the task will be.

The *process model construct* of Fig. 10.1 refers to assumptions about the proper procedure for solving the task. We propose that in order to arrive at a successful outcome, team members not only need a shared understanding of the task itself, but also need a shared understanding about the process for approaching its achievement. In domains in which variations of the process are associated with high risks, standard operating procedures have usually been developed to ensure a uniform procedure of how to perform certain critical tasks (Grote et al. 2004; National Transportation Safety Board 1994). However, standard operating procedures are not suitable for non-routine tasks, which are typically ill defined in the beginning, and the requirements may also change during the problem-solving process. In such cases, the manner in which team members coordinate their processes cannot be predefined, but has to instead be decided based upon the characteristics of the task at hand (Edmondson and Nembhard 2009). Three aspects of process work can be distinguished (1) the planning of when to do what; (2) the procedure of how to solve a problem and which strategies or methods to use; and (3) reflection as meta-analysis of the process (Wetzstein and Hacker, 2004). Reflection (aspect 3) has been identified as a powerful mechanism of self-correction to adapt the process of responding to changing conditions or to unsuccessful results (Hackman and Wageman 2005; West 2002).

The *team model construct* of Fig. 10.1 refers to representations about the other members' individual abilities, knowledge, skills, and experience, as well as the roles and responsibilities each team member holds. As team members discuss the allocation of roles, they establish collective action regulation (Weber 1997), which is pivotal in helping a team to reach its goals effectively. Role allocation can be more or less centralised: In teams that have a formal or informal leader, it will be this person who verbalises the most coordination issues. If several members contribute to role allocation, responsibility and authority are likely to be more shared in the group.

The *climate model construct* of Fig. 10.1 describes the extent to which members feel that they belong to a particular group (Goodman et al. 1987). We define three communication processes that contribute to a cohesive climate: informal talk, appreciation, and mutual confirmation. Informal talk is non-task-related communication, which, according to Goodman et al. (1987), increases mutual knowledge about individual preferences and characteristics beyond the task. Appreciation refers to explicit positive statements concerning another team member or his or her contributions, and thus has an emotional component. Confirmation is another form of feedback that also indicates a positive 'evaluation' or a simple signal reinforcing the continuation of the conversational flow. These signals seem to be akin to the vocalisation in non-human primates to coordinate group movements (for details, see Chap. 13).

10.2.3 Development of Shared Mental Models

The concept of mental models implies that it is initially an individual representation: Each person develops his or her own mental models about the task, process, team, and climate. These mental models of individual team members can become shared mental models when communicating with other team members. In order to successfully work on a task as a team, the individual mental models of the members need to be shared to a certain extent – the general consensus among small group researchers being that the degree of sharedness and successful taskwork are generally positively correlated (e.g. Mathieu et al. 2008; Mesmer-Magnus and DeChurch 2009). Therefore, information exchange between team members is crucial, because when information is exchanged, the shared knowledge at the team level increases (see Fig. 10.1).

The main assumption of our model is that shared mental models develop from explicit, mainly verbal information exchange about the task, process, and team. This explicit verbalisation enables the development of shared team mental models, which in turn allows the team members to draw on similar representations. Teams with shared mental models are assumed to operate in a coordinated fashion without the continuous need for explicit verbal exchange: They coordinate their actions implicitly. We propose the following two stages in the development of shared mental models (see Fig. 10.1):

1. The first phase consists of explicit coordination; this is the moment of measurement t_1. By means of verbal communication, the team develops shared mental models. The verbal communication is the observable behaviour, which is assessed for further analysis.
2. The second phase starts after a certain degree of sharedness has been achieved, which is characterised by less explicit and more implicit coordination.

The following points in time (t_2, t_3, t_4) indicate that the exact moment of these two stages cannot be known in advance; therefore, more than two moments of measurement are necessary to analyse changes in the process. This conceptualisation draws on the idea of stage models of team development (Tuckman 1965), which purports that teams first need to form group norms before they can achieve their maximum effectiveness. Partly in accordance with this assumption is also Gersick's punctuated equilibrium model (Gersick 1988), which presupposes that groups develop through a sudden revision of a 'framework for performance' from phase 1 to phase 2 as outlined in Fig. 10.1. It is also in line with the idea that teams first establish coordination and then enact it (Arrow et al. 2000).

Our model of how shared mental models develop (Fig. 10.1) visualises the following six main assumptions:

1. Explicit exchange of information and knowledge enables the team to build shared mental models on the task, process, team, and climate.
2. After a period of explication, the team will achieve a degree of sharedness for each of the different types of mental models. The time required for explicit coordination depends on the characteristics of the task, the team members, the history of the team, etc.
3. For knowledge developed during the explication process, three different types can be distinguished: task-related knowledge, strategic knowledge, and transactive knowledge. Task-related knowledge comprises the factual knowledge about the task at hand. Strategic knowledge refers to knowledge about possible ways in which to approach a task. Compared to task-specific knowledge, strategic knowledge contributes to the task as well as the process and is augmented through exposure to different problems over time. Transactive knowledge comprises the knowledge about other team members: It contributes to transactive memory, which refers to the finding that team members memorise who knows what and whom to ask rather than remembering the knowledge content itself (Wegner 1986; Wegner et al. 1985).
4. Task-related knowledge is dependent on the specific contextual environment. Thus, any change in the problem or sub-problem necessitates explicit information exchange in order to arrive at a shared understanding. In contrast to interaction on the task, information exchange on how to collaborate on the given task can be valid for longer periods of work. Sharedness of the individual mental models on procedural aspects allows for implicit coordination; as long as there is no need for change, no further explications are needed.
5. Mental models about the team are assumed to serve two functions: a motivational function and a coordinative function. The motivational aspect refers to

maintaining a team climate that ensures sufficient cohesiveness so that the team members want to stay in the team together. The coordinative function encompasses the assignment of tasks, roles, and responsibilities according to the team members' preferences, capabilities, and experience (Stempfle et al. 2001). Explicit coordination is important for building shared mental models of the task, team, process, and climate; a shared mental model allows the behaviour of team members to be predicted. Thus, the need for further repetition of this knowledge becomes obsolete, and the frequency of explicit coordination after such a period of information exchange will decrease.

6. Individual mental models are not necessarily shared equally among all team members. Individuals might have similar background knowledge, share working experiences, or develop similar solution ideas. Those with the same disciplinary background often also share the same jargon, which in turn makes it even more difficult for team members with a diverse background to reach a shared understanding. Earlier research has pointed out that dyads in a team often develop a shared mental model, and through this shared mental model teams achieve a better performance (Bierhals et al. 2007). The frequency of interactions in the team will influence which team members share their mental models more than others.

10.3   Observational Approach to Measuring the Development of Shared Mental Models

The following section explains our methodological approach, which will be illustrated in an analysis of a case study of two team meetings. The definition of shared mental models as being the degree of convergence among team members with regard to the content of known elements, as well as the structure between elements (Mohammed et al. 2000), is purely descriptive. It does not provide any information about how to measure these elements nor the relationships between the elements. For assessing the components of mental models of a person working on a given task, questionnaires can be used to directly request knowledge of the participants, or a thorough team task analysis can be conducted (see Chap. 9). Common methods such as Pathfinder (Schvaneveldt 1990) or concept mapping (Tergan and Keller 2005) attempt to reveal a snapshot of the relationships between key elements in an individual's mind, which these methodologies often depict graphically. For an overview of measurement techniques, see Langan-Fox et al. (2000) and Mohammed et al. (2000), as well as Sect. 9.3.2.1.

For complex tasks such as design problems, no predefined routines exist to solve such a problem. Thus, key elements with regard to the task are dependent on the problem definition and solution ideas of the designer, and therefore cannot be identified easily beforehand; the solution space for most problems in design is by far too big. Continuous observation over time provides a possible solution to the challenge of measuring the development and adaptation of shared mental models during such problem solution processes. In the absence of predefined criteria, observations rely on overt behaviour such as team communication as a valuable way to obtain access to the development of team mental models. The underlying argument is that team members themselves also cannot read each others' minds and therefore have to rely on what is being said to develop shared understanding.

Our methodological approach therefore analyses explicit verbal communication as a natural observational angle into team mental models. All verbal utterances are coded according to the categories of the proposed model, from which we, in turn, infer the development of team mental models.

10.3.1 Coding Scheme

We developed a categorisation system of verbal communication using the above-mentioned four types of team mental models: task, process, team, and climate (see Table 10.1). If more than one topic was addressed within one utterance, the statement was split so that each part could be assigned a single category per categorisation system. Verbal activities were coded by three members of the research team, and Fleiss' kappa (Fleiss 1971) was found to be 0.72 on the level of the subcategories.

10.3.2 Measuring Dissociate Versus Comprehensive Sharedness

The second measure addresses whether sharedness is established in a *dissociate manner* based on dyadic interactions, or in a *comprehensive manner* that encompasses all members of the team. In order to calculate this aspect of sharedness, a contingency analysis is conducted to determine how often each person follows up on another person's utterance: Every time person B follows an utterance of person A, this is counted as an instance of B replying to A. As this relation is unidirectional, the instance of A replying to B is added as additional information regarding the explicit communication between A and B in the network. The results are mapped as a network of all individual contributions. If sharedness is dissociate, only a subgroup of the team actively shares its mental models. The other team members may interpret the behaviour of the active team members as being more knowledgeable or more powerful. It remains unclear from the verbal interaction if those who are only listening are also incorporating what has been said into their mental models. A more comprehensive form of sharedness is likely to arise in teams with equal status and an even spread of knowledge, or it could be the result of standard operating procedures that require read-back behaviour for mission-critical situations.

Table 10.1 Categorisation system for verbal activities (explicit coordination) in teams

Task

Problem definition and elaboration	Defining the problem, elaborating and analysing the constraints, the requirements, and the goal of the task: e.g. "The main focus is that style pen where you'll create a set of patterns by moving on plain white paper."
New solution idea or new solution aspect	Stating a new product-idea or a new solution for an earlier defined problem or sub-problem or new aspects building on an earlier mentioned solution idea: e.g. "...maybe like a flat base with a sort of universal joint like a windsurf mast."
Solution analysis	Analysis of properties and the feasibility of a solution idea; analysis of the usage of a product idea and its potential applications, e.g. by referring to similar products; and evaluation of a solution idea by appraising its feasibility or analysing failure and safety aspects: e.g. "With this feature you get both thick and thin lines."
Explanation	Clarification of questions and explanations about specific (technical) issues, e.g. by referring to specialised knowledge: e.g. "How's that achieved? Is that achieved by a plastic that's bendable?"
Solution decision	Making a solution definitive by accepting it in the whole team: e.g. "So we stick to the second option, the barcode thing."

Process

Statements on organisation re when to do what (planning), how to approach the task, e.g. how to apply a method (procedure), and utterances about what and how the team is doing (reflection)

Planning	Statements on the organisation when to do what: e.g. "We're going to try to deal with that a fair bit on Monday."
Procedure	Statements about how to approach the task, e.g. how to apply a method: e.g. "I think we should concentrate on what the mechanism might look like."
Reflection	Statements about what and how the team is doing: e.g. "I didn't realise what this was. This is a prototype."
Team	Role allocation to team members and references to personal abilities, knowledge, skills, or experience: e.g. "Chris, could you work this out in detail [...], you know best how to do it."
Climate	All aspects that indicate signals about group coherence are included in this category: appreciations about a solution idea or a problem definition, confirmation, and informal communication (e.g. joking).
Appreciation	Statements about a solution idea or a problem definition. All aspects that indicate signals about group coherence: e.g. "It's an interesting idea."
Confirmation	Positive statement which confirms the other member(s) of the team: e.g. "Yeah."
Informal communication	Statements which do not refer the actual task: e.g. "Your housemates are not going to be pleased."
Other	All utterances that are not defined in another category

10.4 Illustration of the Methodological Approach in a Case Study

In order to illustrate the application of our methodological approach, we report a case study that was conducted as part of the Design Thinking and Research Symposium held in London in July 2007 (see McDonnell and Lloyd 2009). The organisers had recorded meetings of design teams in companies and provided participants with video recordings as well as verbal transcripts of the meetings. Each research team analysed the data set (or parts thereof) according to their research interests and methodology. The data set consisted of two meetings, during which a team of engineers and other professionals generated ideas for a new product. All verbal data were categorised following these steps: The transcribed group videotapes were segmented into utterances, and then coded according to two distinct categorisation systems. The analysis focused on how the verbal communication in the defined categories developed over time in terms of task, process, team, and climate.

For coding, the software programme Mangold InterAct (version 7.0, http://www.mangold.de) was used. This programme enables the coding of many types of behavioural data per time unit, and the coded results can be easily transferred into statistical programmes.

10.4.1 Sample

Seven members from a technology development company attended both meetings. Not unusual for projects in industry, there was some fluctuation of attendees between Meetings 1 and 2, and there were no clear boundaries for what constituted membership of this design team. The first meeting included a business consultant acting as the moderator, an electronics engineer, a business developer, three mechanical engineers, an expert for ergonomics and usability issues, and an industrial design student who was doing an internship with the company. The intern also functioned as project leader. In the second meeting, two electronics/software experts and one electrical engineer participated instead of the business consultant and the two mechanical engineers. No further information about team tenure or experience was collected by McDonnell & Lloyd.

10.4.2 Procedure

The task was to develop a print head mounting for a thermal printing pen. The team was instructed to brainstorm in both meetings. Prior to the first meeting, the meeting participants had been asked to consider analogies or possible solutions for the assignment and were informed about the major topics to be discussed during each session.

The frequency of verbal activities (and thus the development of mental models) over time was analysed by comparing the two meetings, as well as the first and second halves of each meeting. For purposes of structuring our coding, a timeline of the frequency of codes was established for segments of the meeting. In order to compare frequencies, the segments were based on duration rather than content, as they had to be of equal length and not too short for statistical analysis of frequencies. Consequently, each meeting was divided into five segments.

10.4.3 Hypotheses

The following section explains our hypotheses regarding the development from unshared to shared mental models derived from the framework described above.

H1 Task. At the beginning of the meeting, verbal utterances should be related to problem definition, explanation, solution generation, and evaluation. Assuming that the group works together for the first time, shared understanding would be low, and thus a high number of explanations is expected. As taskwork progresses, knowledge about the task and possible solutions is expected to increase, and thus fewer problem definition utterances and explanations should occur. The team is expected to continue generating new ideas and analyse solutions, so these task-related utterances should increase over time.

H2 Process. A shared mental model related to the process will be developed by an increase in planning and procedural aspects in the first phase. In the second phase, only rescheduling and minor adaptations should be necessary, while reflection is likely to be useful throughout the whole process. Once explicit coordination has led to a shared mental model about how and when to do what, implicit coordination can occur. The frequency of process utterances should therefore decrease over time.

H3 Team. Transactive knowledge exchanges, such as role allocation utterances and references to each other's knowledge and skills, are expected to be more frequent in the first phase and should decrease after shared understanding on these issues has been achieved.

H4 Climate. Team climate utterances are expected to be initially high when team members strive to gain mutual acceptance, and to then decline once a certain level of team cohesion has been established. If the climate is negatively affected by conflicts or inappropriate group member behaviour, an increase in climate-related utterances will be needed once again to regain a positive climate in the team.

H5 Dissociate versus comprehensive sharedness. A more equal distribution of interactions between members will be related to a more comprehensive sharedness within the whole team. If dissociated patterns emerge in terms of increased levels of interaction in dyads or subgroups, this will be associated with the team's mental models being shaped and shared by only those members.

10.4.4 Findings on the Development of Sharedness in the Case Study

In the following sections, the results from the case study are discussed in terms of our proposed hypotheses.

10.4.4.1 Task: Extent of Sharedness Indicated by Degree of Lower Number of Problem Definition and Solution Analysis and Explanation

Figure 10.2 shows the frequencies of utterances in the team along five moments (M) in time in both meetings (M1.1–M1.5 and M2.1–M2.5).

There are two results of further interest: those for *problem definition* and those for *explanations and solution analysis*. According to our hypotheses, sharedness in the team should be indicated by a decreased frequency in problem definition utterances. The results show that within each meeting, the amount of problem definition significantly decreased from the first to the second half of the first meeting ($\chi^2 = 10.14$; $p < 0.01$). While this result confirms our hypothesis, the comparison between the two meetings tells a different story. The frequencies of problem definition utterances increased significantly from the first to the second meeting ($\chi^2 = 24.26$; $p < 0.01$), and even in the second half of Meeting 2, the frequency of problem definition utterances was significantly higher.

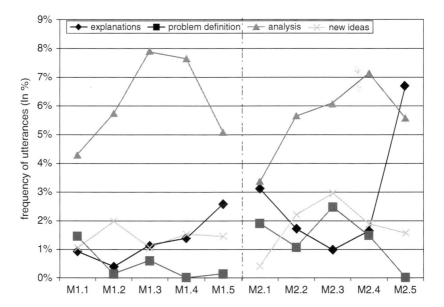

Fig. 10.2 Task communication: percentage of problem definition, solution analysis and explanations for both meetings, each divided into five equal parts

Explanations and solution analysis were expected to increase during the first meeting, but decrease in the second meeting when a shared mental model had been established. This was not found for the data presented here (see Fig. 10.2). The frequencies of explanations increased significantly from the first to the second meeting ($\chi^2 = 23.8; p < 0.01$); furthermore, the frequency also increased suddenly towards the end of each meeting.

The frequencies of solution analysis decreased, albeit not significantly. However, it can clearly be seen that analysis utterances increased during each meeting as more detailed, unshared aspects that had not been mentioned before were discussed. Also, explanations became more frequent towards the end. Utterances categorised as new ideas did not significantly decrease between the meetings. However, analysis utterances became more frequent after new ideas had increased in frequency, thereby showing that once a new, unshared topic was introduced, explicit communication about this topic was required.

These findings suggest that the team still felt the need to discuss the problem definition and to clarify the problem in the second meeting, which would suggest that sharedness had not been developed to a sufficient degree until then. However, a qualitative analysis of the latter half of Meeting 2 showed that a more likely explanation is that the group began to tackle more intricate problems and discussed issues at a deeper level. The problem definition was not actually redefined – the content issues were the same, referring to the powering, charging, cost, and heat of the product. Some members reiterated these issues to reinforce the importance of the issues, and new solution ideas were judged against the already agreed-upon requirements.

10.4.4.2 Process: Sharedness Indicated by a Decrease in Planning Utterances

According to our model, we expected that after a phase of explicit coordination in terms of planning and procedure, the team would develop a shared understanding on these issues. This increase in sharedness should lead to more implicit coordination. Thus, the number of explicit process utterances should decrease towards the end of a team meeting. Reflections on *what* the team is doing and *how* they are proceeding should not decrease dramatically, because teams should maintain awareness about their process (a 'meta-view') throughout the meeting.

This expected negative correlation between implicit and explicit communication once sharedness has taken root is exactly what happened. There was a significant decrease in the summarised process utterances from Meeting 1 to Meeting 2. Process utterances decreased most dramatically between the first and second parts of Meeting 1, whereas the frequency of process utterances stayed the same during the second meeting. In analysing the three process categories of planning, procedure, and reflection in detail (Fig. 10.3), the distribution of these categories shows that there were different developments for 'planning' and 'procedure' on the one hand, and 'reflection' on the other. Reflection should lead to a better meta-view regarding one's own procedures and planning and can overcome content and causal

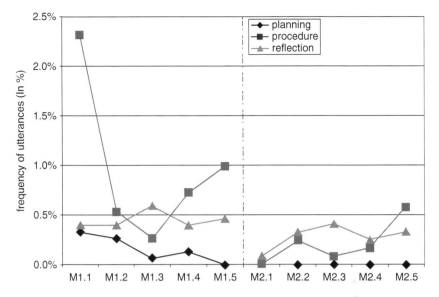

Fig. 10.3 Process communication: frequencies of planning, procedure and reflection utterances for both meetings, each divided into five equal sections

gaps in mutual understanding when sharedness is missing or lost. Reflection was therefore not expected to decrease during the meeting, which is what was found.

10.4.4.3 Team: Allocation of Tasks, Roles, and Responsibilities

Explicit team coordination was expected to be relatively high, especially in the first phase, and expected to then decrease in the second phase if the team had achieved a shared understanding on coordination. Our data of the design team were very clear: There was hardly any explicit team coordination, with only 17 utterances made in both meetings (Fig. 10.4). This surprising result can be attributed to the nature of the meetings, which were set up as brainstorming sessions. Thus, both meetings focused on gathering and developing new ideas; task allocation beyond the meeting was not discussed.

10.4.4.4 Climate: Sharedness Indicated by Continuous Backing Up

Although mainly observed as signals of attentive listening (e.g. 'Yeah'), the cohesion utterances made this category the highest frequency in both meetings. Following Owen (1985), who found that members of cohesive groups are more likely to engage in active communication, climate utterances are very likely to be a prerequisite for the development of team mental models. Based on our two-stage

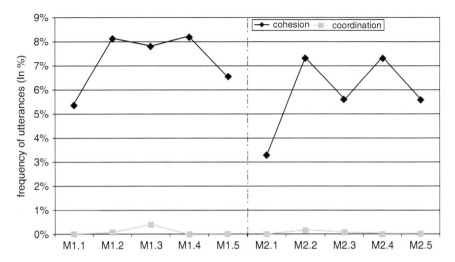

Fig. 10.4 Team and climate communication: percentage of utterances of team coordination and climate for both meetings, each divided into five equal parts

model, a decline of climate utterances over time was expected (see Fig. 10.4). There was no significant difference between the two meetings, but within each meeting climate utterances initially increased and then decreased.

10.5 Analysis of Dissociate Versus Comprehensive Sharedness

The extent to which sharedness was established in a dissociate vs. comprehensive manner was measured by analysing which group member most actively communicated and with whom. Individual contributions of the team members and their interactions were explored using a communication network analysis. A contingency analysis for all speakers was applied to determine who talked to whom and how often, thereby identifying dyads within the team. Figure 10.5 shows a network based on the contingency analysis of the individual contributions (the names are fictitious as used in McDonnell and Lloyd 2009). The thickness of the lines indicates the amount of individual talking in a given dyad, thus indicating how the information was transferred within the team. Individuals with many thick lines connected to them were more central in the team; those with thin lines were not highly involved. The most obvious dyad was Tommy and Todd, who talked to each other more than to any other team member. Both of them were experts in the field that was most relevant to the task, and therefore presumably had the greatest need, as well as capacity, to create a shared mental model about the topic. It is also noticeable that Alan, the facilitator, spoke considerably more to Tommy, Todd, and Jack than to the other members. They, in turn, communicated more with Alan. Again, the most likely reason is that these people had the most relevant knowledge

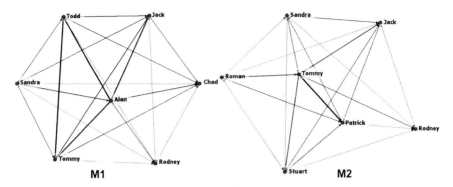

Fig. 10.5 Network based on the contingency analysis of the individual contributions, providing information about the inclusiveness of sharedness in the team

for that task, thereby making sure they shared an understanding about these issues. This is in line with the idea that in some teams it is sufficient for dyads of experts to share their knowledge.

The second meeting revealed a clear dyad between Tommy and Patrick, with Tommy replying to Patrick with 34.5% of his utterances and Patrick replying to Tommy with 55.8% of his. Again, these two team members were the experts for the task at hand. All other members spoke mainly to Tommy, who could be seen as an informal group leader. Additionally, Patrick could be seen as a second facilitator in the team, as he led the discussion about another topic with Roman and Sandra.

Interestingly, this analysis of the inclusiveness of sharedness can also help to reveal why some individuals are less involved in a meeting. For example, Rodney had very weak links to all members except to Tommy. Rodney was a trained industrial designer who could not contribute much to the actual meeting apart from aiding Tommy, who was the facilitator. A network analysis can thus help to investigate dyads in a team and, in turn, to understand the development of sharedness within these dyads.

10.6 Discussion

This chapter focused on the question of how far sharedness can be measured on the basis of observed data, how it develops over time, and to what extent sharedness encompasses the entire group vs. concentrates in dyads. According to our theoretical concept of the role of shared mental models in the transition from explicit to implicit coordination, hypotheses were derived and tested. Our approach presents a methodology for studying the development of shared mental models based on the analysis of overt, explicit behaviour. As such, the approach could also be applied in other domains beyond designing by – if necessary – adapting the category system to the task context, although the task categories of this categorisation system should be sufficiently generic for most areas of complex problem solving. For studying non-human primates, the methodology would obviously need to be adapted, but the transition from explicit to implicit coordination may also be found in non-human primates, for instance in the initiation of offspring into the habits of the group, or in newly formed conspecific or mixed-species groups.

Our specific emphasis in this chapter was the development of shared mental models over time. Based on the proposed model (Fig. 10.1), the frequency of certain verbal utterances in our study sample was expected to initially increase in order to achieve sharedness and then decline over the course of a meeting as a result of increased sharedness. This prediction refers to the two phases of sharedness for coordination of task, process, team, and climate aspects.

In accordance with our assumptions, the process of coordination showed the transition from explicit to implicit coordination for the planning activity, considered to be the most important part of process coordination. However, in this sample, no explicit team coordination could be observed in either of the meetings. In our previous field research, we also found only a limited amount of coordination (Stempfle and Badke-Schaub 2002), but typically in these previous studies there were at least some utterances regarding task allocation and additional post-process project planning beyond the meeting. However, the team in this study did none of these, which presumably is related to the specific situation: The team already shared a common history, as there had been previous team meetings. The two observed meetings in this sample were designated as brainstorming sessions where decisions are typically deferred; additionally, only some members had direct responsibilities for execution of work beyond the meeting.

The results of the meetings further showed that the utterances regarding problem definition increased significantly from the first to the second meeting, which would indicate that there was no sufficient sharedness related to problem definition – which makes sense given the change in membership composition from one meeting to the next. The same was found for the number of explanations, which also increased significantly from Meeting 1 to Meeting 2. According to the hypotheses, both results would indicate a low sharedness in the team related to task-specific knowledge. However, the qualitative analysis revealed that these two findings were related to a more in-depth discussion of a new sub-problem. The findings, therefore, do not violate the assumptions of the model but rather indicate that the process of creating sharedness started again in Meeting 2 at a more detailed level – as is perfectly normal for the progression of design work. This could be integrated in the model by adding a further dimension referring to the resolution level of the observed behaviour.

Overall, this two-phase model of the development of team mental models provides a framework for measuring the development of shared mental models in teams by addressing the temporal, dynamic dimension of the development of shared mental models. Its predictions and the results from the case study are also in line with models of group development such as Tuckman (1965) and results of empirical investigations such as the observations of Gersick (1988), who found that project groups worked with a muddling-through strategy until halftime of the task completion, or until critical events occurred that necessitated a (re-)structuring process in the group. We propose that the group needs the first part of its time to explicate the individual representations (transactive, task, and strategic knowledge) and through this process eventually develops shared mental models.

A limitation of the approach is that the content of mental models is only inferred from the verbal communication. Unless the content of mental models is simultaneously measured using another approach (such as Pathfinder), it is impossible to distinguish between low-coordination utterances as a sign of high sharedness and implicit coordination, or as a sign of a lack of coordination because team members inappropriately assume that they share the same mental model. The rules with regard to what is explicated may also vary between domains and teams. In the case study presented here, we had no control over how the data were collected and could therefore not measure the content of mental models at the start and end of the two meetings. However, we would still advise scholars to use this model and methodological approach in combination with other approaches of measurement of shared mental models, as the constructs of the development of shared mental models were confirmed.

References

Arrow H, McGrath JE, Berdahl JL (2000) Small groups as complex systems: formation, coordination, development and adaptation. Sage, Thousand Oaks, CA

Badke-Schaub P, Neumann A, Lauche K, Mohammed S (2007) Mental models in design teams: a valid approach to performance in design collaboration? CoDesign 3:5–20

Bierhals R, Schuster I, Kohler P, Badke-Schaub P (2007) Shared mental models – Linking team cognition and performance. CoDesign 3:75–94

Cannon-Bowers JA, Salas E, Converse S (1993) Shared mental models in expert team decision making. In: Castellan NJ Jr (ed) Individual and group decision making: current issues. Lawrence Erlbaum, Hillsdale, NJ, pp 221–246

Cooke NJ, Salas E, Cannon-Bowers JA, Stout RJ (2000) Measuring team knowledge. Hum Factors 42:151–173

Craik KJW (1943) The nature of explanation. Cambridge University Press, Cambridge

Eccles DW, Tenenbaum G (2004) Why an expert team is more than a team of experts: a social-cognitive conceptualization of team coordination and communication in sport. J Sport Exercise Psy 26:542–560

Edmondson AC, Nembhard IM (2009) Product development and learning in project teams: the challenges are the benefits. J Product Innovat Manage 26:123–138

Entin EE, Serfaty D (1999) Adaptive team coordination. Hum Factors 41:312–325

Espinosa JA, Lerch FJ, Kraut RE (2004) Explicit versus implicit coordination mechanisms and task dependencies: one size does not fit all. In: Salas E, Fiore SM (eds) Team cognition: understanding the factors that drive process and performance. American Psychological Association, Washington, DC, pp 107–129

Espinosa JA, Slaughter SA, Kraut RE, Herbsleb JD (2007) Team knowledge and coordination in geographically distributed software development. J Manage Inform Syst 24:135–169

Fleiss JL (1971) Measuring nominal scale agreement among many raters. Psychol Bull 76: 378–382

Gentner DA, Stevens AL (1983) Mental models. Lawrence Erlbaum, Hillsdale, NJ

Gersick CJG (1988) Time and transition in work teams: toward a new model of group development. Acad Manage J 32:9–41

Goodman PS, Ravlin E, Schminke M (1987) Understanding groups in organizations. In: Staw BM, Cummings LL (eds) Research in organizational behavior. JAI, Greenwich, CT, pp 121–175

Grote G, Zala-Mezö E, Grommes P (2004) The effects of different forms of coordination on coping with workload. In: Dietrich R, Childress TM (eds) Group interaction in high risk environments. Ashgate, Aldershot, UK, pp 39–55

Hackman JR, Wageman R (2005) A theory of team coaching. Acad Manage Rev 30:269–287

Hoegl M, Gemuenden HG (2001) Teamwork quality and the success of innovative projects: a theoretical concept and empirical evidence. Organ Sci 12:435–449

Johnson-Laird PN (1980) Mental models in cognitive science. Cogn Sci 4:71

Kieser A, Kubicek H (1992) Organisation. de Gruyter, Berlin

Klimoski R, Mohammed S (1994) Team mental model – Construct or metaphor? J Manage 20: 403–437

Langan-Fox J, Code S, Langfield-Smith K (2000) Team mental models: techniques, methods, and analytic approaches. Hum Factors 42:242–271

Lim B-C, Klein KJ (2006) Team mental models and team performance: a field study of the effects of team mental model similarity and accuracy. J Organ Behav 27:403–418

Mathieu J, Maynard MT, Rapp T, Gilson L (2008) Team effectiveness 1997–2007: a review of recent advancements and a glimpse into the future. J Manage 34:410–476

Mathieu JE, Heffner TS, Goodwin GF, Cannon-Bowers JA, Salas E (2005) Scaling the quality of teammates' mental models: Equifinality and normative comparisons. J Organ Behav 26:37–56

McDonnell J, Lloyd P (2009) About: designing – analysing design meetings. Taylor and Francis, London

Mesmer-Magnus JR, DeChurch LA (2009) Information sharing and team performance: a meta-analysis. J Appl Psychol 94:535–546

Mohammed S, Klimoski R, Rentsch JR (2000) The measurement of team mental models: we have no shared schema. Organ Res Methods 3:123–165

National Transportation Safety Board (1994) A review of flightcrew-involved major accidents of U.S. air carriers, 1978 through 1990. Safety study, NTSB/SS-94/01. Washington, DC

Orasanu J (1994) Shared problem models and flight crew performance. In: Johnston N, McDonald N, Ruller R (eds) Aviation psychology in practice. Avebury Technical, Aldershot, UK, pp 255–285

Owen WF (1985) Metaphor analysis of cohesiveness in small discussion groups. Small Group Res 16:415–424

Rentsch JR, Hall RJ (1994) Members of great team think alike: a model of the effectiveness and schema similarity among team members. Adv Interdisc Stud Work Teams 1:223–261

Rousseau V, Aube C, Savoie A (2006) Teamwork behaviors–a review and an integration of frameworks. Small Group Res 37:540–570

Salas E, Cooke NJ, Rosen MA (2008) On teams, teamwork, and team performance: discoveries and developments. Hum Factors 50:540–547

Schvaneveldt RW (ed) (1990) Pathfinder associative networks: studies in knowledge organization. Ablex, Norwood, NJ

Stempfle J, Badke-Schaub P (2002) Thinking in design teams – an analysis of team communication. Design Stud 23:473–496

Stempfle J, Hubner O, Badke-Schaub P (2001) A functional theory of task role distribution in work groups. Group Proc Intergroup Relat 4:138–159

Stewart GL (2006) A meta-analytic review of relationships between team design features and team performance. J Manage 32:29–55

Stout RJ, Cannon-Bowers JA, Salas E, Milanovich DM (1999) Planning, shared mental models, and coordinated performance: an empirical link is established. Hum Factors 41:61–71

Tergan SO, Keller T (eds) (2005) Knowledge and information visualization: searching for synergies. Springer lecture notes in computer science. Springer, Heidelberg/NY

Tuckman BW (1965) Developmental sequence in small groups. Psychol Bull 63:384–399 van de Ven AH, Delbecq AL, Koenig RJ (1976) Determinants of coordination modes within organizations. Am Sociol Rev 41:322–338

Weber WG (1997) Analyse von Gruppenarbeit. Kollektive Handlungsregulation in sozio-technischen Systemen [Group task analysis. Collective action regulation in socio-technical systems]. Huber, Bern

Wegner DM (1986) Transactive memory: a contemporary analysis of the group mind. In: Mullen B, Goethals GR (eds) Theories of group behavior. Springer, New York, pp 185–208

Wegner DM, Giuliano T, Hertel P (1985) Cognitive interdependence in close relationships. In: Ickes WJ (ed) Compatible and incompatible relationships. Springer, New York, pp 253–276

West MA (2002) Sparkling fountains or stagnant ponds: an integrative model of creativity and innovation implementation in work groups. Appl Psychol 51:355–424

Wetzstein A, Hacker W (2004) Reflective verbalization improves solutions – the effects of question-based reflection in design problem solving. Appl Cogn Psychol 18:145–156

Wittenbaum GM, Vaughan SI, Stasser G (1998) Coordination in task-performing groups. In: Tindale RS, Heath L, Edwards J, Posvoc EJ, Bryant FB, Suarez-Balcazar Y, Henderson-King E, Myers J (eds) Social psychological applications to social issues: applications of theory and research on groups. Plenum, New York, pp 177–204

Chapter 11
Effective Coordination in Human Group Decision Making: MICRO-CO: A Micro-analytical Taxonomy for Analysing Explicit Coordination Mechanisms in Decision-Making Groups

Michaela Kolbe, Micha Strack, Alexandra Stein, and Margarete Boos

Abstract In this chapter we present a taxonomy we have developed for assessing coordination mechanisms during group decision-making discussions (MICRO-CO). Since there is a convincing number of findings on poor-quality outcomes of human group decisions and tragic examples found in politics (e.g. Bay of Pig invasion of Cuba), there is an escalating need to foster quality group decision making, particularly with regard to group coordination. Especially for ordinary, daily work-group decision processes (e.g. in project teams; during personnel selection), the current state of scientific research does not offer conclusive explanations of how group members communicate in order to coordinate information exchange and decision making. This research question seems interesting given the growing number of decision-making guidebooks for practical use. In recognition of this need, we have developed MICRO-CO, applying theoretical as well as data-driven methods in order to more decisively study the effectiveness of coordination mechanisms for group decision making. It consists of 30 categories organised in three main and four medium levels, with inter-rater reliability testing resulting in

---

M. Kolbe (✉)
Department of Management, Technology, and Economics, Organisation, Work, Technology Group, ETH Zürich, Kreuzplatz 5, KPL G 14, 8032 Zürich, Switzerland
e-mail: mkolbe@ethz.ch

M. Strack and M. Boos
Georg-Elias-Müller-Institute of Psychology, Georg-August-University Göttingen, Goßlerstrasse 14, 37075 Göttingen, Germany
e-mail: mstrack@uni-goettingen.de, mboos@uni-goettingen.de

A. Stein
Georg-Elias-Müller-Institute of Psychology, Georg-August-University of Göttingen, Gosslerstrasse 14, 37075 Göttingen
e-mail: Alexa7@gmx.de

M. Boos et al. (eds.), *Coordination in Human and Primate Groups*,
DOI 10.1007/978-3-642-15355-6_11, © Springer-Verlag Berlin Heidelberg 2011

substantial to very good agreement. We also report initial experiences using MICRO-CO and discuss its limitations and benefits.

11.1 Introduction

Do you remember your last board or project meeting where you had to come to a decision within your group? Unfortunately, as you may confirm, the process of joint decision making seems to be a challenging endeavour and human group decisions are far from perfect (Kerr and Tindale 2004; Stasser and Titus 1985). Reasons for poor human group decision quality stem from (1) an inadequate exchange of information relevant to the decision (Larson et al. 1998a; Mesmer-Magnus and DeChurch 2009; Stasser and Titus 1985), (2) an insufficient evaluation of the possible negative consequences of ego-based or predetermined decision preferences (Gigone and Hastie 1993; Greitemeyer and Schulz-Hardt 2003; Kauffeld 2007a, b; Schauenburg 2004), and (3) an inappropriate integration of different information, leading to a lack of consensus and delaying possible decisions (Nijstad 2006). These findings represent Steiner's (1972) notion postulating that actual group productivity is a function of both the potential group productivity and process losses occurring during the group interaction. Process losses emerge through a malfunction of motivation and coordination (Stroebe and Frey 1982). By concentrating on the latter, the question arises as to how human groups can be effectively coordinated during their decision process in order to minimise process losses and to optimise decision quality simultaneously. Even if we knew the precise demands during group decision making and the potential mechanisms to meet those demands, we still need to study their effectiveness. Studying coordination in human decision-making groups requires measurement tools that allow for assessing the *quality* of coordination processes during group decision-making discussions.

In this chapter we present a micro-analytical taxonomy for analysis of coordination mechanisms in decision-making groups (MICRO-CO). It allows us to (1) measure coordination mechanisms used by group members during decision-making discussions, and thus also to (2) compare effective and ineffective decision-making groups with regard to their explicit coordination behaviour. As will be outlined, effective group decision making requires a high degree of explicitness; MICRO-CO therefore particularly focuses on explicit coordination mechanisms. Compared to existing taxonomies of group processes, MICRO-CO permits a detailed analysis of the coordinative function of statements made during group discussion, for example, by distinguishing among seven types of steering questions. The coding system operates on the micro-level of verbal interaction behaviour based on the premise that, especially in tasks of high complexity, coordination is performed via communication (Reimer et al. 1997).

This chapter is organised as follows: In a first step, we briefly explain the coordination demands of human group decision tasks. Afterwards we present the taxonomy for group decision coordination mechanisms, and we then close with a discussion of both the advantages and challenges of MICRO-CO plus further research needs.

11.2 Coordination Requirements During Group Decision Making

Why and how does group decision making involve coordination? We will focus on the particular characteristics of decision-making tasks (e.g. structuring the process, information requirements, evaluation demands), as this seems a promising method for predicting teamwork requirements (see Chap. 6).

11.2.1 The Nature of Group Decision Tasks

We propose that the conflictive nature of human group decisions and their opaque structure (McGrath 1984), the high information and evaluation demands, as well as social, affiliative, and hierarchical sources of information (Gouran and Hirokawa 1996) lead to very high requirements for coordination. As suggested by Hirokawa (1990), human group tasks can be analysed with regard to three characteristics: structure, information requirements, and evaluation demand. By applying these characteristics to the group decision task, it becomes apparent why human group decision making must be coordinated (Boos and Sassenberg 2001; Kolbe and Boos 2009):

1. Group decision making is a complexly structured process because its goals and means of goal achievement are often part of the decision-making task itself. Establishing a consensus between individual and group goals and matching individual task representations to a shared mental model of the decision task must be achieved as a basis for joint work.
2. The inherent information requirements of group decisions are very high in most cases, because initial information is typically unequally distributed between group members, making a final high-quality decision possible only via shared and integrated information. Without appropriate coordination, these inherent clarification, reconciliation, and information integration qualities of human group decision making tend to result in poor information processing. For example, relevant information often either is not mentioned (Stasser and Titus 2006) or gets lost during discussion due to not being repeated, summarised, or otherwise stored (e.g. Kolbe 2007). Given that most human decision-making groups consist of different experts, and therefore of different views of problems or standards, simply sharing information is not sufficient. In addition, the meaning of the shared information often needs to be reconciled (e.g. Waller and Uitdewilligen 2008).

3. The evaluation demands of group decision tasks are set very high, because the correctness of most human group decisions cannot be determined objectively. This requires that diverse individual opinions, preferences, and evaluation criteria need to be discussed (Boos and Sassenberg 2001) and that the initial ambiguity of information needs to be clarified (Poole and Hirokawa 1996).

Given these high coordination demands of human group decision tasks, we will outline in the next section mechanisms of human group coordination and reveal how they led to the creation of MICRO-CO.

11.2.2 Coordination of Group Decision-Making Discussions

We regard human group coordination as the task-dependent management of inter-dependencies of group tasks, members, and resources by regulating action and information flow (see Chap. 2).

Wittenbaum et al. (1998) considered coordination in task-performing groups a concept with two dimensions. The first dimension refers to the point of time when the coordination mechanism is applied (prior to vs. during the actual interaction). The second dimension refers to the degree of explicitness (see Chap. 4 for a full description). Explicit coordination is mainly used for coordination purposes and, by definition, is expressed in a definitive and unambiguous manner. Statements of explicit coordination leave almost no doubt regarding their underlying purpose, and the coordination intention of an explicitly coordinating group member is often recognised as such by other group members. In instances of implicit coordination, human group members anticipate the actions and needs of the other group members and adjust their own behaviour accordingly (Rico et al. 2008; Wittenbaum et al. 1996). Contrary to explicit coordination, implicit coordination mechanisms typically do not use clear and conclusive behaviours. Instead, coordination is reached tacitly through anticipation and adjustment.

Considering the extremes of these coordination dimensions leads to four simplified modes of group coordination (1) preplans (pre-interaction and explicit, e.g. 'time scheduled for group discussion'), (2) in-process planning (interaction and explicit, e.g. 'summarising opinions'), (3) tacit pre-coordination (pre-interaction and implicit, e.g. 'unspoken expectations and behavioural norms'), and (4) in-process tacit coordination (interaction and implicit, e.g. 'providing task-relevant information without being requested to do so') (Wittenbaum et al. 1998).

Given the process character of human group decision making and the coordination losses that occur during the decision process (Steiner 1972), we will now focus on explicit and implicit in-process coordination mechanisms used during the decision-making process. How can both be effective for group-decision making? The potential effectiveness of implicit in-process coordination lies in (1) its time-saving manner, and (2) its strategic potential for elegantly steering the process by circumnavigating recurrent orders or requests that could result in feelings of inappropriateness or redundancy, as professional group members generally do not wish to be 'directed' or feel that their intelligence or know-how is discounted or disrespected. However, effective implicit in-process coordination requires the participating group members to have an accurate and shared idea of the decision task, procedure, and interaction. Such shared mental models have been defined as group members' knowledge structures enabling them to form accurate explanations and expectations of the task (Cannon-Bowers et al. 1993) and have been classified as an implicit, pre-process coordination mechanism based on the four-cell Coordination Mechanism Circumplex Model emerging from Wittenbaum's mechanism concept (Wittenbaum et al. 1998; see also Chap. 4). If an accurate and appropriately shared mental model does not exist among the involved decision partners, relying on mere implicit in-process coordination can fall short of coordination needs (e.g. van Dijk et al. 2009). In line with this, Wittenbaum et al. (1998) postulated that implicit coordination alone would be ineffective in complex and interdependent tasks. They suggested that divergent goals and intentions, unequal information distribution, and ambiguity of opinions and preferences require – as is typically the case during group decision making – increased levels of explicit in-process coordination. As such, explicit coordination can be considered most important at the beginning of the interaction, as it facilitates the development of shared mental models and thus facilitates later implicit coordination (Orasanu 1993) (see also Chap. 10). The advantages of explicit in-process coordination are its directness and clarity. Even though being explicit requires communicative effort and time and sometimes courage, the trade-off is that it enhances comprehensibility, transparency, and unambiguousness. An interview study focusing on subjective coordination theories of experienced group leaders and facilitators has shown that explicitness is typically used for instructions (e.g. suggesting a procedure for decision making; asking somebody to provide information), process structuring (e.g. goal definition, making and using notes), and fostering shared cognition via clarification questions, solution questions, and procedural questions. It is also considered vital for 'setting the tone' in terms of defining communication rules (Kolbe and Boos 2009). Taken together, it appears that the demands of the decision-making task benefit from a certain amount of explicit in-process coordination.

The relationship among task demands, explicit coordination, and group performance has been well investigated in high-risk work environments. For example, Grote et al. (2010) have found that in cockpit crews, explicit in-process coordination and crew performance were positively correlated. Similarly, medical research has recently focused on the role of group coordination in ensuring patient safety (Künzle et al. 2010; Manser et al. 2008; Rosen et al. 2008; Zala-Mezö et al. 2009) (see also Chaps. 5 and 6). For example, the lack of explicit in-process coordination in the form of questioning decisions and/or by notifying other group members of critical events has been generally found to be a main source of error (Greenberg et al. 2007; Hyey and Wickens 1993).

For ordinary, daily work-group decision processes (e.g. in project teams; during personnel selection), however, the current state of research does not allow conclusions to be drawn about how group members communicate in order to explicitly coordinate information exchange and decision making, which is interesting given the growing number of guidebooks for practical use (e.g. Bens 2005; Edmüller and Wilhelm 2005; Hartmann et al. 2000; Hunter et al. 1995; Kanitz 2004; Seifert 2005; Wikner 2002). We do know that group moderators apply their knowledge about group functioning and their attitudes towards coordination mechanisms when using explicit coordination mechanisms according to perceived task requirements (Kolbe and Boos 2009). But we do not yet know whether this is relevant for optimal group decision performance. Drawing on the functional perspective of groups (Hackman and Morris 1975; Wittenbaum et al. 2004), there appears to be an escalating need to study the effectiveness of using explicit coordination mechanisms during group decision making within the process and for the overall decision outcome, and in turn to investigate whether explicit coordination mechanisms help to avoid common mistakes such as not mentioning, not repeating, or failing to store decision-relevant information (Kolbe 2007; Stasser and Titus 2006).

In addressing these research needs, we present a micro-analytical taxonomy for the analysis of coordination mechanisms in decision-making groups (MICRO-CO). Given the relevance of explicitness during group decision making as outlined earlier, MICRO-CO focuses primarily on explicit rather than implicit in-process coordination mechanisms.

11.3 MICRO-CO: A Micro-analytical Taxonomy for Analysis of Explicit Coordination Mechanisms in Decision-Making Groups

In this section we explain the analysis of group coordination by means of interaction analysis and describe the taxonomy as well as the related coding procedure.

11.3.1 Micro-analytical Interaction Analysis

The goal of the coordination taxonomy is to assess mechanisms used for coordination during group decision-making processes. A coordination mechanism is defined as a statement or action by which group coordination is executed during interaction, whereby the interdependencies of tasks, members, and resources by regulating action and information flow are managed (see Chap. 2). As stated earlier, the focus of MICRO-CO is the assessment of explicit in-process coordination.

Group coordination in decision making can be analysed by means of interaction analysis (see Becker-Beck 1994, 1997; Becker-Beck et al. 2005; Boos 1996; Boos et al. 1990; Brauner 1998; Brauner and Orth 2002; Hirokawa 1982; Kerr et al. 2000; Marks et al. 2001; McGrath et al. 2000; Nägele 2004; Tschan 2000; Weingart et al. 2004; Wittenbaum et al. 2004). We developed this taxonomy based on (1) findings of an explorative study (Kolbe and Boos 2009), (2) videotaped group decision-making discussions of an experimental study (Kolbe 2007), and (3) the formal model of group coordination (see Chap. 2). This model suggests that, based on a coordination occasion, a specific coordination mechanism is used, which in turn is followed by a certain consequence. Applying the formal coordination model requires the detailed description of coordination actions, as well as their prerequisites (coordination occasion) and their proximate consequences. This can be done by micro-analytically coding the coordination utterances of the group members. The term 'micro-analytical' refers to the level of fine-grained analysis of statements of individual group members during interactions.

11.3.2 Taxonomy of MICRO-CO

The taxonomy of MICRO-CO consists of three main categories: explicit in-process coordination, content-related statements, and additional categories (Fig. 11.1).

As suggested in the literature on coding system development (e.g. Brosius and Koschel 2001; Früh 2004; Weingart 1997), we developed the taxonomy in a theoretical as well as data-driven way (see Table 11.1). With regard to theory, we referred to (1) the formal model of group coordination (see Chap. 2), (2) the literature on group coordination and group interaction analysis (Beck and Fisch 2000; Gottman 1979; Grote et al. 2003; Hirokawa 1982; Kauffeld 2007a, b; Larson et al. 1998b; Simon 1997; Yukl 2002) and (3) findings of a study on subjective coordination theories (Kolbe and Boos 2009). This led us to an initial taxonomy, which we tested for usability and reliability, and we then adapted it in a subsequent iterative procedure using five transcribed group decision-making discussions of a previous study (Boos 1996).

The main category of explicit in-process coordination includes four medium-level categories with respective categories whose source will be explained in Table 11.1: 'addressing' (personally and by name); 'instructions' ('asking sb. to do sth.', 'assigning tasks or responsibilities', 'suggesting procedure', 'asking sb. to clarify sth.', 'asking sb. to suggest sth.', 'reminding sb.'), 'structuring' ('summarising', 'repeating', 'goal setting', 'goal indicating', 'deciding', 'explaining own behaviour'), and 'questions' ('requesting information', 'requesting opinion', 'requesting clarification', 'procedural questioning', 'requesting solution', 'requesting agreements', 'requesting decision'). The content-related statements include 'declaring', 'providing information', 'providing opinion', 'agreeing', 'disagreeing', 'content-related suggesting', and 'suggesting solution'. Finally, there are two additional categories: 'interrupting' and 'one-word focusing statements' (Fig. 11.1). A detailed description of these categories, including examples, can be found in Table 11.1. For example, the category 'explaining own behaviour' belongs to the medium-level categories structuring (see first column), which in turn belongs to the main-level category 'explicit in-process coordination' (see vertical row). It is defined as a person who makes his or her manner or attitude (in which he or she behaves) clear or comprehensible. An example would be, 'I will now tell you the pros and cons of this procedure'. The last column of Table 11.1 shows that this category was

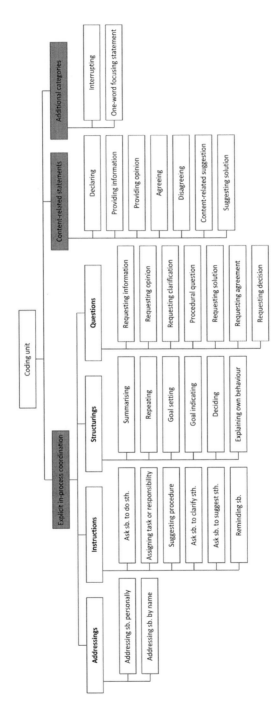

Fig. 11.1 Micro-analytical taxonomy for analysis of coordination mechanisms in decision-making groups (MICRO-CO)

11 Effective Coordination in Human Group Decision Making

Table 11.1 Definitions, examples, and sources of MICRO-CO categories

Main- and medium-level category	Category	Definition	Example	Source
Explicit in-process coordination				
Addressing	Addressing sb. Personally	The person who talks speaks to someone directly in order to give him or her instructions	'This falls into your area of expertise'.	Data-based
	Addressing sb. by name	The person who talks approaches someone directly by using his or her name in order to get his or her attention (e.g. by using the first or last name)	'Steve, would you please explain your point of view?'	Data-based
Instructions	Ask sb. to do sth.	Someone directs a request for something at someone	'Would you please hand out the write-ups?'	Grote et al. (2003), Yukl (2002), Kolbe and Boos (2009)
	Assigning tasks or responsibilities	Someone allocates or gives out tasks or delegates responsibilities to someone.	'It's your job to make sure the process runs smoothly'.	Yukl (2002), Kauffeld (2007a)
	Suggesting procedure	Someone proposes a series of steps which should help to accomplish a task	'Let's first listen to everyone's opinion before coming to a conclusion'.	Fisch (1998), Simon (1997), Kauffeld (2007a), Kolbe and Boos (2009)
	Ask sb. to clarify sth.	A person requests that someone makes his or her statement clearer and easier to understand in order to eliminate confusion or uncertainty	'Please illustrate your suggestion'.	Simon (1997), Yukl (2002)
	Ask sb. to suggest sth.	A person tries to induce someone to come up with an idea in order to start a process by which one thought leads to another.	'Could someone offer an idea of how to solve this problem?'	Simon (1997), Yukl (2002)

(continued)

Table 11.1 (continued)

Main- and medium-level category	Category	Definition	Example	Source
	Reminding sb.	Someone causes another to remember something. He or she puts something back into his or her mind	'We must keep in mind that we have to solve this problem by the end of the day'.	Yukl (2002), Kolbe and Boos (2009)
Structuring	Summarising	Someone gives a recapitulation of salient facts	'To sum up the situation it has been said that...'	Grote et al. (2003), Kauffeld (2007a)
	Repeating	Someone repeats or reinforces something already said. Therefore he or she expresses him- or herself in the same way or with the same words	Person A: 'We should take a look at our law enforcement!' Person B: 'Yeah, we should [take a look at the law enforcement]!'	Kolbe and Boos (2009), Larson et al. (1998b)
	Goal setting	A technique leading to the assigning and choosing of specific, objective, concrete targets or goals which one or more members of the group strive to achieve	'We have to figure out who is going to do what in order to solve this case!'	Kolbe and Boos (2009), inductive
	Goal indicating	Someone points out or shows the way to or the direction of where the discussion is supposed to end and what results they want to achieve	'I think if we find some clues to such-and-such we'll be able to solve the puzzle'.	Data-based
	Deciding	Someone is able to determine the outcome and therefore reaches a decision	'I've reached a decision on how we go from here...'	Kolbe and Boos (2009), Simon (1997)
	Explaining own behaviour	A person makes his or her manner/ attitude (in which he or she behaves) plain or comprehensible	'I will now tell you the pros and cons of this procedure'.	Kolbe and Boos (2009), data-based

Questions	Requesting information	A person asks someone to share their knowledge or facts and data they've collected on the subject	'Haven't you read something about this lately? What was it about exactly?'	Grote et al. (2003), Kauffeld (2007a), Fisch (1998), Simon (1997)
	Requesting opinion	A person asks someone for his or her belief/judgment based on his or her special knowledge	'What do you think about this suggestion?'	Kauffeld (2007a)
	Requesting clarification	A person asks someone to explain or clarify a certain aspect of the discussion	'What exactly are you suggesting?'	Simon (1997), Kolbe and Boos (2009), Yukl (2002)
	Procedural question	A person asks someone to specify exactly what has to be done and in what order so everyone knows what series of steps have to be taken to accomplish an end	'How do you suggest we proceed from here?'	Kauffeld (2007a), Fisch (1998), Kolbe and Boos (2009)
	Requesting solution	A person requests an answer to the current problem	'Would someone like to suggest a solution?'	Kolbe and Boos (2009), data-based
	Requesting agreements	A person suggests reaching a harmonious consensus in order to come to terms on the subject	'Are you all OK with this decision?'	Grote et al. (2003), Simon (1997), Kolbe and Boos (2009)
	Requesting decision	A person asks someone else to reach a conclusion or make up one's mind	'Would someone please share the conclusion they've reached?'	Simon (1997)
Content-related statements			–	
	Declaring sth.	A person makes a full statement.	'Never underestimate anybody!'	Fisch (1998)
	Providing information	Someone presents/lays out his or her knowledge or a collection of facts or data	'This law is about the rights of all citizens'.	Fisch (1998), Kolbe and Boos (2009), Grote et al. (2003)
	Providing opinion	Someone reveals/gives away his or her beliefs or judgments	'I don't think that's such a good idea because we shouldn't	Fisch (1998)

(continued)

Table 11.1 (continued)

Main- and medium-level category	Category	Definition	Example	Source
			forget to include everyone on this matter'.	
	Agreeing	Someone comes to terms about a thing and reaches a mutual understanding	'We are on the same page when it comes to how to solve this'./'I absolutely agree with everything you've said'.	Fisch (1998), Kauffeld (2007a), Gottman (1979)
	Disagreeing	Someone has a different opinion about something	'Unfortunately I have other priorities!'/'I don't quite agree with you on that one'.	Fisch (1998), Kauffeld (2007a), Gottman (1979)
	Content-related suggestion	Someone makes a content wise proposition that helps solving a problem	'We should take a look at the casebooks on this matter'.	Fisch (1998), Kauffeld (2007a), Gottman (1979)
Suggesting solution	A person provides an answer to the current problem		'Why don't we try it like this. . .'	Fisch (1998)
Additional categories				–
	Interrupting	A person stops someone else in mid-sentence	Person A: 'I assure that. . .' Person B: 'You can say what you want, we still should do...'	Kauffeld, (2007a), Kolbe and Boos (2009)
	One-word focusing statement	Someone makes a declaration that consists of only one word	'Absolutely!'/'Done!'	Data-based

developed inductively as well as based on the exploratory study (Kolbe and Boos 2009).

Since MICRO-CO is designed for the analysis of decision-making discussions, which are usually of a verbal character, it focuses on verbal communication used as coordination mechanisms. It does not include non-verbal coordination behaviour (e.g. proving task-relevant action without requests, such as holding up the correct surgical instrument after the attending surgeon has announced a change in procedure), as is the case in other coordination taxonomies that have been designed for analysing group coordination in high-risk, time-pressed work environments (e.g. Grote et al. 2003; Kolbe et al. 2009b; Manser et al. 2009).

11.3.3 Coding Procedure

The analysis of group decision processes requires the definition of the sampling rule (which subjects are to be observed and when) and the recording rule (how behaviour is recorded). Regarding the sampling rule, the most satisfactory approach to studying groups using means of observation is the so-called focal sampling method (Martin and Bateson 1993). Thereby, the whole group is observed for a specified period of time (e.g. duration of group discussion). In the case of MICRO-CO, all occurring behaviour is coded by applying the above MICRO-CO categories and by indicating who is communicating to whom. The recording rule indicates the way of coding, typically either continuous sampling (all occurrences are coded) or time sampling (behaviour is sampled periodically, e.g. every 10 s) (Martin and Bateson 1993). Despite the fact that periodic time sampling is usually considered more reliable for reasons explained ahead, we recommend recording the coordination behaviour continuously, which allows for assessing the 'true' frequencies and durations of events (Martin and Bateson 1993). The literature on observation sampling contends that in order to analyse the dynamic coordination process and to determine whether a certain coordination act is followed by another certain act as well as how long each act lasts, continuous coding is necessary, as it facilitates appropriate data analysis methods – for example, lag sequential analysis (Bakeman 2000; Bakeman and Gottman 1986). However, continuous coding challenges the proper definition of coding units, especially when defining the amount of communication behaviour that is coded into one category vs. another (McGrath and Altermatt 2002). We therefore recommend a technique for the systematic definition of coding units using ten segmentation rules[1] based on grammar (SYNSEG; Kolbe et al. 2007). This technique can be used for preparing transcribed or merely videotaped group decision-making discussions that can subsequently be coded with the subcategories of MICRO-CO.

---

[1]A description of the ten segmentation rules can be requested from the first author.

11.3.4 Reliability of MICRO-CO

In an ongoing study, we are investigating the impact of explicit in-process group coordination mechanisms on group decision quality. Within this study we tested MICRO-CO for inter-rater reliability. Two trained coders independently coded a group decision-making discussion of 25 min' duration. Faced with a personnel selection task, four group members had to choose one of four candidates. Coding units were defined using the above-mentioned grammar-based technique suggested by Kolbe et al. (2007), which resulted in 585 units. Analysis of Cohen's kappa to assess inter-rater agreement showed a mean value of $\kappa = 0.89$. The MICRO-CO categories 'suggesting solution', 'deciding', and 'interrupting' were especially reliable (each $\kappa = 0.99$), whereas the categories 'ask sb. to do sth'. ($\kappa = 0.66$) and 'suggesting procedure' ($\kappa = 0.79$) were the least reliable. Table 11.2 shows the mean kappa values for the six medium-level categories ranging between 'substantial' and 'almost perfect' reliability (Landis and Koch 1977, p 165).

11.3.5 First Experience for MICRO-CO Category Occurrence

In exemplifying the usage of MICRO-CO, we will first show its sensitivity in assessing the coordination character of statements made during group discussions. We will then refer to the validity of the three-level organisation of the taxonomy (main level, medium level, category level) and explain how these features help indicate the quality of the decision and also the usefulness of explicit coordination.

Applying the German version of MICRO-CO to group decision-making discussions showed the occurrence of a considerable proportion of explicit in-process coordination. A sample of 32 group discussions (duration 11–45 min, MD = 23 min, SD = 7.6 min) of experimental four-person groups was segmented as suggested by Kolbe et al. (2007), resulting in 22,920 units. They were coded using a slightly simplified version of MICRO-CO, which differs from Fig. 11.1 in only minor aspects (the two addressing categories were combined; only three types of instructions were differentiated; 'goal setting' and 'goal indicating' were combined; and five instead of seven types of questions were discriminated).

Table 11.2 Cohen's kappa for MICRO-CO medium-level categories (two coders, 585 units)

Medium-level category	Cohen's kappa values
Instructions	0.73
Structuring	0.91
Questions	0.95
Content-related statements	0.89
Additional categories	0.97
Addressing	*Did not occur*

11 Effective Coordination in Human Group Decision Making 213

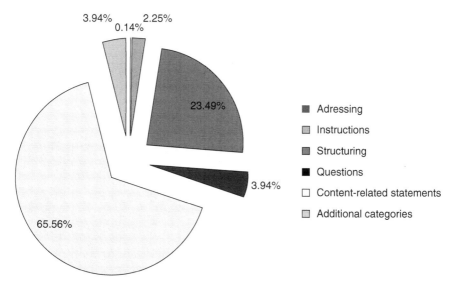

Fig. 11.2 Occurrence of medium-level categories of MICRO-CO (32 four-person groups, 22,920 units)

Figure 11.2 depicts the occurrence of medium-level categories: About one third of the discussions served coordination purposes, mainly by structuring statements. These results indicate that MICRO-CO serves its purpose in being very sensitive to the explicit coordination character of statements made during group decision discussion.

The 32 groups analysed in the above-mentioned study are part of an ongoing experimental series on the effectiveness of coordination mechanisms. Assuming that repeating already-mentioned information facilitates group cognition during decision making and thus contributes to the quality of group decision, Kolbe (2007) instructed one of the four members per group to facilitate the group discussion, instructing half of these lay facilitators to specifically 'repeat' important information others had mentioned. 'Repeating' is a subcategory of the 'structuring' category on the medium level which belongs to the 'coordination' super-ordinate category of MICRO-CO (see Fig. 11.1). According to the manipulation, groups varied in the amount of 'repeating' information from 6–27% of their units, leading to a range of 15–37% 'structuring' behaviour, and 23–46% 'coordinating' behaviour, respectively. The group decision served as the main dependent measure: A correct decision represented a solved hidden profile. Forty percent of the groups were correct. Figure 11.3 illustrates the logistic regression of correct decisions on the amount of coordinating behaviour on the three levels of MICRO-CO (coordinating, structuring, repeating). As Fig. 11.3 shows, prediction of decisions is possible on each of the three levels of the taxonomy (Nagelkerke's $R^2 = 0.41$ for 'repeating', $R^2 = 0.23$ for 'structuring', and $R^2 = 0.21$ for 'coordinating' on the highest level). This means that decision-making quality in this study can be

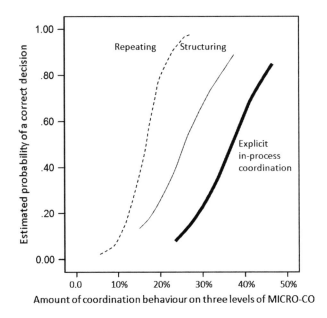

Fig. 11.3 Logistic regression of the correct decision on coordinating behaviour on the three levels of MICRO-CO (32 groups)

attributed to a large degree to sound explicit coordination, to structuring, and especially to repeating information. Thus, using MICRO-CO allowed for these multi-layer findings on the usefulness of explicitness during group decision making to be obtained.

11.4 Discussion

The MICRO-CO taxonomy presented here allows for micro-analytically analysing coordination mechanisms in human decision-making groups. Developed on the basis of both coordination theory (see Chap. 2) and empirical data from group decision-making discussions, it provides a reliable and manageable set of coding categories applicable to group decision-making discussions. Compared to existing taxonomies of group processes, MICRO-CO permits a detailed analysis of the coordination character of statements made during group discussion by precisely distinguishing among a variety of explicit mechanisms in a hierarchical framework. MICRO-CO is designed to permit group researchers to assess the occurrence and duration of both explicit in-process coordination mechanisms as well as content-oriented utterances. The resulting codes can subsequently be integrated in lag sequential analysis (Bakeman and Gottman 1986), revealing insights into the proximate functions of the explicit mechanisms for the ongoing group discussion.

This analysis could, for example, investigate whether content-oriented statements occur only in response to explicit 'request' categories (see Sect. 11.1.2.3). An unsolicited occurrence of task-relevant content-oriented statements could be regarded as an indicator of implicit in-process coordination (anticipation ratio; see Serfaty et al. 1993; Toups and Kerne 2007). Due to these built-in tools enabling the micro-analysis of content occurrence, duration, source, and especially the ability to assess the degree of explicit vs. implicit coordination MICRO-CO is also useful for comparing effective and ineffective decision-making groups with regard to their coordination behaviour. One example was given in Fig. 11.3, where the solution of a hidden profile was regressed to coordination on the three hierarchical levels of MICRO-CO. With regard to the distal effects of coordination behaviour, we consider this micro-process analysis an important contribution to increasing our understanding of the characteristics of effective human group decision making. Furthermore, MICRO-CO can be used to analyse the coordination behaviour of group facilitators or leaders and contribute to the training of the effectiveness of their behaviour.

Further research should address the issue of validity of MICRO-CO. Dickinson and McIntyre (1997) have pointed out that construct validity of the observation method requires the ability to discriminate between groups. An important advantage of micro-coding systems such as MICRO-CO is that they allow for identifying task-relevant and rather fine-grained, subconsciously occurring behaviour – group behaviour not detectable on a more aggregated level. On the other hand, in micro-coding systems, category membership is frequently based on the meaning or structure of single-member statements, resulting in a lack of synthesis of individual utterances into interaction or group process indices. In this sense, Marks and colleagues (Marks et al. 2001, p. 364) have discussed that 'detecting processes often requires more macro observation of the verbal exchanges and behaviours that take place during a particular episode'. We suggest that lag sequential analysis might serve as a tool for analysing proximate functions of individual coordination behaviour, empirically clustering micro-level findings on macro-level patterns. As such, lag sequential analysis can be used to investigate the antecedents and functions of single coordination behaviours within and between group members (e.g. Grote et al. 2010; Kolbe et al. 2009a). The resulting coordination patterns can be utilised as dynamic group-level indices and can be compared between different groups. Lag sequential analysis can also be applied to classify behavioural codes by means of their functional similarity (Jacobs and Krahn 1987). The said application addresses the issue of appropriate aggregation, which is particularly important when we are interested in the distal consequences of coordination behaviour such as its long-term effectiveness (Dickinson and McIntyre 1997). Further observational research on group processes such as coordination might profit enormously from an integrative comparison of the methods applied in the domain of observing human and non-human primate groups. In particular, important methodical issues such as sampling, aggregating codes, editing, and analysing observational group process data are worth comparing vis-à-vis their construct validity. There may not be the silver bullet, but some standards would be beneficial.

References

Bakeman R (2000) Behavioral observation and coding. In: Reis HT, Judd CM (eds) Handbook of research methods in social and personality psychology. Cambridge University Press, New York, pp. 138–159

Bakeman R, Gottman JM (1986) Observing interaction: an introduction to sequential analysis. Cambridge University Press, Cambridge

Beck D, Fisch R (2000) Argumentation and emotional processes in group decision-making: Illustration of a multi-level interaction process analysis approach. Group Process Intergroup Relat 3:183–201

Becker-Beck U (1994) Strukturanalyse des Interaktionsverhaltens in Diskussionsgruppen [in German]. Gruppendynamik 25:95–106

Becker-Beck U (1997) Soziale Interaktion in Gruppen: Struktur- und Prozessanalyse [in German]. Westdeutscher Verlag, Opladen

Becker-Beck U, Wintermantel M, Borg A (2005) Principles of regulating interaction in teams practicing face-to-face communication versus teams practicing computer-mediated communication. Small Group Res 36:499–536

Bens I (2005) Facilitating with ease! Core skills for facilitators, team leaders and members, managers, consultants, and trainers. Wiley, San Francisco, CA

Boos M (1996) Entscheidungsfindung in Gruppen: Eine Prozessanalyse [in German]. Huber, Bern

Boos M, Morguet M, Meier F, Fisch R (1990) Zeitreihenanalysen von Interaktionsprozessen bei der Bearbeitung komplexer Probleme in Expertengruppen [in German]. Z Sozialpsychol 21: 53–64

Boos M, Sassenberg K (2001) Koordination in verteilten Arbeitsgruppen [in German]. In: Witte EH (ed) Leistungsverbesserungen in aufgabenorientierten Kleingruppen: Beiträge des 15. Hamburger Symposiums zur Methodologie der Sozialpsychologie. Papst, Lengerich, pp. 198–216

Brauner E (1998) Die Qual der Wahl am Methodenbuffet oder: Wie der Gegenstand nach der passenden Methode sucht [in German]. In: Ardelt-Gattinger E, Lechner H, Schlögl W (eds) Gruppendynamik. Anspruch und Wirklichkeit der Arbeit in Gruppen. Verlag für Angewandte Psychologie, Göttingen, pp. 176–193

Brauner E, Orth B (2002) Strukturen von Argumentationssequenzen in Gruppen [in German]. Z Sozialpsychol 33:65–81

Brosius H, Koschel F (2001) Methoden der empirischen Kommunikationsforschung - Eine Einführung [in German]. Westdeutscher Verlag, Wiesbaden

Cannon-Bowers JA, Salas E, Converse S (1993) Shared mental models in expert team decision making. In: Castellan JN (ed) Individual and group decision making. Current issues. Lawrence Erlbaum, Hillsdale, NJ

Dickinson TL, McIntyre RM (1997) A conceptual framework for teamwork measurement. In: Brannick MT, Salas E, Prince C (eds) Team performance assessment and measurement. Lawrence Erlbaum, Mahwah, NJ

Edmüller A, Wilhelm T (2005) Moderation in German. Haufe, Planegg, Germany

Fisch R (1998) Konferenzkodierung: Eine Methode zur Analyse von Problemlöseprozessen in Gruppen [in German]. In: Ardelt-Gattinger E, Lechner H, Schlögl W (eds) Gruppendynamik: Anspruch und Wirklichkeit der Arbeit in Gruppen. Verlag für Angewandte Psychologie, Göttingen, pp. 194–206

Früh W (2004) Inhaltsanalyse [in German]. UVK, Konstanz, Germany

Gigone D, Hastie R (1993) The common knowledge effect: information sharing and group judgement. J Pers Soc Psychol 65:959–974

Gottman JM (1979) Marital interactions. Experimental investigations. Academic, New York

Gouran DS, Hirokawa RY (1996) Functional theory and communication in decision-making and problem-solving groups. An expanded view. In: Hirokawa RY, Poole MS (eds) Communication and group decision making. Sage, Thousand Oaks, CA, pp. 55–80

11 Effective Coordination in Human Group Decision Making

Greenberg CC, Regenbogen SE, Studdert DM, Lipsitz SR, Rogers SO, Zinner MJ, Gawande AA (2007) Patterns of communication breakdowns resulting in injury to surgical patients. J Am Coll Surg 204:533–540

Greitemeyer T, Schulz-Hardt S (2003) Preference-consistent evaluation of information in the hidden profile paradigm: Beyond group-level explanations for the dominance of shared information in group decisions. J Pers Soc Psychol 84:322–339

Grote G, Zala-Mezö E, Grommes P (2003) Effects of standardization on coordination and communication in high workload situations. Linguistische Berichte, Sonderheft 12:127–155

Grote G, Kolbe M, Zala-Mezö E, Bienefeld-Seall N, Künzle B (2010) Adaptive coordination and heedfulness make better cockpit crews. Ergonomics 52:211–228

Hackman JR, Morris CG (1975) Group tasks, group interaction process, and group performance effectiveness: a review and proposed integration. In: Berkowitz L (ed) Advances in experimental social psychology. Academic, New York, pp. 44–99

Hartmann M, Funk R, Wittkuhn KD (2000) Gekonnt moderieren [in German]. Teamsitzung, Besprechung und Meeting: Zielgerichtet und ergebnisorientiert. Beltz, Weinheim, Germany

Hirokawa RY (1982) Group communication and problem-solving effectiveness I: a critical review of inconsistent findings. Commun Quart 30:134–141

Hirokawa RY (1990) The role of communication in group decision-making efficacy. A task-contingency perspective. Small Group Res 21:190–204

Hunter D, Bailey A, Taylor B (1995) The art of facilitation: how to create group synergy. Fisher, Tucson, AZ

Hyey BM, Wickens CD (1993) Team leadership and crew coordination. In: Hyey BM, Wickens CD (eds) Workload transition: implications for individual and team performance. National Academy, Washington, DC, pp. 229–247

Jacobs T, Krahn G (1987) The classification of behavioral observation codes in studies of family interaction. J Marriage Fam 49:677–687

Kanitz VA (2004) Gesprächstechniken [in German]. Haufe, Planegg, Germany

Kauffeld S (2007a) Kompetenzen in der betrieblichen Praxis: Das Kasseler-Kompetenz-Raster [in German]. Berufsbildung, 103/104, pp 24–26

Kauffeld S (2007b) Jammern oder Lösungsexploration? Eine sequenzanalytische Betrachtung des Interaktionsprozesses in betrieblichen Gruppen bei der Bewältigung von Optimierungsaufgaben [in German]. Z Arb Organ 51:55–67

Kerr N, Tindale RS (2004) Group performance and decision-making. Annu Rev Psychol 55:623–655

Kerr NL, Aronoff J, Messé LA (2000) Methods of small group research. In: Reis HT, Judd CM (eds) Handbook of research methods in social and personality psychology. Cambridge University Press, New York, pp. 160–189

Kolbe M (2007) Koordination von Entscheidungsprozessen in Gruppen Die Bedeutung expliziter Koordinationsmechanismen [in German]. VDM, Saarbrücken

Kolbe M, Boos M (2009) Facilitating group decision-making: Facilitator's subjective theories on group coordination. Forum Qualitative Sozialforschung/Forum: Qualitative Social Research 10, http://nbn-resolving.de/urn:nbn:de:0114-fqs0901287

Kolbe M, Stein A, Boos M, Strack M (2007) SYNSEG – a method for developing coding units in verbal interaction processes. University of Goettingen, Göttingen

Kolbe M, Künzle B, Zala-Mezö E, Wacker J, Grote G (2009a) Adaptive coordination in anaesthesia crews. Micro-analytical analysis of coordination behaviour in live clinical settings. In: 14th European Congress of Work and Organizational Psychology, Santiago de Compostela, Spain

Kolbe M, Künzle B, Zala-Mezö E, Wacker J, Grote G (2009b) Measuring coordination behaviour in anaesthesia teams during induction of general anaesthetics. In: Flin R, Mitchell L (eds) Safer surgery. Analysing behaviour in the operating theatre. Ashgate, Aldershot, UK, pp. 203–221

Künzle B, Kolbe M, Grote G (2010) Ensuring patient safety through effective leadership behaviour: a literature review. Safety Sci 48:1–17

Landis JR, Koch GG (1977) The measurement of observers agreement for categorial data. Biometrics 33:159–174

Larson JR, Christensen C, Franz TM, Abbott AS (1998a) Diagnosing groups: the pooling, management, and impact of shared and unshared case information in team-based medical decision making. J Pers Soc Psychol 75:93–108

Larson JR, Foster-Fishman PG, Franz TM (1998b) Leadership style and the discussion of shared and unshared information in decision-making groups. Pers Soc Psychol B 24:482–495

Manser T, Howard SK, Gaba DM (2008) Adaptive coordination in cardiac anaesthesia: a study of situational changes in coordination patterns using a new observation system. Ergonomics 51: 1153–1178

Manser T, Howard SK, Gaba DM (2009) Identifying characteristics of effective teamwork in complex medical work environments: adaptive crew coordination in anaesthesia. In: Flin R, Mitchell L (eds) Safer surgery: analysing behaviour in the operating theatre. Ashgate, Aldershot, UK, pp. 223–239

Marks MA, Mathieu JE, Zaccaro SJ (2001) A temporally based framework and taxonomy of team processes. Acad Manage Rev 26:356–376

Martin P, Bateson P (1993) Measuring behavior. An introductory guide. Cambridge University Press, Cambridge

McGrath JE (1984) Groups, interaction and performance. Prentice Hall, Englewood Cliffs, NJ

McGrath JE, Altermatt TW (2002) Observation and analysis of group interaction over time: some methodological and strategic choices. In: Hogg MA, Tinsdale S (eds) Blackwell handbook of social psychology: group processes. Blackwell, Boston, MA, pp. 525–556

McGrath JE, Arrow H, Berdahl JL (2000) The study of groups: past, present, and future. Pers Soc Psychol Rev 4:95–105

Mesmer-Magnus JR, Dechurch LA (2009) Information sharing and team performance: a meta-analysis. J Appl Psychol 94:535–546

Nägele C (2004) Dynamic group processes in co-acting virtual teams. Université de Neuchâtel, Neuchâtel

Nijstad BA (2006) Decision refusal in groups: the role of time pressure and leadership. 1st Annual Conference of INGRoup 2006, Pittsburgh

Orasanu JM (1993) Decision-making in the cockpit. In: Wiener EL, Kanki BG, Helmreich RL (eds) Cockpit ressource management. Academic, San Diego, CA, pp. 137–172

Poole MS, Hirokawa RY (1996) Introduction. Communication and group decision making. In: Hirokawa RY, Poole MS (eds) Communication and group decision making. Sage, Thousand Oaks, CA, pp. 3–18

Reimer T, Neuser A, Schmitt C (1997) Unter welchen Bedingungen erhöht die Kommunikation zwischen Gruppenmitgliedern die Koordinationsleistung in einer Kleingruppe [in German]? Z Exp Psychol 44:495–518

Rico R, Sánchez-Manzanares M, Gil F, Gibson C (2008) Team implicit coordination processes: a team knowledge-based approach. Acad Manage Rev 33:163–184

Rosen MA, Salas E, Wilson KA, King HB, Salisbury ML, Augenstein JS, Robinson DW, Birnbach DJ (2008) Measuring team performance in simulation-based training: adopting best practices for healthcare. Simul Healthcare 3:33–41

Schauenburg B (2004) Motivierter Informationsaustausch in Gruppen: Der Einfluss individueller Ziele und Gruppenziele [Motivated information sampling in groups: The influence of individual and group goals]. Dissertation. University of Göttingen, Göttingen. Available at http://webdoc.sub.gwdg.de/diss/2004/schauenburg/

Seifert JW (2005) 30 Minuten für professionelles Moderieren [in German]. Gabal, Offenbach, Germany

Serfaty D, Entin EE, Volpe C (1993) Adaptation to stress in team decision-making and coordination. In: Proceedings of the human factors and ergonomics society 37th annual meeting. human factors and ergonomics society, Santa Monica, CA

Simon P (1997) SYNPRO. Interaktions-Beobachtungssystem [in German]. Manual zur Kodierung, Unveröffentlichtes Manuskript. Universität Regensburg, Regensburg

Stasser G, Titus W (1985) Pooling of unshared information in group decision making: Biased information sampling during discussion. J Pers Soc Psychol 48:1467–1578

Stasser G, Titus W (2006) Pooling of unshared information in group decision making: Biased information sampling during discussion. In: Levine JM, Moreland R (eds) Small groups. Psychology, New York, pp. 227–239

Steiner ID (1972) Group processes and productivity. Academic, New York

Stroebe W, Frey BS (1982) Self-interest and collective action: the economics and psychology of public goods. Brit J Soc Psychol 21:121–137

Toups ZO, Kerne A (2007) Implicit coordination in firefighting practice: design implications for teaching fire emergency responders. In: Proceedings of the SIGCHI conference on human factors in computing systems. ACM, New York

Tschan F (2000) Produktivität in Kleingruppen. Was machen produktive Gruppen anders und besser [in German]? Huber, Bern.

Van Dijk E, De Kwaadsteniet EW, De Cremer D (2009) Tacit coordination in social dilemmas: The importance of having a common understanding. J Pers Soc Psychol 96:665–678

Waller MJ, Uitdewilligen S (2008) Talking to the room. Collective sensemaking during crisis situations. In: Roe RA, Waller MJ, Clegg SR (eds) Time in organizational research. Routledge, Oxford, pp. 186–203

Weingart LR (1997) How did they do that? The ways and means of studying group process. Res Organ Behav 19:189–239

Weingart LR, Olekalns M, Smith PK (2004) Quantitative coding of negation behavior. Int Negot 9:441–455

Wikner U (2002) Besprechungen moderieren. Top-Tools für effiziente Meetings. Financial Times/ Prentice Hall, München

Wittenbaum GM, Hollingshead AB, Paulus PB, Hirokawa RY, Ancona DG, Peterson RS, Jehn KA, Yoon K (2004) The functional perspective as a lens for understanding groups. Small Group Res 35:17–43

Wittenbaum GM, Stasser G, Merry CJ (1996) Tacit coordination in anticipation of small group task completion. J Exp Soc Psychol 32:129–152

Wittenbaum GM, Vaughan SI, Stasser G (1998) Coordination in task-performing groups. In: Tindale RS, Heath L, Edwards J, Posavac EJ, Bryant FB, Suarez-Balcazar Y, Henderson-King E, Myers J (eds) Theory and research on small groups. Plenum, New York, pp. 177–204

Yukl G (2002) Leadership in organizations. Prentice Hall, Upper Saddle River, NJ

Zala-Mezö E, Wacker J, Künzle B, Brüesch M, Grote G (2009) The influence of standardisation and task load on team coordination patterns during anaesthesia inductions. Qual Saf Health Care 18:127–130

Part III
Primatological Approaches to the Conceptualisation and Measurement of Group Coordination

Chapter 12
Primatological Approaches to the Study of Group Coordination

Peter M. Kappeler

Abstract This chapter outlines why non-human primates provide some of the best comparative models for students of coordination in small human groups. It then summarises what and why non-human primates need to coordinate at the group level. From this review, group movements emerge as the major paradigm of primatologists in this study context. In this integrating chapter, the content of the contributions to Part III is placed within the broader context of this book on coordination in human and non-human primates.

12.1 Introduction

Many animals live in groups where a number of decisions need to be made at the group level on a regular basis. For example, many social insects collectively choose new nest sites, migrating birds agree on a common migration route, some carnivores and primates hunt cooperatively, and virtually all group-living species need to coordinate their daily activities and movements with each other (Conradt and Roper 2005; Kerth 2010). Human groups stand out from those of other animals in that human groups exhibit more diversity, complexity, and social dynamics than those of any other species, including those of non-human primates (i.e. the more than 400 species of lemurs, lorises, tarsiers, monkeys, and apes). Because human societies have their biological roots in the primate order (Chapais 2010), it is evident to evolutionary biologists that many aspects of human behaviour – especially those related to successful survival, reproduction, and parenting – have a more or less pronounced biological legacy (see, e.g. Kappeler et al. 2010). Comparative studies across different species can therefore help identify common principles in behavioural

---

P.M. Kappeler
Department of Behavioral Ecology and Sociobiology, German Primate Center, Kellnerweg 6, 37077 Göttingen, Germany
e-mail: pkappel@gwdg.de

M. Boos et al. (eds.), *Coordination in Human and Primate Groups*,
DOI 10.1007/978-3-642-15355-6_12, © Springer-Verlag Berlin Heidelberg 2011

evolution, as well as unique, species-specific solutions. In the context of group decisions and coordination, recent broad comparisons across animal species have begun integrating relevant research on humans, revealing several fundamental insights regarding important variables that structure collective decision making (Conradt and List 2009).

At a more fine-grained level beyond our biological legacy, non-human primates are the obvious outgroup for obtaining comparative insights into fundamental mechanisms of group coordination that have also shaped human behaviour, specifically because humans and primates share many more features of their life history and socioecology than humans share with social insects or fish, for example. For instance, because human groups and those of other primates are on average much smaller than those of many other taxa, individuals recognise each other and they establish social relationships that last years or even decades (e.g. Jolly and Pride 1999; Silk et al. 2003). The main aims of this integrating chapter are to detail some of the reasons for the suitability of primates as a point of reference for students of human behaviour, to identify obvious limitations of this comparative approach, to explain the dominant research paradigm in primate group coordination studies, and to place the subsequent chapters on various aspects of primate group coordination within this overall framework.

Non-human primates not only exhibit great variation in social systems suitable for interesting comparative studies of adaptation across species, but they also share several traits that predestine them for comparisons with humans beyond the obvious close phylogenetic relationship. First, the majority of extant primates live in groups in which several males and females of different ages are permanently associated, a feature that distinguishes them from groups in the majority of other mammalian orders (van Schaik and Kappeler 1997). This type of social organisation creates conditions for a maximum of inter-individual conflicts of interest based on sex, age, rank, and reproductive status, leading to divergent interests regarding optimal behaviour that require coordination of activities. Second, non-human primate groups are also socially structured, with a variety of individualised competitive, affiliative, and mating relationships that can exacerbate conflicting interests. Third, non-human primates have some of the relatively largest brains among mammals, providing opportunities for more complex behavioural mechanisms and cognitive solutions of dyadic and group-level problems that resemble some of those observed among humans (Conradt and List 2009). Comparative studies among primates indeed support the hypothesis that large brain size has evolved to cope with the social problems created by increasingly large and complex groups (Dunbar and Shultz 2007; Shultz and Dunbar 2007). Fourth, non-human primates dispose of a great variety of communicative signals in different modalities (especially gestures and vocalisations), which are useful for the proximate regulation of group coordination. Finally, great apes (orangutans, gorillas, chimpanzees, and bonobos), uniquely among non-human primates, exhibit rudimentary forms of shared intentionality (Tomasello et al. 2005; see below), which underlies most cases of human group coordination. Thus, non-human primates share several key features of their social systems and life histories with humans that distinguish them from group-living insects, fish, or birds, which tend to rely on a limited set of mechanisms to achieve group coordination (Couzin et al. 2005).

What and why non-human primates need to coordinate at the group level can be explained primarily with respect to ecological factors. This may constitute a major difference to post-modern humans, where most coordinated group action has a social goal and function (see Chap. 7). Ecological factors that make group coordination in primates (as well as in other animals) advantageous can be related to the fundamental advantages of group living per se. Most primatologists agree that groups evolved from solitary or pair-living ancestors along with an evolutionary transition to diurnality, because living in groups confers several fundamental benefits that reduce per-capita predation risk (van Schaik 1983). Primates that became diurnal were faced with a new suite of predators, notably visually hunting raptors and large carnivores. Under these circumstances, aggregating into permanent groups (preferably with relatives to minimise the concomitant costs of sociality such as increased feeding competition) reduces the individual risk of being taken by a predator because of the dilution effect (Hamilton 1971): A single individual has a probability of one of being attacked by a predator it encounters; in a group of ten, this risk is reduced to 1/10. Animals in groups are also better at detecting an approaching predator and at confusing an attacking predator, and they can share vigilance, resulting in an overall increase in vigilance levels for the group while at the same time reducing the frequency of this costly behaviour for the individuals (Bertram 1978). Presumably secondary benefits of grouping include improved territorial defence against neighbours and new opportunities for cooperation, especially with kin. Only some primates, including our closest-living relatives (chimpanzees and bonobos), adopt a less cohesive, so-called fission-fusion lifestyle in response to relaxed predation risks and/or other selective pressures related to their feeding ecology (Aureli et al. 2008). Thus, the primary selfish aim of individuals in virtually all group-living primates is to maximise and maintain tight group cohesion.

The preservation imperative of group cohesion explains why non-human primates need to coordinate themselves at the group level. Inter-individual conflicts of interest, which are ultimately related to various costs of group living, exert a centrifugal force on group cohesion. These conflicts need to be reconciled at the group level to maintain cohesion. Selfish interests are therefore constrained by the trade-off that individuals face with respect to group cohesion. Because the potential costs of abandoning the safe harbour of a group will always outweigh the costs of foregoing a particular individual interest, group coordination is expected to be widespread (Conradt and Roper 2005; Conradt and List 2009).

Because non-human primates are relatively active, medium-sized mammals with relatively large energy-demanding brains, they need to spend most of their time searching and processing food. The size of a group's home range increases with group size and depends on their primary type of food, but all primate groups move several hundred metres in search of food, water, and suitable resting sites every day (Clutton-Brock and Harvey 1977). Because different classes of individuals within the group are thought to have diverging nutritional needs – compare, for example, a lactating female with a recently weaned juvenile – group movements provide the ecologically most salient context in which group coordination is required. Given that there are alternative, largely incompatible options for the subsequent activity (feeding, drinking, resting, socialising), and that for each subsequent activity there are alternative sites available (water hole a or b, feeding tree x or y), a consensus about the next destination must be reached. In addition, there may be conflicts about how much time to spend in each patch, such as when to move on, as some individuals need to drink or eat less or more than others. The overwhelming ecological significance of these decisions, along with their quantitative dominance, explains why primatologists have studied group coordination phenomena predominantly in the context of group movements (Petit and Bon 2010; see also Chap. 3).

In Chap. 13, Julia Fischer and Dietmar Zinner discuss communicative and cognitive aspects of group coordination within a single species. Also focusing on collective group movements, they briefly review different processes giving rise to collective activities in non-human primates and other animal groups. Their review outlines variations across species along several axes. Similar outcomes in different species can be characterised, for example, according to the level of self-organisation vs. explicit decision making, the level of the group (local vs. global) at which a decision is taken, whether individual interests overlap or diverge, or whether decisions are shared or unshared (see also Conradt and List 2009). Moving on to information transmission, they point out that both signals and cues can be important in bringing about a collective decision. They illustrate these different processes and mechanisms by reviewing recent studies of group movements in baboons. Because these animals live and move in an open, primarily two-dimensional habitat, group movements are comparatively easy to study, rendering a correspondingly large amount of detail for this taxon. Their chapter closes with a discussion of the cognitive underpinnings of collective decisions, and emphasises the lack of theory of mind and intentionality in non-human primates, which, in turn, compromise comparisons with many examples of human group coordination.

In Chap. 14, Juliane Kaminski highlights an experimental paradigm in great ape research that touches upon a critical difference between human and non-human primate group coordination: shared intentionality and its communication. Apart from language, humans also dispose of the ability to understand other individuals' mental states (theory of mind), both of which may facilitate group decisions and coordination on an explicit as well as implicit level (see Chap. 4 for a thorough treatment of this subject). The combination of the group members' individual brains in constructing shared goals is an aspect that uniquely characterises and facilitates human group coordination. The psychological mechanism underlying this effect, in combination with a prosocial disposition, defines shared intentionality (Tomasello et al. 2005). Kaminski discusses how the proximate behavioural mechanisms facilitating shared intentionality include gaze following, pointing, and other triadic gestural interactions. The existence of these behavioural building blocks of human uniqueness has been studied in several primate species, especially in great apes. Kaminski reviews these studies (and similar ones conducted with small children and dogs) in her chapter, emphasising the cognitive gap that separates *Homo sapiens* from other primates in this context. Differences at this level may explain the many unique functions and mechanisms of human group coordination discussed at length in the other contributions to this volume.

In the final chapter of Part III, Eckhard Heymann focuses on groups of non-human primates formed by members of two or more species within the broader context of safety and foraging. Such mixed-species groups are rare among primates, but they provide an interesting opportunity to study the mechanisms of group coordination because communication among different species is required. For humans, communication and coordination with members of other species may be relevant in the contexts of hunting, domestication, and animal training, but certainly to a less significant degree than among animals. Chapter 15 provides a summary of all known cases of regular heterospecific associations between non-human primates. His analysis of the potential costs and benefits confirms insights from similar studies in other taxa (e.g. Fitzgibbon 1990): By associating with a group of another species, a given group of primates can effectively double the anti-predator benefits without a pari passu increase to their ecological costs, because the cooperative species, by definition, exploit different feeding niches. Given these net benefits, Heymann goes on to explore with which behavioural mechanisms inter-specific coordination is achieved. It turns out that the exchange of loud calls appears to provide the main mechanism used by members of different species to establish and maintain spatial proximity and cohesion. Upon closer inspection, beyond the loose semblance of loud call exchange between inter-specific non-human primates and verbal communication among humans, more differences than similarities with human group coordination become apparent. These differences have to do primarily with the fact that the formation of shared mental states among non-human primates is unlikely.

The study of collective decision making and group coordination in non-human primates is still in a phase where inductive approaches predominate. Despite some early pioneering studies (e.g. Kummer 1968), this topic entered mainstream primatology only after stimulation provided by the publication of *On the Move* (Boinski and Garber 2000), an eye-opener to a fundamental problem in the behavioural ecology of all group-living animals, including non-human primates. The last decade has seen a diversification of study species and a sophistication of methods (see the summary provided in Chap. 3), but a predictive theoretical framework is still lacking. As exemplified by the non-human primate-oriented contributions to this volume, the communicative and cognitive mechanisms underlying group coordination have since been recognised as interesting topics. It is in these areas where comparative studies of non-human primates in particular can inform and inspire corresponding studies of humans. At the moment, the functional contexts of group coordination that are being studied differ too widely between humans and non-human primates for meaningful integration, although analytical models are being developed to do just that (see, for instance, Chaps. 2 and 4). One interesting approach towards bridging this gap could be the study of traditional human forager societies, who live in mobile camps with a median population of 26 individuals and migrate about seven times per year (Marlowe 2005). Thus, both collective group movements as well as all other communal decisions in the social domain could be profitably studied with a broad comparative perspective.

References

Aureli F, Schaffner CM, Boesch C, Bearder SK, Call J, Chapman CA, Connor R, Di Fiore A, Dunbar RIM, Henzi SP, Holekamp K, Korstjens AH, Layton R, Lee P, Lehmann J, Manson JH, Ramos-Fernandez G, Strier KB, van Schaik CP (2008) Fission-fusion dynamics: new research frameworks. Curr Anthropol 49:627–654

Bertram BCR (1978) Living in groups: predators and prey. In: Krebs JR, Davies NB (eds) Behavioural ecology. Blackwell, Oxford, pp 64–97

Boinski S, Garber PA (2000) On the move: how and why animals travel in groups. University of Chicago Press, Chicago

Chapais B (2010) The deep structure of human society: primate origins and evolution. In: Kappeler PM, Silk JB (eds) Mind the gap: tracing the origins of human universals. Springer, Heidelberg, pp 19–51

Clutton-Brock TH, Harvey PH (1977) Species differences in feeding and ranging behaviour in primates. In: Clutton-Brock TH (ed) Primate ecology: studies of feeding and ranging behaviour in lemurs, monkeys, and apes. Academic, London, pp 557–584

Conradt L, List C (2009) Group decisions in humans and animals: a survey. Philos Trans Roy Soc Lond B: Biol Sci 364:719–742

Conradt L, Roper TJ (2005) Consensus decision making in animals. Trends Ecol Evol 20:449–456

Couzin ID, Krause J, Franks NR, Levin SA (2005) Effective leadership and decision-making in animal groups on the move. Nature 433:513–516

Dunbar RIM, Shultz S (2007) Evolution in the social brain. Science 317:1344–1347

Fitzgibbon CD (1990) Mixed-species grouping in Thomson's and Grant's gazelles: the antipredator benefits. Anim Behav 39:1116–1126

Hamilton WD (1971) Geometry for the selfish herd. J Theor Biol 31:295–311

Jolly A, Pride E (1999) Troop histories and range inertia of *Lemur catta* at Berenty, Madagascar: a 33-year perspective. Int J Primatol 20:359–374

Kappeler PM, Silk JS, Burkart JM, von Schaik CP (2010) Primate behavior and human universals: exploring the gap. In: Kappeler PM, Silk JB (eds) Mind the gap: tracing the origins of human universals. Springer, Heidelberg, pp 3–15

Kerth G (2010) Group decision-making in animal societies. In: Kappeler PM (ed) Animal behaviour: evolution and mechanisms. Springer, Heidelberg, pp 241–265

Kummer H (1968) Social organization of hamadryas baboons. A field study. University of Chicago Press, Chicago, IL

Marlowe FW (2005) Hunter-gatherers and human evolution. Evol Anthropol 14:54–67

Petit O, Bon R (2010) Decision-making processes: the case of collective movements. Behav Process 84:635–647.

Shultz S, Dunbar RIM (2007) The evolution of the social brain: Anthropoid primates contrast with other vertebrates. Proc Roy Soc Lond B: Biol Sci 274:2429–2436

Silk JB, Alberts SC, Altmann J (2003) Social bonds of female baboons enhance infant survival. Science 302:1231–1234

Tomasello M, Carpenter M, Call J, Behne T, Moll H (2005) Understanding and sharing intentions: the origins of cultural cognition. Behav Brain Sci 28:675–691

van Schaik CP (1983) Why are diurnal primates living in groups? Behaviour 87:120–144

van Schaik CP, Kappeler PM (1997) Infanticide risk and the evolution of male-female association in primates. Proc Roy Soc Lond B 264:1687–1694

Chapter 13
Communicative and Cognitive Underpinnings of Animal Group Movement

Julia Fischer and Dietmar Zinner

Abstract The topic of collective animal behaviour has seen a surge of interest in recent years, with the diversity of organisms under study ranging from bacteria to humans in crowds. A large part of this research has been devoted to the identification of the mechanisms underlying decision making in the context of collective movement. In this chapter, we provide an overview of different processes that have been invoked to explain group coordination. Using baboons as a model, we illustrate the importance of signalling behaviour and behaviour-reading to achieve group movement, and we discuss the cognitive processes associated with collective action. We conclude by evaluating the differences in human collective action compared to collective action in other animals, with particular regard for the intentional structure of human communication.

13.1 Introduction

It is the wee hours of an African morning. The air is still cool and the birds have just begun to sing. Up in their sleeping trees by the Gambia River, a group of Guinea baboons is showing the first signs of activity. While some animals are still stretching their limbs, the dry leaves of the palm trees rustle as others begin the descent from the trees to linger in the area. Some animals huddle, others bask in the first sun rays. Eventually, as if on command, the entire group walks off in one direction to begin their daily travel routine.

What are the rules underlying this apparently coordinated behaviour? Do different animals take different roles in initiating the descent from the sleeping trees or the onset of travel? How does a group of baboons reach a decision about when to leave and where to go? Which signals – if any – do these animals exchange when they

---

J. Fischer (✉) and D. Zinner
Cognitive Ethology, German Primate Center, Kellnerweg 4, 37077 Göttingen, Germany
e-mail: fischer@cog-ethol.de; dzinner@gwdg.de

M. Boos et al. (eds.), *Coordination in Human and Primate Groups*,
DOI 10.1007/978-3-642-15355-6_13, © Springer-Verlag Berlin Heidelberg 2011

behave collectively? And what are the cognitive underpinnings underlying group movement? These are the questions that we will address in the present chapter.

The topic of collective animal behaviour has received increasing attention in recent years, with the diversity of organisms under study ranging from bacteria to humans. The movement patterns of ants, honey bees, and locusts, as well as of fish schools and bird flocks, have become the topic of experimental studies as well as mathematical modelling (reviewed in Sumpter 2009). The central question in collective animal behaviour concerns the process of how individual 'decision making' gives rise to collective behaviour. Collective behaviour encompasses all instances where aggregations of subjects engage in one type of activity. Coordinated behaviour, or group coordination, refers to a subset of these instances where some form of regulation of group activity appears to be needed to achieve collective action. 'Group decision making' is frequently used to describe the process that eventually leads to coordinated action. However, it should be made clear that such 'group decisions' are based on the behaviour of individuals. 'Group decision making' is of interest because individual decisions strongly depend on the behaviour of others and are thus not independent of each other, resulting in specific dynamics of certain behaviours at the group level.

The fact that individual subjects make specific decisions points to one of the major factors, namely, the question of whether individual interests overlap or diverge. Further, it raises the questions of in which way the behaviour of others may influence individual decisions and how subjects integrate environmental and behavioural information to make their choices. In the following sections, we will focus on the mechanisms that have been invoked to explain collective behaviour, with particular regard for potential conflicts of interest, the flow of information, and the cognitive processes (see Chap. 15 for an additional discussion on collective primate behaviour). Conflict of interest refers to the fact that individuals have to pit the costs associated with a given decision against the benefits of group living. Before we turn to the decision-making processes in more detail, we briefly review the advantages of group living. We will then illustrate some of the processes outlined beforehand with a review of studies on baboons (*Papio* spp.), whose coordination of group travel has been studied in some detail. We conclude with a discussion of the differences and similarities in animal and human group coordination.

13.2 Living in Groups

Many species form aggregations or groups, and the vast majority of primates, including humans, live in groups. Whenever the advantages of group living for individuals outweigh the costs incurred by the disadvantages, such as increased transmission of pathogens or competition with group mates for food or mating partners, group living should be selected (see Chap. 2; Krause and Ruxton 2002). Once group living has evolved, animals are often faced with the need to choose, collectively, between mutually exclusive actions.

The benefits of group living have been explored in a large range of species, and a thorough review of the respective literature is far beyond the scope of this chapter. Here we will just list the main benefits that may have facilitated the evolution of group living. However, the various advantages do not apply to all species in the same way, and data to test hypotheses regarding the specific benefits are difficult to obtain, in particular for large and long-lived vertebrates such as non-human primates (Krause and Ruxton 2002).

Reducing the risk of predation is believed to be one of the major driving forces of group living in most species (Alexander 1974; Anderson 1986; van Schaik 1983). In addition, group living allows animals to defend their home range or feeding grounds cooperatively: A group can drive other animals out of its home range or away from a food resource, and larger groups can outcompete smaller ones (van Schaik 1983; Wrangham 1980). Group hunting is an important strategy in a number of large carnivores such as wild dogs, lions, and hyenas (Bertram 1978). Although non-human primates are not carnivores, group hunting also occurs in the primate order, albeit rarely. Only some populations of chimpanzees, *Pan troglodytes,* are known to hunt in groups (e.g. Boesch 1994). Finally, behavioural thermoregulation may favour group living, as huddling together to conserve energy and share body heat during cold periods can be an advantage. This applies mainly to small-bodied species and appears to be relevant only for a small number of primate species (Hanya et al. 2007; Ostner 2002).

In order to take advantage of the benefits of group living, animals face the challenge to stay together, coordinate group movement, and synchronise their behaviour within a group. These processes may pose additional costs on group-living animals that are related to decision making. Decision making about the direction of the morning departure from a sleeping site in hamadryas or chacma baboons, for example, can take up to 1 h (Stolba 1979; Stückle and Zinner 2008).

13.3 Concepts in Group Coordination

13.3.1 A Terminological Conundrum

Research topics that transcend disciplines require a careful clarification of terminology. Depending on the discipline, 'decision making' may refer to the outcome of a statistical process, such as the firing of neurons (Beck et al. 2008). In the behavioural sciences, 'decision making' is used to describe situations in which subjects can adopt one of two (or more) alternative behaviours. In human research in particular, 'decision making' is frequently associated with reasoning, that is, a mental simulation of the outcomes of different scenarios, such as the next move in a chess game (Koechlin and Hyafil 2007). Moreover, decision making in humans is typically linked to intentionality (e.g. Sutter 2007). For the time being, we will be agnostic with regard to the underlying cognitive processes and simply use the term

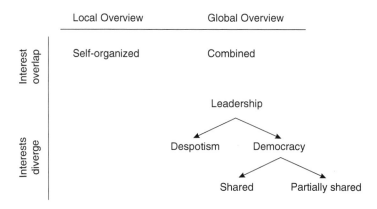

Fig. 13.1 Conceptions of the processes underlying group decision making. Two of the major determinants are the distribution of information (local vs. global overview) and the distribution of interests (overlapping vs. divergent). Within this framework, more detailed characterisations are possible. Note also that local and global processes may interact

'decision making' broadly, reserving the discussion of the cognitive requirements for the final section of this chapter.

In terms of the processes giving rise to collective action, there are unfortunately a number of different, partly overlapping, and not fully compatible concepts. To make matters worse, the terms are not always used consistently. In addition, a number of terms are adopted that describe highly complex human social institutions (see Conradt and List 2009 for an overview of terms used in the natural and social sciences, respectively). In some cases, such a transfer of terms appears to have larger metaphorical than explanatory power.

There are two major axes explaining the dynamics of decision making in animal or human groups or aggregations: One dimension is whether the behaviour is organised at the local level, where subjects simply pay attention to what their neighbours are doing, or whether global information is also factored in. The second dimension is whether or not the individual constituents of the group have overlapping or diverging interests (see Fig. 13.1). These processes are embedded in the respective physiological and ecological conditions, which may pose specific constraints (Conradt and List 2009).

13.3.2 From Rule-Governed Behaviour to Complex Processing of Information

Many people may know the example of starlings when they perform their spectacular acrobatic swarm flights on their way to their autumn sleeping sites. In fact, such seemingly complex group behaviours can be generated by rules that govern the behaviour locally, without any information at the individual level about the global movement pattern. These processes have been described as 'self-organisation' (Camazine et al. 2001). In principle, the behaviour of swarms can be generated by modelling individuals as self-propelled particles linked to their neighbours through attraction, repulsion, and alignment. Directed behaviour of large groups, as well as the formation of more complex behaviours such as the confusion of predators, can be explained by these simple rules (Gregoire et al. 2003).

The issue becomes more complex when individuals possess both local knowledge about the behaviours of those around them as well as global knowledge about some aspect of the environment. When both types of information are not in agreement, subjects need to make decisions. Take, for instance, a situation at the airport where a crowd is exiting the arrival hall through one of two doors, although the second door is also clearly marked with the word 'Exit'. You may be inclined to factor in that global information and leave through the second door, or you may be influenced by other people's behaviour and walk with the crowd, possibly musing that some of them must know that the other door is locked. In such cases, conflicts of interest may occur and individuals have to consider the costs (checking if the door is locked requires the risk of wasted time and energy) and benefits (less inhibited exiting) of choosing one alternative over another.

It is important to note that a subject's probability of exhibiting a given behaviour can take the form of a sharply non-linear function of the number of other individuals already performing this behaviour. This phenomenon has been described as *quorum response* (Sumpter 2006). One form of quorom responses, namely quorum sensing has also been invoked to explain the collective action of bacteria. For instance, individual bacteria are able to detect the number and strain identity of other bacteria around them and vary their behaviour (e.g. biofilm production) accordingly (Nadell et al. 2008). In bacteria, quorum sensing is based on the secretion and detection of autoinducer molecules. The concentration of these molecules gives some indication of the cell density in that particular area. Once concentrations of these molecules have reached a certain threshold, it affects the behaviour of quorum-sensing cells. Quorum sensing has been linked to a number of processes in bacteria, including sporulation and bioluminescence (Miller and Bassler 2001), and it constitutes one of the simplest means of achieving collective action in social organisms.

A more complex example where quorum sensing and decision making both play a role is the emigration of an insect society from an old nest to a new one. First of all, this requires the assessment of different sites, a comparison between them, and an eventual choice of the new nest site (Pratt et al. 2002). Once a suitable new nest site has been found, the entire colony needs to be moved there. Several studies have shown that honey bee (*Apis mellifera*) and ant colonies are able to achieve this goal without any central control. Furthermore, these cases constitute intriguing examples for studying the link between behavioural rules and the flow of information between individual members. The ant *Leptothorax albipennis*, for instance, achieves the selection of a new site on the basis of a process in which several individuals function as scouts. Once a scout has identified a suitable site, it returns home. The initiation of recruitment behaviour is inversely proportional to the quality of the site, thus providing some indirect information about site quality (Pratt et al. 2002). The recruitment occurs in the form of so-called tandem runs. Once a certain number of animals are found at the new site (quorum sensing), tandem runs are abandoned in favour of transport runs in which passive members are carried to the new site. In sum, a combination of relatively simple rules is sufficient to achieve such a complex collective action as the move of a colony to a new nest site.

13.3.3 Overlapping and Diverging Interests

When interests overlap, collective behaviour may arise through 'combined' decision making (Conradt and Roper 2005). In a combined decision, there is no conflict of interest and external stimuli direct individual decisions in the same direction. An example is provided by a herd of thirsty animals walking to the nearest water source. When interests diverge, subjects need to reach 'consensus' decisions. Such scenarios raise the question of whether specific individuals exert a disproportionate influence on the outcome of behaviour, for instance, because of their social role in the group or because of an uneven distribution of information. In such cases, animals can be characterised as leaders and followers (King et al. 2009). Two principal forms of leadership have been proposed (1) *personal leadership* (*despotism*), where a single individual uses its high-dominance status or experience to lead the group, resulting in an unshared decision; (2) *distributed leadership* (*democracy*), where either all group members reach an equally shared decision, or a subgroup of individuals reaches a partially shared decision (Conradt and Roper 2005; Leca et al. 2003). In cases of divergent interests, decision making also requires some form of interaction among the involved individuals, whereas a combined decision can be achieved without any such interaction. Hence, consensus decisions are also described as *aggregate* or *collective* decisions, whereas combined decisions are described as *interactive* decisions (Conradt & List 2009).

One of the most famous examples for despotism is found in mountain gorillas, where the silverback male directs the group by heading in his preferred direction (Watts 2000). In this example, the apparently coordinated behaviour is based on the fact that other group members have no choice but to follow the leader if group cohesion is to be maintained. Similar processes have also been observed in mongooses (Rasa 1983) and wolves (Mech 1970). Shared decisions are generally thought to be more profitable for group members than accepting unshared decisions made by a single leader (Conradt and Roper 2007), because on average there is a greater overlap of interest. Distributed leadership has been observed in such diverse species as honey bees (Seeley and Buhrman 2001), coatis (Gompper 1996), and red deer (Conradt and Roper 2003). Note that in relation to group movements, leadership often refers to individuals initiating movements or changes in direction that are followed by the rest of the group (e.g. Leca et al. 2003; Trillmich et al. 2004). These animals are not necessarily the decisive subjects when it comes to agreeing on a certain direction (see ahead). In a recent study, Conradt and colleagues showed that the assertiveness with which animals opt for one option over another depends on the cost of splitting, compared to the benefits of going to the leader's preferred target. The assertiveness, in turn, can be thought of as a function of meeting an individual's need. Animals that are highly motivated to direct the travel to a water source because they are thirsty may have a disproportionate effect on the group's eventual travel direction than less needy individuals (Conradt et al. 2009). Conradt and Roper explored the conditions under which unshared decisions are – in evolutionary terms – more successful than shared ones. Using a combination of self-organising systems and game theory, they revealed that shared as well as unshared decision types can evolve without invoking global knowledge in the individual members of a group. They found that unshared decisions are favoured when conflicts are high compared to grouping benefits due to the inherent expediency and simplicity of unshared decisions (Conradt and Roper 2009).

13.3.4 Information Transmission

As we have seen earlier, the transmission of information among individuals can be an important determinant of collective action. Information can be transmitted through communication, that is, the usage of signals that are emitted by a sender and received and processed by one or many recipients (McGregor and Peake 2000; Skyrms 2009). Signals are defined as structures or behaviours that predominantly serve information transmission, having little survival value otherwise. In contrast to signals, cues are considered to reflect more directly the physiological or morphological state of an individual. Thus, cues encompass all sorts of features or behaviours that may influence a specific animal's actions (Maynard Smith and Harper 2003). Note that not all authors use these terms consistently (e.g. Hauser 1996), but it is generally agreed that signals evolve from cues, which in turn are a subset of an animal's behavioural features.

Importantly, from the recipient's point of view, signals are not necessarily more informative than cues or the simple behaviour of another subject. In other words, if one animal watches another animal move in a certain direction, there is no doubt that the animal is doing just that – moving in a certain direction. In contrast, signals typically predict imminent behaviour, but not fully reliably. Nevertheless, signalling is important because it constitutes a means of information transfer that is generally less costly than the proposed action itself, and, depending on the circumstances, the use of signals may be the only way of transmitting information. The signalling systems involved in collective action may take a variety of forms, including the secretion of chemicals in bacteria that lead to quorum sensing, as well as the waggle dance of bees or the usage of vocal signals.

Animals gather information about others' locations not only by attending to their signals, but also by generally attending to their activities. Among socially living animals, knowledge about food or water sources is frequently transmitted socially without invoking explicit instruction. Research during the last two decades has identified different social learning mechanisms that encompass a range of different forms with varying degrees of cognitive complexity. The common denominator is that the behaviour of one subject facilitates or influences the behaviour of another subject. Of particular importance in the current context are social facilitation, stimulus enhancement, and local enhancement (see Fischer 2008 for a full review). Social facilitation is invoked when an individual's learning is affected by the activity of another animal. Animals typically pay a lot of attention to what others, particularly their group mates, are doing. This may lead to stimulus enhancement, such as an increase in the salience of stimuli others are paying attention to, as well as local enhancement, such as when the subject learns something about a specific situation simply because it is near an individual who does something particular. Social facilitation may lead to quorum responses, with a non-linear spread of specific behaviours, and it may even influence the dynamics of information transmission in a given group.

13.4 Baboons as Models to Study Animal Group Coordination

13.4.1 Social Organisation and Decision Making

The initiation of group movements and the coordination of travel have been studied extensively in baboons. Baboons range all over sub-Saharan Africa and the Arabian Peninsula. They inhabit a large range of different habitats and exhibit very different social systems. Hamadryas baboons, for instance, live in a multi-layered society that consists of small one-male units (OMU) of one male with a few females. Several OMUs may form a clan, which forage together and which are believed to be connected by close kin relationships among the male leaders of the respective OMUs. Several such clans form a band, which shares a common home range and can include more than 100 individuals. Bands are seen as the ecological units of the hamadryas society, whereas OMUs are seen as reproductive units. Individuals rarely change from one band to another. Several bands may aggregate at certain rare sleeping cliffs where they spend the night and form troops of up to several hundred individuals.

Typically, members of a band leave their sleeping cliff together at the same time and in the same direction. Each morning before leaving the sleeping site, a significant amount of time is devoted to agreeing on a joint direction of travel. Hans Kummer, who pioneered the study of group movement initiation, distinguished between initiative individuals (I) and decision-making individuals (D-ID-System). The initiative and decision-making individuals are males, mainly older ones, with leaders of OMUs having apparently the greatest influence in the decision-making process (Kummer 1968). Kummer concluded that 'leading' individuals do not need to determine group movement from the vanguard position since it was also possible to 'lead the troop from the back' (Kummer 1968; Stolba 1979). Thus, hamadryas baboons provide an example of partially shared leadership (Kummer 1968; Leca et al. 2003).

Geladas (*Theropithecus gelada*), close relatives of baboons, live in a superficially similar social organisation as hamadryas baboons. However, here the dominant female of the OMU fulfils a pivotal role in maintaining coordination between the male and his other females. Progressions are usually initiated by lactating females, but decisions whether to follow or not are shared by the dominant female and the male. Females monitor the behaviour of the male but try at the same time to stay in proximity to their preferred social partners (Dunbar 1983).

In contrast to hamadryas baboons and geladas, savannah baboons typically live in female-bonded multi-male–multi-female societies. Savannah baboons encompass the traditionally recognised morphotypes of chacma, yellow, and olive baboons. Because of the importance of females in these groups, it became of interest to elucidate their role in initiating group movements. In a study in South Africa, Stückle and Zinner (2008) examined whether a group would reveal evidence for distributed leadership that included females. Before taking off for their daily march, these baboons rested below their sleeping trees. After some time, one animal (the so-called initiator) moved away from the rest of the group. In such instances, others either followed (successful attempt) or stayed behind (unsuccessful attempt). The minimum number of consenting adult individuals that normally guaranteed that the entire group would depart was five. Therefore, six adult individuals (initiator and five followers) seem to be sufficient to pull the entire group of 39 into the pursued direction, making the initiative successful. If not enough animals followed, the initiator normally moved back to the group until a second attempt was made, either by the first initiator or by another animal. Approximately 75% of all adult animals successfully initiated a collective move, with 67% of initiations being made by males. The relative success of an initiation was equally distributed among adult group members, with almost two thirds of the initiatives being successful. If a successful initiation of a collective move is regarded as equivalent to leadership, then these baboons show a system of distributed leadership. However, although the probability of being successful was similar for males and females when initiating a move, males had more influence on the morning departure process by initiating more start attempts, thereby making appreciably more successful initiations. Among males, there was a trend for higher-ranking (more dominant) individuals to make a higher total number of initiation attempts as well as a higher number of successful attempts than lower-ranking ones. A similar trend was not obvious in females, most likely because a possible effect of dominance was masked by a confounding effect of females' reproductive state. High-ranking females were the ones with dependent offspring, which most likely forced them to stay in the centre of the group or in close vicinity of a male protector instead of taking the lead when leaving the sleeping site (Stückle and Zinner 2008).

In contrast, King and colleagues (King et al. 2008) reported that in another chacma baboon population, the decision-making process was despotic and not distributed. In this study, an artificial clumped food source was offered in the home range of a baboon group. The baboons had to decide whether to visit the feeding site or to go elsewhere to forage. When visiting, the dominant male obtained the largest share of the resource, while others received less or nothing at all. Nevertheless, the group's foraging decisions were consistently made by the dominant male. Subordinate group members followed the leader in the interest of staying together despite considerable consensus costs (no food) to these subordinate members. King and colleagues interpreted the behaviour of the subordinates by the value of their social bonds with the dominant male. These baboons seem to face a dilemma: either to maintain close proximity to the dominant male by following him to his preferred feeding site even though they do not benefit from the food, or to leave the dominant male and forage on their own, thus jeopardising their social relationship with the dominant male (King et al. 2008).

13.4.2 Signalling Behaviour: Non-vocal Signals

As in hamadryas baboons and other primate species, particular behaviours or postures may be used to communicate during the decision-making process. Hamadryas males seem to communicate intensively during the decision-making process and 'negotiate' the direction of travel (Kummer 1968; Stolba 1979). Males present each other in a particular way, called *notifying*. Notifying can be a complete behavioural sequence entailing the approach of one male to within arm's reach of another, turning the body, presenting the hindquarter, and looking back to the first male, who can then touch the penis of the presenting male. The mildest form of this behaviour is just a short glance of one male to a second, followed by an abrupt turn of his head. Notifying is normally accompanied by vocalisations of both males. When proposing a certain direction, males may walk in a particular manner in the preferred direction or they may stand on outstretched arms and legs, as stiff as a sawhorse, and will then advance in the direction of the body axis. A male can also vote against a proposed direction by sitting down and abruptly lowering his head on the chest without moving within the next 2 min. Possible equivalent behavioural patterns in chacma baboons also include the presence or absence of a 'back glance' when the initiator is looking back to the rest of the troop while walking away, 'pauses' of the initiator during locomotion, and 'walking speed' of the initiator – either trotting or walking fast away from the troop. Such behaviours have been interpreted as intentional signals to recruit group members (Meunier et al. 2008). This view is also in line with the classical ethological concept that these signals have evolved from intentional movements, which are viewed as expressions of the motivation to move.

13.4.3 Signalling Behaviour: Vocal Signals

Vocal signals play an important role in baboon group movements. It has been assumed that calls indicate the motivational state of the sender and provide some information about the sender's propensity to take a certain action. Moreover, such signals can facilitate emotional contagion, thus raising the probability of collective action such as a group departure. Baboons emit soft tonal calls, called grunts, to facilitate social interactions with others or when they are travelling. Moreover, animals utter grunts when they are moving through high grass, but also when the group is about to initiate a group movement either in the morning or after the group has taken a rest during the day. A pilot study on a Namibian population of baboons indicated that an increase in the call rate reflects an increased probability that the group will start moving (Fig. 13.2). Although this system deserves further detailed study, it suggests that each animal's calling can be viewed as the expression of its motivation to get going. Once a certain number of animals are grunting at the same time, this predicts that others are willing to follow if the first animal sets itself in motion. This system would also lend itself to experimental testing, as the grunt rate could be experimentally augmented. Similarly to baboons, vervet monkeys utter so-called move into open grunts (MIO) when they sit near trees and before entering the open grassland (Cheney and Seyfarth 1982), while Barbary macaques emit soft tonal calls termed 'girneys' when they initiate group movements (Fischer and Hammerschmidt 2002).

In the context of group travel, baboons also frequently emit so-called barks. These barks have a tonal structure (Fischer et al. 2001, 2002) and are emitted when subjects have lost contact with specific individuals – for instance, their offspring – or with the rest of the group. Typically, several animals can be heard calling at the same time, giving rise to the notion that these animals may in fact 'answer' each other's calls. Cheney and colleagues set out to test this assumption in a group of baboons in the Okavango delta in Botswana (Cheney and Seyfarth 1996). They followed the adult females of the group for 2.5 h after the departure from the sleeping site and noted for each call the identity of the caller, the context, and the

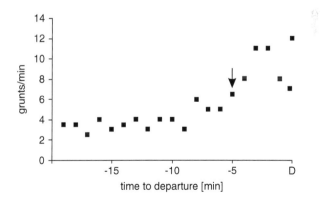

Fig. 13.2 Baboon grunts as cues to imminent group departure. Average grunt rate per individual and minute recorded at the sleeping site of baboons at Tsaobis (Namibia) before departure (D). The *arrow* indicates a significant increase in call rate (determined by means of a change-point analysis) about 5 min prior to the first animal leaving. (Data were kindly made available by Kristine Meise, Christina Keller, and Guy Cowlishaw)

relative location of the animal compared to the rest of the group (front, middle, or rear). For more than 1,600 calls, the observers were able to ascertain the identity of the caller and her location. More than 80% of calls were given by individuals in the rear third of the group. A sequential analysis revealed that these calls typically occurred in a clumped fashion, with one signaller emitting several calls in succession, and several animals calling at the same time. In fact, 92% of calls occurred within 5 min of another call, either following another subject's call, or because the animal was emitting a series of calls in a bout. A more detailed inspection revealed that calling was primarily driven by the caller's risk of becoming 'lost' (hence, these calls have also been termed 'lost calls'). To examine this finding systematically, Cheney and colleagues initiated a set of playback experiments where they played back the clear barks of females to close female relatives, and then checked whether they would 'answer' their kin. In 7 of 36 trials, the subject indeed responded to the playback with a vocal response within 5 min. However, as in the previous observational study, these females were themselves lagging behind the group, suggesting that the immediate context and the risk of getting lost drive the calling behaviour, and not the intention to inform other group members about one's own location (Cheney and Seyfarth 1996).

Contact-calling may be influenced by the visibility (or lack thereof) and, hence, the visual control of the other group members' locations. Accordingly, one would predict that animals call more frequently when visibility is poor. Indeed, in Botswana, clear calls occurred more frequently in woodland habitat than on open flood plains, or when the group gathered near their sleeping trees (Cheney and Seyfarth 1996). Similarly, the rate of grunts given by olive baboons studied in Uganda varied with the habitat type: Baboons uttered grunts at a significantly higher rate in forest than in open habitat. In a population of olive baboons in Nigeria, in contrast, the grunt rate did not vary with the habitat quality and the visibility conditions. However, in both cases, the calls were significantly longer in the forest compared to the open habitat (Ey et al. 2009). Moreover, the rate of loud calls in Nigeria did not differ in relationship to the habitat. This suggests that calling rates are not simply determined by visibility. Moreover, this observation raises the question of whether the differential call rate in Botswana could be indirectly mediated by the fact that in this group, animals lose contact with others less frequently in open habitats, while the much smaller group in Nigeria is generally more cohesive.

13.5 Cognition and Collective Behaviour

Humans like to think of themselves as being highly cognitive creatures who deliberately consider different outcomes of specific actions before making a decision. Frequently, humans believe that they are approaching an optimum when doing so. Yet, they often rely on simple heuristics (Gigerenzer 2001). Morgan pointed out that, 'In no case is an animal activity to be interpreted in terms of higher psychological processes, if it can be fairly interpreted in terms of processes which stand lower in the scale of psychological evolution and development' (Morgan 1903, p. 59). This is meant to also apply to the human animal. Therefore, considering coordinated behaviour in putatively more simple organisms is useful in identifying how minimal sets of rules can give rise to collective action without the need to invoke cognitively demanding operations. For instance, a number of studies have shown how simple local processes govern the movement patterns of human crowds (e.g. Dyer et al. 2008).

At the same time, it is also of interest to clarify how cognitively elaborate group decision-making processes in animals can be. One important issue here is the question of whether the usage of signals (or other forms of behaviour) can be considered as intentional. In the domain of animal communication and cognition, researchers frequently invoke the definition of Dennett (1971). Dennett described different stages of intentionality, where zero-order intentionality would apply to simple expressions of emotion or fixed action patterns given in response to sign stimuli. First-order intentionality describes communicative acts employed in order to alter the behaviour of the recipient. This does not necessarily imply that the signaller is conscious of her own behaviour or mental state (Bruner 1981) in the sense that the sender is aware that she does have such an intention. To identify (first-order) intentionality, Tomasello and Call (1997) proposed, as a diagnostic, the observation that animals may have different means to achieve a specific goal. Second-order intentionality would apply to cases where the sender intends to alter the knowledge state of the other but not necessarily his or her immediate behaviour. For second-order intentionality to apply, the sender must know that the receiver's mental state can be different from his own mental state. So far, there is no convincing evidence for second- (or higher-) order intentional communication in animals (Seyfarth and Cheney 2003). Thus, most of the studies ask whether primates use signals with the intention to alter their group mates' behaviour.

For our specific purposes, Dennett's scheme is powerful because it clearly distinguishes between the intent to alter the mental state vs. the behaviour of another subject. This research falls under the umbrella of so-called Theory of Mind research, a field that investigates the attribution of beliefs, desires, and knowledge to others. The term was introduced by Premack and Woodruff (1978) in a paper entitled 'Does the Chimpanzee have a Theory of Mind?' To date, there is some evidence that non-human primates indeed understand something about the link between seeing and knowing (Hare et al. 2000; Kaminski et al. 2008; Tomasello and Call 2006; also see Cheney and Seyfarth 1990), but they appear to lack a full-blown attribution of mental states. Thus, it seems fairly safe to conclude that non-human primates use signals with the possible intention to alter the movement patterns of others (see Hesler and Fischer 2007), but they do not call with the intention to provide others with the information that they are about to leave (Fischer 2008; Seyfarth and Cheney 2003). Accordingly, Couzin (2009) suggested that the field of animal collective action may profit from studying the properties of neural assemblies and the information flow in such systems instead of invoking higher-level cognitive processes.

The differences in mental state attribution have important implications for the conceptualisation of collective action and mechanisms underlying decision making. According to Bratman (1992), shared collective action refers to situations where (human) subjects have a concept that a given activity is shared, and that each subject intends that the group performs the collective action by coordinating 'subplans that mesh' (Bratman 1992). For instance, each member in a team of surgeons in an operating room will assume that all the other members have the same goal, namely, to successfully perform the operation, and also assume that the others are aware of the communality of that goal (see Chap. 5). Along these lines, Tomasello pointed out that human communication is grounded in fundamentally cooperative and shared intentions (Tomasello 2008). In contrast, no such cognitive assumptions are made in the animal domain.

13.6 Conclusion

Animal models are valuable for investigating the mechanisms that lead to collective action without invoking attribution of mental states. As we have seen, simpler models are useful to understand the minimal requirements to achieve a certain type of collective behaviour and at the same time identify the differential needed to achieve a more complex form of coordination, especially in the identification of the mechanisms underlying decision making in the context of collective movement. Thus, bird flocks are useful to understand the behaviour of human crowds, while bacteria help to illuminate the selective pressures that play a role in the evolution of sociality in general. However, animal models are often less useful for gaining insights into the psychological, cognitive processes at work when humans engage in shared collective behaviour.

Acknowledgements We thank Elisabeth Scheiner, Guy Cowlishaw, Peter M. Kappeler, and Michaela Kolbe for helpful comments, Margarita Neff-Heinrich for her careful editing of the manuscript, as well as Kristine Meise, Christina Keller, and Guy Cowlishaw for making unpublished data available.

References

Alexander R (1974) The evolution of social behaviour. Ann Rev Ecol Syst 5:325–383
Anderson C (1986) Predation and primate evolution. Primates 27:15–39
Beck JM, Ma WJ, Kiani R, Hanks T, Churchland AK, Roitman J, Shadlen MN, Latham PE, Pouget A (2008) Probabilistic population codes for Bayesian decision making. Neuron 60:1142–1152
Bertram B (1978) Living in groups: predators and prey. In: Krebs J, Davies N (eds) Behavioural ecology: an evolutionary approach. Blackwell, Oxford, pp 64–96
Boesch C (1994) Cooperative hunting in wild chimpanzees. Anim Behav 48:653–667
Bratman ME (1992) Shared cooperative activity. Philos Rev 101:327–341

13 Communicative and Cognitive Underpinnings of Animal Group Movement 243

Bruner JS (1981) Intention in the structure of action and interaction. Adv Infancy Res 1:41–56
Camazine S, Deneubourg J, Franks N, Sneyd J, Theraulaz G, Bonabeau E (2001) Self-organization in biological systems. Princeton University Press, Princeton, NJ
Cheney DL, Seyfarth RM (1982) How vervet monkeys perceive their grunts: field playback experiments. Anim Behav 30:739–751
Cheney DL, Seyfarth RM (1990) Attending to behaviour versus attending to knowledge: examining monkeys' attribution of mental states. Anim Behav 40:742–753
Cheney DL, Seyfarth RM (1996) The function and mechanisms underlying baboon "contact" barks. Anim Behav 52:507–518
Conradt L, List C (2009) Group decisions in humans and animals: a survey introduction. Phil Trans Roy Soc Lond B: Biol Sci 364:719–742
Conradt L, Roper T (2003) Group decision-making in animals. Nature 421:155–158
Conradt L, Roper T (2005) Consensus decision making in animals. Trends Ecol Evol 20:449–456
Conradt L, Roper T (2007) Democracy in animals: the evolution of shared group decisions. Proc Roy Soc B: Biol Sci 274:2317–2326
Conradt L, Roper TJ (2009) Conflicts of interest and the evolution of decision sharing. Phil Trans Roy Soc B: Biol Sci 364:807–819
Conradt L, Krause J, Couzin ID, Roper TJ (2009) "Leading According to Need" in self-organizing groups. Am Nat 173:304–312
Couzin ID (2009) Collective cognition in animal groups. Trends Cogn Sci 13:36–43
Dennett DC (1971) Intentional systems. J Philos 68:68–87
Dunbar RIM (1983) Structure of gelada baboon reproductive unit. 4. Integration at group level. Z Tierpsychol 63:265–282
Dyer JRG, Ioannou CC, Morrell LJ, Croft DP, Couzin ID, Waters DA, Krause J (2008) Consensus decision making in human crowds. Anim Behav 75:461–470
Ey E, Rahn C, Hammerschmidt K, Fischer J (2009) Wild female olive baboons adapt their grunt vocalisations to environmental conditions. Ethology 115:493–503
Fischer J (2008) Transmission of acquired information in nonhuman primates. In: Byrne JH (ed) Learning and memory: a comprehensive reference. Elsevier, Oxford, pp 299–313
Fischer J, Hammerschmidt K (2002) An overview of the Barbary macaque, *Macaca sylvanus*, vocal repertoire. Folia Primatol 73:32–45
Fischer J, Hammerschmidt K, Cheney D, Seyfarth R (2001) Acoustic features of female chacma baboon barks. Ethology 107:33–54
Fischer J, Hammerschmidt K, Cheney DL, Seyfarth RM (2002) Acoustic features of male baboon loud calls: influences of context, age, and individuality. J Acoust Soc Am 111:1465–1474
Gigerenzer G (2001) Simple heuristics that make us smart. Oxford University Press, Oxford
Gompper ME (1996) Sociality and asociality in white-nosed coatis (*Nasua narica*): foraging costs and benefits. Behav Ecol 7:254–263
Gregoire G, Chate H, Tu YH (2003) Moving and staying together without a leader. Physica D 181:157–170
Hanya G, Kiyono M, Hayaishi S (2007) Behavioral thermoregulation of wild Japanese macaques: comparisons between two subpopulations. Am J Primatol 69:802–815
Hare B, Call J, Agnetta B, Tomasello M (2000) Chimpanzees know what conspecifics do and do not see. Anim Behav 59:771–785
Hauser MD (1996) The evolution of communication. MIT, Cambridge, MA
Hesler N, Fischer J (2007) Gestural communication in Barbary macaques (*Macaca sylvanus*): An overview. In: Tomasello M, Call J (eds) The gestural communication of apes and monkeys. Lawrence Earlbaum, Mahwah, NJ, pp 159–195
Kaminski J, Call J, Tomasello M (2008) Chimpanzees know what others know, but not what they believe. Cognition 109:224–234
King AJ, Douglas CMS, Huchard E, Isaac NJB, Cowlishaw G (2008) Dominance and affiliation mediate despotism in a social primate. Curr Biol 18:1833–1838
King AJ, Johnson DDP, Van Vugt M (2009) The origins and evolution of leadership. Curr Biol 19: R911–R919

Koechlin E, Hyafil A (2007) Anterior prefrontal function and the limits of human decision-making. Science 318:594–598

Krause J, Ruxton G (2002) Living in groups. Oxford University Press, Oxford

Kummer H (1968) Social organization of hamadryas baboons. A field study. University of Chicago Press, Chicago, IL

Leca J, Gunst N, Thierry B, Petit O (2003) Distributed leadership in semifree-ranging white-faced capuchin monkeys. Anim Behav 66:1045–1052

Maynard Smith J, Harper D (2003) Animal signals. Oxford University Press, Oxford

McGregor PK, Peake TM (2000) Communication networks: social environments for receiving and signalling behaviour. Acta Ethol 2:7181

Mech L (1970) The wolf. The ecology and behavior of an endangered species. Doubleday, New York

Meunier H, Deneubourg J, Petit O (2008) How many for dinner? Recruitment and monitoring by glances in capuchins. Primates 49:26–31

Miller MB, Bassler BL (2001) Quorum sensing in bacteria. Annu Rev Microbiol 55:165–199

Morgan CL (1903) An introduction to comparative psychology. Walter Scott, London

Nadell CD, Xavier JB, Levin SA, Foster KR (2008) The evolution of quorum sensing in bacterial biofilms. PLoS Biol 6:0171–0179

Ostner J (2002) Social thermoregulation in redfronted lemurs (Eulemur fulvus rufus). Folia Primatol 73:175–180

Pratt SC, Mallon EB, Sumpter DJT, Franks NR (2002) Quorum sensing, recruitment, and collective decision-making during colony emigration by the ant *Leptothorax albipennis*. Behav Ecol Sociobiol 52:117–127

Premack D, Woodruff G (1978) Does the chimpanzee have a Theory of Mind? Behav Brain Sci 1:515–526

Rasa OAE (1983) Dwarf mongoose and hornbill mutualism in the Taru desert, Kenya. Behav Ecol Sociobiol 12:181–190

van Schaik C (1983) Why are diurnal primates living in groups? Behaviour 87:120–144

Seeley T, Buhrman S (2001) Nest-site selection in honey bees: how well do swarms implement the "best-of-*N*" decision rule? Behav Ecol Sociobiol 49:416–427

Seyfarth RM, Cheney DL (2003) Signalers and receivers in animal communication. Annu Rev Psychol 54:145–173

Skyrms B (2009) Evolution of signalling systems with multiple senders and receivers. Phil Trans Roy Soc B: Biol Sci 364:771–779

Stolba A (1979) Entscheidungsfindung in Verbänden von Papio hamadryas [in German]. Dissertation. Universität Zürich, Zürich

Stückle S, Zinner D (2008) To follow or not to follow: decision making and leadership during the morning departure in chacma baboons (*Papio hamadryas ursinus*). Anim Behav 75:1995–2004

Sumpter DJT (2006) The principles of collective animal behaviour. Phil T Roy Soc B 361:5–22

Sumpter DJT (2009) Collective animal behavior. Princeton University Press, Princeton, NJ

Sutter M (2007) Outcomes versus intentions: on the nature of fair behavior and its development with age. J Econ Psychol 28:69–78

Tomasello M (2008) Origins of human communication. MIT, Cambridge, MA

Tomasello M, Call J (1997) Primate cognition. Oxford University Press, New York, Oxford

Tomasello M, Call J (2006) Do chimpanzees know what others see – Or only what they are looking at? In: Hurley S, Nudds M (eds) Rational animals. Oxford University Press, Oxford, pp 371–384

Trillmich J, Fichtel C, Kappeler P (2004) Coordination of group movements in wild Verreaux's sifakas (*Propithecus verreauxi*). Behaviour 141:1103–1120

Watts D (2000) Mountain gorilla habitat use strategies and group movements. In: Boinski S, Garber P (eds) On the move: how and why animals travel in groups. University of Chicago Press, Chicago, IL, pp 351–374

Wrangham R (1980) An ecological model of female-bonded primate groups. Behaviour 75:262–300

Chapter 14
Communicative Cues Among and Between Human and Non-human Primates: Attending to Specificity in Triadic Gestural Interactions

Juliane Kaminski

Abstract Humans have a sophisticated understanding of other individuals' mental states. But differences between humans and others species are already apparent when focusing on more basic social cognitive capacities. This chapter focuses on a very basic social cognitive skill: gaze following. A rich set of data supports the hypothesis that the ability to interpret others' gaze shift as an indicator that others see things in their environment seems to be widespread among primates (and other species). However, this is in contrast to another set of data that shows that non-human primates, unlike humans, seem to have difficulties interpreting others' gaze shift as indicators that they communicate about things in the environment. This chapter argues that non-human primates (and probably other species) may lack certain cognitive and motivational components, which help to identify the relevance and specificity of a triadic gestural communicative interaction.

14.1 Introduction

The ability to coordinate complex collaborative actions such as those involving making collective decisions and coordinating group member actions requires certain social cognitive and communicative skills. Human social cognition seems to be unique compared to non-human primates: Humans have developed a highly flexible form of communication, language, and also a sophisticated understanding of other individuals' mental states. But differences between humans and other species are already apparent when focusing on more basic social cognitive capacities.

This chapter focuses on very basic social communicative cues among human and non-human primates: gaze-shift following, pointing, and ostensive cues (e.g. eye contact, tone of voice, etc.). A rich set of data supports the hypothesis that the ability to interpret others' gaze shift as an indicator that others see things in their environment seems to be widespread among primates (and other species) (for an overview, see

---

J. Kaminski

Max Planck Institute for Evolutionary Anthropology, Deutscher Platz 6, 04103 Leipzig, Germany
e-mail: kaminski@eva.mpg.de

M. Boos et al. (eds.), *Coordination in Human and Primate Groups*,
DOI 10.1007/978-3-642-15355-6_14, © Springer-Verlag Berlin Heidelberg 2011

Rosati and Hare 2009). However, this is in contrast to another set of data that shows that non-human primates, unlike humans, seem to have difficulties interpreting others' gaze shift as indicators that they communicate about things in the environment (for an overview, see Call and Tomasello 2005). Human children from an early age seem to be able and highly motivated to interpret gaze shifts as communicative acts performed by others in order to inform them about certain entities in their environment. In contrast to that, even our closest living relatives, the chimpanzees, have problems when it comes to interpreting communicative gaze shifts provided by others, for instance to help them find something interesting such as a piece of desirable food.

14.2 Following Gaze to See What Others See

A very basic social cognitive skill is gaze following. Humans demonstrate very early in life that faces, especially the eyes, are an important visual stimulus that as newborns they attend to specifically (Morton and Johnson 1991; Johnson et al. 2005). By the age of 6–9 months, infants begin to sense the gaze direction of others (Scaife and Bruner 1975) and follow the gaze of others to visible targets (D'Entremont et al. 1997; Csibra and Volein 2008; Flom and Pick 2003). By the age of 9–12 months, infants spontaneously follow others' gaze direction to specific targets that are not directly in view (Corkum and Moore 1995; Carpenter et al. 1998). Gaze following seems to be widespread in the animal kingdom as well.

Different primate species have been observed to follow the gaze of a human experimenter or a conspecific to an outside entity (Bräuer et al. 2005; Burkart and Heschl 2006; Emery et al. 1997; Goossens et al. 2008; Povinelli and Eddy 1996; Tomasello et al. 1998, 1999). In the classic paradigm, a human experimenter sits opposite a subject and at a predetermined moment looks up as if she has seen something interesting above the subject. Several primate species (including all great ape species) respond to this behaviour by also looking up. They do so significantly more often than in a control condition in which the human stares at the opposite side of the room (Bräuer et al. 2005). This provides good evidence that another individual's gaze direction is taken as a meaningful cue and potentially as information about some outside entity. Chimpanzees and several monkey species also follow the gaze direction of their conspecifics. Tomasello et al. (1998) tested chimpanzees and various monkey species for their ability to follow the gaze of their group members. An experimenter, situated above the group, attracted the attention of one individual by presenting food to her. Once this individual had shifted her gaze towards the food, it was recorded whether a nearby subject that had not seen the food would respond with co-orientation to the conspecific's gaze shift. All species tested in this setting followed the gaze direction of their conspecific. This finding supported earlier findings from a computerized task in which Emery and colleagues showed that rhesus macaques were able to locate an object according to the gaze direction of a conspecific depicted on a TV monitor (Emery et al. 1997; see also Deaner and Platt 2003).

Recent research suggests that the gaze-following behaviour of great apes is based on some understanding of others' perspective. Different primate species seem to have some understanding of others' line of sight and others' visual perspective (Bräuer et al. 2005; Burkart and Heschl 2006; Povinelli and Eddy 1996; Tomasello et al. 1999). If an individual interprets gaze as an indicator of another individual's line of sight, the gaze-follower should not just automatically shift its head but should, if necessary, relocate to a position where it can see what the other animal is looking at. All great ape species follow the gaze direction of a human experimenter around a barrier situated between the target of the gaze and the ape. Bräuer et al. (2005) compared all great ape species for their ability to track a human experimenter's line of sight. In that study, the apes watched a human look behind different types of barriers. To track the human's gaze direction and be able to see what the human was looking at, the subjects had to move to a different corner of the room. Subjects indeed moved more often in this experimental condition compared to a control condition in which the human looked in another direction (Bräuer et al. 2005).

Okamoto-Barth et al. (2007) investigated whether the great apes' gaze following around obstacles is based on an understanding that the other individual is seeing something different, or alternatively that their movement is based on a simpler mechanism such as the motivation to simply co-orient with others – a mechanism conceptualized as operating geometrically in space (Okamoto-Barth et al. 2007). The authors therefore conducted a study in which a human experimenter gazed in the direction of a target object and the line of sight was either blocked by an opaque barrier or by a barrier with a large window in it, such that the experimenter could potentially see what was behind it. If the subjects understand something about the human experimenter's visual perspective in that situation, they would expect the target of the human's attention to be behind the barrier if the barrier had a window or in front of the barrier if the barrier was opaque. Chimpanzees and bonobos looked behind the barrier more frequently when it had a window and they looked in front of the barrier when it was opaque. This suggests that chimpanzees and bonobos take the perspective of the experimenter when following the gaze and deduce that the experimenter is seeing something (Okamoto-Barth et al. 2007).

Taken together, this evidence suggests that non-human primate species use the gaze direction of others flexibly as a cue to scan the environment for something interesting. There is also evidence that in some species this behaviour seems to be based on an understanding of others' line of sight and visual perspective and not just on some kind of automatic reflex (see also Hare et al. 2000, 2001; Kaminski et al. 2008; Melis et al. 2006a; Santos et al. 2006).

14.3 Following Gaze to See What Others Communicate About

Gaze following in primates also presents us with a conundrum, however, because a slight alteration in the procedure leads to a significant drop in the subjects' performance. Take, for example, the situation in which the experimenter sits opposite the subject and whose gaze is directed to one specific target object from a number of potential objects, all visible to the subject. If the experimenter tries to inform the subject through gaze about the location of hidden food from a number of potential targets, for example, all primate species tested (including chimpanzees) seem to ignore the experimenter's gaze. The paradigm used in this case is the so-called object-choice paradigm (Anderson et al. 1995). Here, a human experimenter hides food in one of several locations out of view of the subject. The human then provides the subject with information about where to find the hidden food by indicating its location, either by pointing (often accompanied by gaze alternation) or by just gazing at the correct cup. Different primate species, including chimpanzees, coming from different labs and tested by different experimenters seem to ignore these communicative gestures while making their choice as they choose randomly between the cups (Anderson et al. 1995, 1996; Call et al. 1998; Herrmann et al. 2007; Itakura et al. 1999; Povinelli et al. 1997; Tomasello et al. 1997). Furthermore, the subjects' performance is not enhanced by interaction with a conspecific. Itakura et al. (1999) trained a chimpanzee to reliably indicate the location of hidden food to other group members. In this setting, the chimpanzee subjects ignored the conspecific's pointing gesture just as they ignored the human experimenter's (Itakura et al. 1999).

One problem with the standard object-choice task could be that it is highly artificial. Typically, the containers and food are on the human's side of a cage or barrier, and, thus, the human does not really need the chimpanzee's help in locating the food – she could easily just lift the containers and look herself. A second issue is that chimpanzees might follow the pointing gesture and assume that the container itself is the target, rather than what might be inside. However, in a recent study, Kirchhoffer et al. (submitted) showed that chimpanzees also either ignore or do not cognitively register a human's gesture if it is directed at one of two objects that are placed within the subjects' reach and that remain visible throughout the entire test. In said study, a transparent box was placed within the subjects' reach. The box had two separate compartments at either end of the box, each containing an object. Then the human indicated which of the two objects was to be delivered by the chimpanzee by gesturing towards it and offering the subject some food in return. All chimpanzees fetched one of the two objects in most of the trials; however, all chimpanzees that were tested seemed not to take into account the human's gesture while doing so. Indeed, none of the 20 chimpanzees tested chose the target object above chance. This suggests that it is not just the potential artificiality of the object-choice paradigm that is affecting the subjects' behaviour, but presumably something more fundamental on a cognitive level (Kirchhoffer et al. submitted).

One difference between the standard gaze-following situation and cases of perceptual co-presence of the referent and the gaze cue is that in the latter, the problem cannot be solved simply by scanning the environment for anything relevant. The referent is already present, which is why, in order to be successful, the subject has to understand the referential aspect of the communicative gesture. To do so, the recipient has to visually track the gesture to a specific referent in the environment, direct the subject's attention to that referent, and then infer why the subject's attention was directed towards it.

14.4 Human Child Development

From an early age, humans seem to be especially adapted to attending to and acting upon communicative cues produced by others (Csibra and Gergely 2009; Senju and Csibra 2008). The use and production of certain communicative gestures (e.g. pointing) by 12-month-olds seem to be underpinned by an interpretation of another's intentions, desires, and some consideration of the context and common ground between both individuals. This is supported by studies showing that children as young as 12-months old use others' gestures as information about where to find a desirable object (Behne et al. 2005; Gräfenhain et al. 2009). Also, children as young as 18 months of age interpret others' pointing gestures based on what experience they have shared with them (Liebal et al. 2009; Saylor and Ganea 2007). Liebal et al. (2009) set up a situation in which children shared one activity with one adult and a second activity with another adult. Later, one of the adults pointed towards a referent, which was appropriate for either activity. The children's response to the adult's gesture depended upon which adult pointed, supporting the view that children took the pointing gesture to be relevant to a shared experience that they had just had with a specific adult some moments before, and therefore a common ground they had established with that adult (Liebal et al. 2009).

Furthermore, children at this age produce communicative gestures in order to inform others about certain entities in the environment (Liszkowski 2005; Liszkowski et al. 2006). Sometimes they do so solely with a helpful motive; in other words, without expecting anything in return for having done so (Liszkowski et al. 2008). But human communication goes beyond communicating about referents that are currently present. From an early age, children base their communicative attempts on the assumption that others can make inferences about absent referents (Liszkowski et al. 2009). Situations where the referent is currently absent are special because here, the only possible way to solve the problem is for the recipient to interpret the larger context and common ground in order to be successful (Clark 1996; Tomasello and Carpenter 2007). As shown above, this could be based on some shared experience or shared knowledge, which then helps both individuals to specify which referent the communication is about and what it has to do with that referent (Clark 1996; Liebal et al. 2009; Aureli et al. 2009; Tomasello and Carpenter 2007).

Imagine going to a friend's home that you have not seen for some time. You ask your friend to show you around because you want to know what has changed over the past few months. During your tour, you suddenly point to an empty wall without saying anything, and your friend replies, "Yes, I've sold it". He knows you are referring to a painting that was hanging there the last time you saw his home. In that moment your knowledge overlaps with your friend's and you therefore have a common ground, and also share the knowledge that you do. Therefore, communicative interactions, especially if they take place non-verbally, are based on both individuals' assumption that both can mentally orient towards a referent. Having this shared perspective helps both individuals to successfully communicate, especially in the absence of a referent (Tomasello et al. 2005; Tomasello and Carpenter 2007).

14.5 Chimpanzees

Chimpanzees' failure to use human-like pointing is even more puzzling given that chimpanzees produce pointing accompanied by gaze alternation regularly when requesting something from a human partner (Leavens and Hopkins 1998; Leavens et al. 1996). There are only very few observed instances in which great apes pointed for conspecifics in the wild (Véa and Sabater-Pi 1998; Inoue-Nakamura and Matsuzawa 1997), but for humans, great apes point in order to indicate the location of food, for example. They do so with strong persistence; that is, they will not stop pointing at a referent until the human hands it over, even if offered alternatives (Leavens et al. 2005; Leavens and Hopkins 1998). This indicates that their gesture is directed at a specific referent and is not just a random begging gesture. There is also anecdotal evidence that chimpanzees will guide a human over a great distance by pointing towards a location where a desirable reward is hidden. These observations are especially impressive, as the chimpanzees in this situation show tremendous persistence and change the direction of their pointing in a flexible way as soon as the human veers in the wrong direction. However, unlike children, chimpanzees do not seem to produce gestures readily in order to help others find objects in which they themselves currently have no interest (Bullinger et al. in press; see also Zimmermann et al. 2009). In the study by Bullinger et al. (in press), the authors compared chimpanzees' and children's motivation to point towards an object in which the subjects themselves were interested or an object in which only the adult was interested. Unlike the children, the chimpanzees' level of pointing dropped dramatically after the context changed from a more selfish (the chimpanzee gets a desired object) to a more helpful (the human gets a desired object) context. Therefore, rather than informing others about certain entities, chimpanzees seem to produce gestures mainly to request self-desired objects.

This insight has led to the hypothesis that chimpanzees use gestures more as imperatives, and suggests that they are directed at humans to make them do something such as make them hand over food that they desire rather than inform the human recipients about a certain entity in the environment (Tomasello 2008; Bullinger et al. in press). Directing the recipient imperatively in order to make the recipient do things is substantially different from drawing the recipient's attention towards something in order to make the recipient know things about the environment (Tomasello et al. 2005). Communicating imperatively does not necessarily require an understanding of others' mental states. Instead, the recipient is more used as a social tool, and the gestures are used in order to request a certain action from the recipient such as to move into a certain direction. Informing individuals about things in the environment is substantially different, as by communicating the sender's aim is to provide or share information (Zimmermann et al. in press). This requires an ability to be motivated beyond the obtaining of direct-reward objects, as well as the cognitive ability to detect when a recipient is ignorant about certain aspects of the environment, plus the understanding and motivation to provide this information (Tomasello 2008).

That chimpanzees regard communication as a tool for changing others' behaviour rather than as a means to make others mentally orient towards a referent would also explain why chimpanzees fail to produce gestures in the absence of a referent. Liszkowski et al. (2009) compared chimpanzees and 12-month-old children in an identical setting. These test subjects watched two individuals, a giver and a requester, communicating with each other. The giver had two types of objects (desired or undesired), which the giver would offer to the requester upon request. During the demonstration phase, the requester would sometimes produce specific or unspecific requests. Specific requests consisted of the requester looking at the giver, nodding, clapping hands, and saying something like, "I want a ball", upon which the giver would offer the desired object. Unspecific requests consisted of the requester looking around, frowning, raising hands, and saying, "Give me something", upon which the giver would offer the undesired object. Desired and undesired objects were placed on or below a table – but always on the same side of the table. Therefore, even if the referent (e.g. the desired object) was gone, pointing towards the empty location where it used to be would produce the desired effect and the giver would hand over the desired object.

When the children were placed in the role of the requester, they made specific requests towards the correct location, irrespective of whether the referent was present but hidden behind the barrier or genuinely absent, meaning that the requester had to point to an empty location. The chimpanzees, on the other hand, made specific requests towards the desired object if it was hidden behind a barrier, but not if it was absent and the table was empty. Thus, it appears that unlike human children, the chimpanzees needed some perceptible referent in order to elicit requesting behaviour. This supports the hypothesis that chimpanzees do not regard and use communication as a means to make the giver mentally orient towards a referent.

From the evidence supporting the view that chimpanzees produce gestures imperatively leads to the hypothesis that they also interpret gestures imperatively and as a request to do things. This context of imperative motivation and interpretation helps explain why they may struggle with the cooperative, helpful nature of the pointing gesture in communicative interactions. An understanding of the pointing gesture in the object-choice paradigm presupposes a more general understanding of others' helpful motives and their motivation to help or inform us about things that they assume are relevant to our purposes (Moll and Tomasello 2007; Tomasello 2008). Chimpanzees may simply lack an understanding of others' cooperative motives, which is why they fail to use a human's pointing gestures in a food-finding situation (Tomasello 2008).

However, Kirchhoffer et al. (submitted) showed that chimpanzees also seem to have difficulties using requests in return for an edible reward. The human in the study did not inform the chimpanzees helpfully, but instead indicated by pointing at which object she wanted in return for the edible reward. The pointing gesture was therefore clearly underlined by an imperative rather than informative motive. Still, none of the chimpanzees used this imperative pointing gesture in order to gain a direct reward. This is difficult to reconcile with the view that it is the cooperative motive of the pointing gesture alone that chimpanzees have difficulties with. An alternative hypothesis could be that chimpanzees simply do not register the gesture, because acting upon it requires a motivation to act cooperatively, which chimpanzees may lack.

Interestingly, chimpanzees do not hesitate in fetching an object upon request if it is the only possible referent in view, even if there is no direct benefit for them doing so (Yamamoto et al. 2009; see also Melis et al. in press for similar evidence). In the Yamamoto et al. study, two chimpanzees sat in adjacent rooms. Both individuals needed a tool located in the other individual's room to retrieve food located in their own room. Both individuals helped each other by handing over the necessary tool requested by the other individual. The same was true if only one individual needed a tool that was placed in the other individual's room and the giver had no direct benefit from handing it over to the requester. Even though the chimpanzees rarely provided the tool voluntarily, they would not hesitate in handing it over upon a specific request produced by the requester such as poking an arm through a hole (Yamamoto et al. 2009). This demonstrates that chimpanzees are generally motivated to act cooperatively upon receiving a communicative request from another individual (see also Melis et al. 2006b; Warneken et al. 2007 for evidence that chimpanzees are generally motivated to act cooperatively and even helpfully).

These findings are difficult to reconcile with the assumption that a general lack of cooperative motivation is the reason for chimpanzees' failure to respond to human-given communicative gestures. Chimpanzees seem to incur difficulties only in those situations where there is more than one possible referent in view or no referent at all. One major difference between a situation with only one perceptible object present vs. two perceptible objects is that the recipient of the gesture will not be successful simply by scanning the environment for just any object, but must attend to the specificity of the gesture or infer the referent from the context such as based on some common ground. If there is no common ground upon which to interpret the context, attending to the specificity of the gesture is the most parsimonious solution.

Imagine you are sitting opposite a stranger and you suddenly find yourself distracted by something behind you. You turn to look, and upon turning back the stranger points and makes a request using no words other than, "Give me". Imagine that you, the recipient, have no further way of requesting additional information. Because the other individual is a stranger, you have absolutely no common ground such as shared past experiences. The only way to solve the problem is by scanning the environment for anything relevant. If there is more than one possible referent, you, the recipient, must invest time and effort figuring out which referent is being requested by closely attending to the specificity of the gesture (e.g. the requester's gaze or pointed finger). If there is only one possible referent, it is likely that the referent is the one in view and therefore time need not be invested attending to the specificity of the gesture. In the event that there is no referent at all in sight, the problem cannot be solved without a tremendous investment of time and effort, such as going through every possible referent in the wider area or identifying the specificity of the gesture. Therefore, finding the correct referent is significantly hindered in the absence of common ground, which serves as a helpful tool to quickly identify the correct object.

The fact that chimpanzees may lack the cognitive abilities allowing them to refer to common ground while communicating with others means that chimpanzees lack an important mechanism to help simplify the identification of a particular referent from several objects. In the absence of an understanding of common ground, chimpanzees must therefore rely solely on identifying the specificity of a gesture. This becomes even more difficult if the possible referents are in close proximity to each other. If the objects are distant from one another, deciphering the reference may be easier, as the specificity of the gesture is easier to identify.

Two studies have been conducted supporting this view: Mulcahy and Call (2009) tested chimpanzees, bonobos, and orangutans in a proximal version and a more distant version of the object-choice test (Mulcahy and Call 2009). The proximal version of the paradigm was comparable to that used in prior studies (Bräuer et al. 2006; Hare et al. 2002; Herrmann et al. 2007). The subject sat in front of two cups positioned on a table located between the experimenter and the subject. While the experimenter pointed to one cup, the subject could also see the other cup, which was in close proximity. As in prior studies, the apes failed this version of the test. The distant version differed in one important respect: The cups were placed at a large distance from one another such that the subject had to move from one room to another in order to make a choice. This meant that while attending to one cup, the other was no longer in view, which made deciphering of the reference easy, as attending to the specificity of the gesture was not necessary. Interestingly, the chimpanzees were successful in this version of the test, supporting the view that it is not the generally cooperative nature of the communicative interaction that is the problem but rather the interpretation of the specificity of the gesture.

The second paradigm corroborating this view is a study by Barth et al. (2006). In this study, the authors also compared two setups of the object-choice paradigm. Both setups had in common that the cups were positioned in close proximity to one another so that they were both in view the entire time the chimpanzees made their choice. But the two setups differed in one important respect. In one setup, the chimpanzees were already in the room, sitting in front of the two cups before the experimenter performed her gesture towards one of them. As the chimpanzees were already attending to both cups when the experimenter gazed at the target cup, the required attention to the specificity of the gesture seemed to be missed, and the chimpanzees chose randomly between both cups. In the alternative setup, the chimpanzees entered the room while the experimenter was already gazing at one of the cups; thus, the first thing the chimpanzees saw was the human gazing in a particular direction, which potentially drew their attention immediately to the specific referent. Chimpanzees probably performed well in this version of the task because when the chimpanzees made their choice, the specificity of the gesture was unambiguous: While entering the room, the chimpanzees most likely saw the gesture before they saw the two cups (Barth et al. 2005).

14.6 Ostensive Signals

Humans have evolved non-costly signals, so-called ostensive cues (e.g. eye contact, high-pitched voice, etc.), which help to indicate that the ensuing interaction is communication directed at the other individual and that the interaction is relevant (Sperber and Wilson 1986; Csibra and Gergely 2009). Human children are sensitive to ostensive cues from a very early age. One example is the special pattern of infant-directed speech (so-called motherese), which can make it manifest that the child is being addressed. It has been observed that newborns prefer infant-directed over adult-directed speech (Cooper and Aslin 1990; Csibra and Gergely 2009). At the age of 9 months, children tend to follow the gaze of others only when they are preceded by an ostensive signal (e.g. eye contact). This suggests that already at this early age, children form in their social environment an expectation of referential communication when following others' gaze direction (Senju et al. 2008; Senju and Csibra 2008).

Chimpanzees, as well as other primates, do not seem to produce or use ostensive signals during their communicative interactions, even though there is evidence that chimpanzees are sensitive to eye contact and also use others' eyes as an important stimulus in communicative interactions with humans (Hostetter et al. 2007; Itakura et al. 1999; Kaminski et al. 2004). There is, as yet, no evidence suggesting that ostensive cues are used in order to manifest relevance or specificity during communicative interactions with conspecifics. Therefore, chimpanzees may lack another important mechanism that, in the absence of a common ground, could help to manifest the relevance of specificity. This may also be why, for chimpanzees, it may simply not be relevant to attend to the specificity of gestures in certain situations, especially if the possible referents are in close proximity and therefore deciphering the reference of the gesture made even more difficult or impossible.

Interestingly, chimpanzees are generally more successful in reading a human's communicative gestures if the context in which the gesture is given is shifted from a cooperative one to a competitive one. When the context is competitive, chimpanzees successfully use human-given gestures, even in situations where the possible referents are in close proximity. In a study by Herrmann and Tomasello (2006), a human experimenter signalled the location of hidden food by extending her arm in the direction of one of two cups. Both cups were in full view of the subject while the human was pointing. If the human had established a competitive context with the chimpanzees (competing with other chimpanzees for food), the extended arm was used to indicate which location to avoid, while in the cooperative setting (engaging in a cooperate interaction with the experimenter vs. competing with other chimpanzees for food), the experimenter indicated informatively where the food was hidden. Chimpanzees were more successful in using the gesture to find the location of the hidden food in the competitive vs. cooperative context (Herrmann and Tomasello 2006; see also Hare and Tomasello 2004).

The reason for this outcome could be that chimpanzees are especially adapted for competitive interactions (Hare 2001). Chimpanzees constantly compete with conspecifics as well as neighbouring groups over resources. It has even been

14 Communicative Cues Among and Between Human and Non-human Primates 255

hypothesized that it was this type of competition that led to certain sophisticated social cognitive skills and that in turn helped individuals to outcompete others (Humphrey 1976; Byrne and Whiten 1988). There is also evidence that chimpanzees show certain social cognitive skills such as understanding others' visual perspective in competitive paradigms (Hare et al. 2000, 2001), but not in more cooperative paradigms (Povinelli et al. 1994). This has led to the hypothesis that chimpanzees may simply find competitive contexts more relevant than cooperative ones, which could also explain why chimpanzees use human-given gestures more frequently when presented in a competitive context (Hare and Tomasello 2004). In line with the framework of our argument, this evidence supports the theory that chimpanzees simply attend to the specificity of the gesture more during competitive situations because it is of greater relevance for them, while in more cooperative contexts the specificity of the gesture is not registered because it is of less relevance.

Interestingly, multiple pieces of evidence suggest that chimpanzees and other primates that have been raised in close proximity to humans generally find it easier to use human-given pointing gestures. This is even the case within a cooperative and informative context and when the potential referents are in close proximity to each other (Itakura and Tanaka 1998; Okamoto-Barth et al. 2008). Call and Tomasello (1996) outlined different mechanisms that might possibly explain differences between enculturated and non-enculturated apes: simple exposure to human life, emulation learning of human actions, explicit training by humans, and being treated by humans as intentional beings (Call and Tomasello 1995; Tomasello and Call 2004). They favour the explanation that being treated as intentional beings may lead to a fundamental change in the social cognition of enculturated apes. However, recent evidence suggests that non-enculturated apes also have some understanding of intentions in others (Call et al. 2004; Buttelmann et al. 2008). Therefore, being treated as intentional beings may not be sufficient to explain why enculturated apes use human-given gestures. An alternative explanation could be that these individuals, through their intensive interactions with human caretakers, consider communicative interactions with humans to be more relevant than their mother-reared counterparts. Rather than learning to associate the gesture with a reward, they may attend to the specificity of the gesture more because they consider the whole interaction to be highly relevant, even if presented in a cooperative context. One could also speculate that those individuals potentially developed some sensitivity to certain ostensive signals, which in turn helps them to identify the relevance and specificity of a communicative interaction.

14.7 Sensitivity to Ostensive Signals in a Non-human Species: The Domestic Dog

Work with domestic dogs shows that sensitivity to ostensive cues alone can lead to a flexible use of gestural communication, even in the absence of a deeper understanding of others' psychological states (a prerequisite for the ability to form a common ground with others). Domestic dogs are extraordinarily flexible in using different human-given communicative gestures. Interestingly, dogs' skills in this domain seem to be influenced by selection processes that occurred during domestication. This hypothesis is supported by two additional facts: First, wolves, dogs' closest living relatives, seem to be less flexible than dogs in using human gestural communication. Even if both species are raised under identical conditions and then tested at the same age, dogs outperform wolves in communication skills (Miklosi et al. 2003; Hare et al. 2002). In addition, major learning events during ontogeny alone cannot account for the dogs' behaviour. This is supported by the fact that puppies from 6 weeks of age are able to use human pointing to find food, even when required to move away from the human's hand to be successful (Riedel et al. 2008). Dogs' behaviour relative to human pointing cannot be explained exclusively by mechanisms such as local or stimulus enhancement, but seems also to be based on some understanding of the triadic nature of the interaction (Hare et al. 1998). In addition, there is evidence that dogs are extremely sensitive to certain ostensive cues such as eye contact or tone of voice (Kaminski et al. submitted). Eye contact seems to be a particularly important signal for dogs to identify when communication is relevant and directed at them (Kaminski et al. submitted; see also Viranyi et al. 2004). Domestic dogs therefore represent an interesting example of how rather simple mechanisms, such as having a certain sensitivity to ostensive cues, may help to identify the referent in triadic communicative interactions despite lacking a complex understanding of others' psychological states.

14.8 Summary and Conclusions

The expectation is that comparative communicative cue studies between humans and non-human primates such as those described in this chapter will reveal relevant information regarding the underpinnings of group cooperation in our species. For instance, in Sects. 14.2 and 14.3, we showed that there is a difference between the ability of non-human primates to interpret gaze shifts as an indicator that others see things in the environment vs. that others are communicating about things – an important cognitive nuance specific to humans. That said, it appears that the gaze-following behaviour of great apes is based on some understanding of another's perspective. Different primate species seem to have some understanding of others' line of sight and others' visual perspective, an amazingly relevant parallel to human cooperative behaviour.

In Sect. 14.4 we also showed that human children as young as 12 months can interpret communicative cues of pointing regarding the pointer's intentions, desires, and even with some consideration to the context and common ground between themselves and the individual pointing. In another six months of age, this cognitive interpretation expands to shared experiences that the child has had with the individual pointing, demonstrating an already-developed grasp of the abstract indicated by the human child's non-reliance on the referent object even being in site – a human-specific trait reinforced by imperative communication experiments comparing human children and chimpanzees' understanding of others' mental states. (Sect. 14.5). What is not specific to humans is the general cooperative nature of communicative interaction, a behaviour trait observed extensively among chimpanzees and limited mostly to their inability to interpret the specificity of the communicative gesture rather than an unwillingness to help or cooperate.

We then talked about the comparative sensitivity of human children and chimpanzees to ostensive cues or signals such as referential speech patterns – chimpanzees neither producing nor responding to such communication except eye contact in some communicative interactions with humans (Sect. 14.6). But what we also know – again, something that could potentially explain certain motivational aspects of human group cooperative behaviour – is that the response among chimpanzees to human-initiated communicative cues is notably heightened when they are issued in a conspecific competitive setting, leading to the hypothesis that the chimpanzees' response to communicative cues (or the lack thereof) may be based more on the relevancy of said cues rather than on the chimpanzee's cognitive understanding of these cues. The last, but certainly not the least, important detail included in this section is the parallel between the levels of understanding in non-human primates (both chimpanzees and apes) of human communicative cues and their level of enculturation/exposure to humans, suggesting that there may be elements of "nurture" in the "nature vs. nurture" group cooperative traits among primates in general – human as well as non-human.

This domestication theme then segues to our final section on comparative studies between dogs and wolves and the extensive understanding that dogs demonstrate regarding human communicative cues. The outperformance dogs demonstrate over their ancestral relatives raised under identical conditions in an experimental setting points to the influences of selection processes that occurred during species domestication. These influences of selection processes, coupled with the certain sensitivity dogs have to human communicative cues, may help enlighten our understanding of triadic communicative interactions even when lacking a complex understanding of others' psychological states.

References

Anderson JR, Sallaberry P, Barbier H (1995) Use of experimenter-given cues during object-choice tasks by capuchin monkeys. Anim Behav 49:201–208

Anderson JR, Montant M, Schmitt D (1996) Rhesus monkeys fail to use gaze direction as an experimenter-given cue in an object-choice task. Behav Proc 37:47–55

Aureli T, Perucchini P, Genco M (2009) Children's understanding of communicative intentions in the middle of the second year of life. Cogn Dev 24:1–12

Barth J, Reaux JE, Povinelli DJ (2005) Chimpanzees' (*Pan troglodytes*) use of gaze cues in object-choice tasks: Different methods yield different results. Anim Cog 8:84–92

Behne T, Carpenter M, Tomasello M (2005) One-year-olds comprehend the communicative intentions behind gestures in a hiding game. Dev Sci 8:492–499

Bräuer J, Call J, Tomasello M (2005) All great ape species follow gaze to distant locations and around barriers. J Comp Psychol 119:145–154

Bräuer J, Kaminski J, Riedel J, Call J, Tomasello M (2006) Making inferences about the location of hidden food: social dog, causal ape. J Comp Psychol 120:38–47

Bullinger A, Zimmermann F, Kaminski J, Tomasello M (in press) Different social motives in the gestural communication of chimpanzees and human children. Dev Sci

Burkart J, Heschl A (2006) Geometrical gaze following in common marmosets (*Callithrix jacchus*). J Comp Psychol 120:120–130

Buttelmann D, Carpenter M, Call J, Tomasello M (2008) Rational tool use and tool choice in human infants and great apes. Child Dev 79:609–626

Byrne RW, Whiten A (1988) Machiavellian intelligence: social expertise and the evolution of intellect in monkeys, apes and humans. In: Byrne RW, Whiten A (eds) Oxford University Press, New York

Call J, Tomasello M (1995) The effect of humans on the cognitive development of apes. In: Russon AE, Bard KA, Parker ST (eds) Reaching into thought. Cambridge University Press, New York, pp 371–403

Call J, Tomasello M (2005) What do chimpanzees know about seeing revisited: an explanation of the third kind. In: Eilan N, Hoerl C, McCormack T, Roessler J (eds) Issues in joint attention. Oxford University Press, Oxford

Call J, Hare BA, Tomasello M (1998) Chimpanzee gaze following in an object-choice task. Anim Cog 1:89–99

Call J, Hare B, Carpenter M, Tomasello M (2004) Unwilling versus unable: Chimpanzees' understanding of human intentional actions. Dev Sci 7:488–498

Carpenter M, Nagell K, Tomasello M (1998) Social cognition, joint attention, and communicative competence from 9 to 15 months of age. Monogr Soc Res Child Dev 63:1–143

Clark HH (1996) Using language. Cambridge University Press, Cambridge

Cooper RP, Aslin RN (1990) Preference for Infant-directed speech in the first month after birth. Child Dev 61:1584–1595

Corkum V, Moore C (1995) Development of joint visual attention in infants. In: Moore C, Dunham PJ (eds) Joint attention: Its origins and role in development. Lawrence Erlbaum, Hillsdale, NJ, pp 61–83

Csibra G, Gergely G (2009) Natural pedagogy. Trends Cogn Sci 13:148–153

Csibra G, Volein A (2008) Infants can infer the presence of hidden objects from referential gaze information. Br J Dev Psychol 26:1–11

D'Entremont B, Hains SMJ, Muir DW (1997) A demonstration of gaze following in 3- to 6-month-olds. Inf Behav Dev 20:569–572

Deaner R, Platt M (2003) Reflexive social attention in monkeys and humans. Curr Biol 13:1609–1613

Emery NJ, Lorincz EN, Perrett DI, Oram MW, Baker CI (1997) Gaze following and joint attention in rhesus monkeys (*Macaca mulatta*). J Comp Psychol 111:286–293

Flom R, Pick AD (2003) Verbal encouragement and joint attention in 18-month-old infants. Inf Behav Dev 26:121–134

Goossens BMA, Dekleva M, Reader SM, Sterck EHM, Bolhuis JJ (2008) Gaze following in monkeys is modulated by observed facial expressions. Anim Behav 75:1673–1681

Gräfenhain M, Behne T, Carpenter M, Tomasello M (2009) One-year-olds' understanding of nonverbal gestures directed to a third person. Cogn Dev 24:23–33

Hare B (2001) Can competitive paradigms increase the validity of experiments on primate social cognition? Anim Cogn 4:269–280

Hare B, Tomasello M (2004) Chimpanzees are more skillful in competitive than in cooperative cognitive tasks. Anim Behav 68:571–581

Hare B, Call J, Tomasello M (1998) Communication of food location between human and dog (*Canis familiaris*). Evol Commun 2:137–159

Hare B, Call J, Agnetta B, Tomasello M (2000) Chimpanzees know what conspecifics do and do not see. Anim Behav 59:771–785

14 Communicative Cues Among and Between Human and Non-human Primates 259

Hare B, Call J, Tomasello M (2001) Do chimpanzees know what conspecifics know? Anim Behav 61:139–151

Hare B, Brown M, Williamson C, Tomasello M (2002) The domestication of social cognition in dogs. Science 298:1634–1636

Herrmann E, Tomasello M (2006) Apes' and childrens' understanding of cooperative and competitive motives in a communicative situation. Dev Sci 9:518–529

Herrmann E, Call J, Hernández-Lloreda MV, Hare B, Tomasello M (2007) Humans have evolved specialized skills of social cognition: the cultural intelligence hypothesis. Science 317: 1360–1366

Hostetter AB, Russell JL, Freeman H, Hopkins WD (2007) Now you see me, now you don't: evidence that chimpanzees understand the role of the eyes in attention. Anim Cogn 10:55–62

Humphrey NK (1976) The social function of intellect. In: Bateson PPG, Hinde RA (eds) Growing points in ethology. Cambridge University Press, Cambridge, pp 303–317

Inoue-Nakamura N, Matsuzawa T (1997) Development of stone tool use by wild chimpanzees (*Pan troglodytes*). J Comp Psychol 111:159–173

Itakura S, Tanaka M (1998) Use of experimenter-given cues during object-choice tasks by chimpanzees (*Pan troglodytes*), an orang-utan (*Pongo pygmaeus*), and human infants (*Homo sapiens*). J Comp Psychol 112:119–126

Itakura S, Agnetta B, Hare B, Tomasello M (1999) Chimpanzee use of human and conspecific social cues to locate hidden food. Dev Sci 2:448–456

Johnson MH, Griffin R, Csibra G, Halit H, Farroni T, de Haan M, Baron-Cohen S, Richards J (2005) The emergence of the social brain network: evidence from typical and atypical development. Dev Psychopathol 17:599–619

Kaminski J, Call J, Tomasello M (2004) Body orientation and face orientation: two factors controlling apes' begging behavior from humans. Anim Cogn 7:216–223

Kaminski J, Call J, Tomasello M (2008) Chimpanzees know what others know, but not what they believe. Cognition 109:224–234

Kaminski J, Schulz L, Tomasello M (submitted) How dogs know when communication is intended for them?

Kirchhofer K, Zimmermann F, Kaminski J, Tomasello M (submitted) Chimpanzees (*Pan troglodytes*) but not dogs (*Canis familiaris*) fail to understand directive pointing gestures

Leavens DA, Hopkins WD (1998) Intentional communication by chimpanzees: a cross-sectional study of the use of referential gestures. Dev Psychol 34:813–822

Leavens DA, Hopkins WD, Bard KA (1996) Indexical and referential pointing in chimpanzees (*Pan troglodytes*). J Comp Psychol 110:346–353

Leavens DA, Russell JL, Hopkins WD (2005) Intentionality as measured in the persistence and elaboration of communication by chimpanzees (*Pan troglodytes*). Child Dev 76:291–306

Liebal K, Behne T, Carpenter M, Tomasello M (2009) Infants use shared experience to interpret pointing gestures. Dev Sci 12:264–271

Liszkowski U (2005) Human twelve-month-olds point cooperatively to share interest with and provide information for a communicative partner. Gesture 5:135–154

Liszkowski U, Carpenter M, Striano T, Tomasello M (2006) 12- and 18-month-olds point to provide information for others. J Cogn Dev 7:173–187

Liszkowski U, Carpenter M, Tomasello M (2008) Twelve-month-olds communicate helpfully and appropriately for knowledgeable and ignorant partners. Cognition 108:732–739

Liszkowski U, Schäfer M, Carpenter M, Tomasello M (2009) Prelinguistic infants, but not chimpanzees, communicate about absent entities. Psychol Sci 20:654–660

Melis AP, Call J, Tomasello M (2006a) Chimpanzees (*Pan troglodytes*) conceal visual and auditory information from others. J Comp Psychol 120:154–162

Melis AP, Hare B, Tomasello M (2006b) Chimpanzees recruit the best collaborators. Science 311:1297–1300

Melis AP, Warneken F, Jensen K, Call J, Schneider AC, Tomasello M (in press) Chimpanzees help conspecifics to obtain food and non-food items. Proceedings of the Royal Society B

Miklosi A, Kubinyi E, Topal J, Gacsi M, Viranyi Z, Csanyi V (2003) A simple reason for a big difference: Wolves do not look back at humans, but dogs do. Curr Biol 13:763–766

Moll H, Tomasello M (2007) Cooperation and human cognition: the Vygotskian intelligence hypothesis. Phil Trans Roy Soc Lond B 362:639–648

Morton J, Johnson MH (1991) CONSPEC and CONLERN: a two-process theory of infant face recognition. Psychol Rev 98:164–181

Mulcahy NJ, Call J (2009) The performance of bonobos (*Pan paniscus*), chimpanzees (*Pan troglodytes*), and orangutans (*Pongo pygmaeus*) in two versions of an object-choice task. J Comp Psychol 123:304–309

Okamoto-Barth S, Call J, Tomasello M (2007) Great apes' understanding of other individuals' line of sight. Psychol Sci 18:462–468

Okamoto-Barth S, Tanaka M, Tomonaga M (2008) Development of using experimenter given cues in infant chimpanzees: behavioral changes in cognitive development. Dev Sci 11:98–108

Povinelli DJ, Eddy TJ (1996) Chimpanzees: joint visual attention. Psychol Sci 7:129–135

Povinelli DJ, Rulf AB, Bierschwale DT (1994) Absence of knowledge attribution and self-recognition in young chimpanzees (*Pan troglodytes*). J Comp Psychol 108:74–80

Povinelli DJ, Reaux JE, Bierschwale DT, Allain AD, Simon BB (1997) Exploitation of pointing as a referential gesture in young children, but not adolescent chimpanzees. Cogn Dev 12:327–365

Riedel J, Schumann K, Kaminski J, Call J, Tomasello M (2008) The early ontogeny of human–dog communication. Anim Behav 75:1003–1014

Rosati AG, Hare B (2009) Looking past the model species: diversity in gaze-following skills across primates. Curr Opin Neurobiol 19:45–51

Santos LR, Nissen AG, Ferrugia JA (2006) Rhesus monkeys, *Macaca mulatta*, know what others can and cannot hear. Anim Behav 71:1175–1181

Saylor MS, Ganea P (2007) Infants interpret ambiguous requests for absent objects. Devel Psychol 43:696–704

Scaife M, Bruner J (1975) The capacity for joint visual attention in the infant. Nature 253:265–266

Senju A, Csibra G (2008) Gaze following in human infants depends on communicative signals. Curr Biol 18:668–671

Senju A, Csibra G, Johnson MH (2008) Understanding the referential nature of looking: Infants' preference for object directed gaze. Cognition 108:303–319

Sperber D, Wilson D (1986) Relevance: Communication and cognition. Blackwell, Oxford; and Harvard Press, Cambridge, MA

Tomasello M (2008) Origins of human communication. MIT, Cambridge, MA

Tomasello M, Call J (2004) The role of humans in the cognitive development of apes revisited. Anim Cogn 7:213–215

Tomasello M, Carpenter M (2007) Shared intentionality. Dev Sci 10:121–125

Tomasello M, Call J, Gluckman A (1997) Comprehension of novel communicative signs by apes and human children. Child Dev 68:1067–1080

Tomasello M, Call J, Hare B (1998) Five primate species follow the visual gaze of conspecifics. Anim Behav 55:1063–1069

Tomasello M, Hare B, Agnetta B (1999) Chimpanzees, *Pan troglodytes*, follow gaze direction geometrically. Anim Behav 58:769–777

Tomasello M, Carpenter M, Call J, Behne T, Moll H (2005) Understanding and sharing intentions: the origins of cultural cognition. Behav Brain Sci 28:721–727

Véa JJ, Sabater-Pi J (1998) Spontaneous pointing behaviour in the wild pygmy chimpanzee (*Pan paniscus*). Folio Primat 69:289–290

Viranyi Z, Topal J, Gacsi M, Miklosi A, Csanyi V (2004) Dogs respond appropriately to cues of human attentional focus. Behav Proc 66:161–172

Warneken F, Hare B, Melis AP, Hanus D, Tomasello M (2007) Spontaneous altruism by chimpanzees and young children. PLoS Biol 5:84

Yamamoto S, Humle T, Tanaka M (2009) Chimpanzees help each other upon request. PLoS One 4:7416

Zimmermann F, Zemke F, Call J, Gómez JC (2009) Orangutans (*Pongo pygmaeus*) and bonobos (*Pan paniscus*) point to inform a human about the location of a tool. Anim Cogn 12:347–358

Zimmermann F, Zemke F, Warneken F, Call J, Gomez J, Tomasello M (in press) Orangutans' (*Pongo pygmaeus*) and bonobos' (*Pan paniscus*) pointing behavior in 2 different motivational contexts. Anim Cogn

Chapter 15
Coordination in Primate Mixed-Species Groups

Eckhard W. Heymann

Abstract Groups formed by individuals from different species (mixed-species groups) are a widespread phenomenon amongst primates. Although the formation and maintenance of such mixed-species groups may incur costs to participating individuals, they render a net benefit, mainly through increased safety from predators and increased foraging efficiency. In contrast to the large number of studies that have examined the benefits and costs of primate mixed-species groups, there are still very few studies that have analysed the mechanisms of group coordination in mixed-species groups. Available evidence suggests that this coordination is mainly through vocal communication, but since the same vocalisations may be employed in intra-specific within-group and between-group communication as well as in inter-specific communication, it is difficult to analytically separate intra- and inter-specific coordination. The need for inter-specific coordination is likely to be highest when asymmetries in benefits from a mixed-species troop's formation are strong. Thus, "goal-dependent management of interdependencies" is necessary to maintain the integrity of mixed-species groups.

15.1 Introduction

Apart from forming groups with conspecifics, many vertebrates habitually associate with *heterospecifics*, that is, members of other species (fish: e.g. Krause et al. 2000; Parrish et al. 2002; birds: Powell 1985; Greenberg 2000; mammals: Stensland et al. 2003; Quérouil et al. 2008). In primates, such mixed-species groups, also known as *inter-specific* or *polyspecific associations*, are widespread amongst Neotropical and African rainforest monkeys but less common or absent in Malagasy lemurs and

---

E.W. Heymann
Behavioral Ecology and Sociobiology Unit, German Primate Center, Kellnerweg 4, 37077 Göttingen, Germany
e-mail: eheyman@gwdg.de

M. Boos et al. (eds.), *Coordination in Human and Primate Groups*,
DOI 10.1007/978-3-642-15355-6_15, © Springer-Verlag Berlin Heidelberg 2011

Asian monkeys (Struhsaker 1981; Waser 1987; Freed 2006; Haugaasen and Peres 2009). Such associations might occur randomly because groups from different species are simultaneously attracted to the same resources and endure for short times only (e.g. Whitesides 1989). However, there is strong evidence for non-randomness of many mixed-species groups (see Sect. 15.2), and many species spend considerable time, if not most of their activity period, in mutual association. Such mixed-species groups give rise to questions concerning their ultimate biological functions as well as the proximate mechanisms for the establishment and maintenance of association. This chapter first presents basic information on the occurrence of mixed-species groups amongst primates and then discusses their biological functions. It then addresses specific questions of coordination in mixed-species groups. Source references will be preferentially made to publications that have emerged since 2000, when the first review on this topic was published (Cords 2000).

15.2 Definition, Non-randomness, and Association Patterns

15.2.1 What Are Mixed-Species Groups?

In order to establish whether primates are found in a mixed-species group paradigm, primatologists usually define a criterion distance. Whenever members of different species are located at or within this criterion distance, the respective single-species groups are considered to be *inter-specific associated*. Criterion distances of 20, 25, and 50 m have been used in different primate studies. It is conceivable that a higher criterion distance results in finding primates more often in association.

15.2.2 Do Mixed-Species Groups Form Randomly?

To examine whether mixed-species groups occur by chance or not, *Waser's gas model* (Waser 1982) is usually employed. This model uses an analogy of primate group movements with the movements of a perfect gas and calculates expected rates of encounters between heterospecific groups with known mean travel speeds and mean group radii. From these variables, the mean duration of association and the expected proportion of time spent in association can be derived and compared to the observed time spent in association (Whitesides 1989; Holenweg et al. 1996).

While the application of Waser's gas model has been instrumental in testing for the randomness of associations, its weaknesses are also immediately apparent. Primate groups do not usually move randomly (as gas molecules do) through their home ranges. Movements can be goal-oriented, with food and water resources, neighbouring groups, shelters, and sleeping sites constituting goals (Garber 2000; Janson 2000). If two species overlap at least partially in the temporal distribution of their activities, if they are attracted to the same goal, if there is a limited set of optimal travel routes between different goals, or if similar mental maps are used for navigation, then groups from different species could meet or associate more often than predicted by chance alone (DiFiore and Suarez 2007). A more conservative test for non-randomness of associations should thus include "attractors" and travel decisions derived from optimal foraging theory. It is likely that for those associations where species spend most or all of their time in association (see Sect. 15.2.3), and where active establishment of association takes place (see Sect. 15.4.1), even more conservative tests will demonstrate nonrandomness of association.

15.2.3 Association Patterns

Primate mixed-species groups are usually composed of members from two or three (rarely more) different species. Participating species may come from the same genus (congeneric mixed-species groups: e.g. *Cercopithecus ascanius – Cercopithecus mitis, Saguinus mystax – Saguinus fuscicollis*) or may stem from different genera (heterogeneric mixed-species groups: e.g. *Cercopithecus diana – Procolobus badius, Saimiri boliviensis – Cebus apella*). The differentiation between congeneric and heterogeneric mixed-species groups is relevant for the comparison of costs for the establishment and maintenance of mixed-species groups (see Sect. 15.3.2).

Mixed-species groups can always be formed by same single-species groups, or a group from one species may associate with various groups of another species at different times. In the former case, for instance, in congeneric mixed-species groups of the genus *Saguinus*, home ranges (i.e. the area in which a group resides) of the participating groups are usually of the same size and overlap completely or almost so. In the latter case, for instance, in heterogeneric mixed-species groups of *S. boliviensis* and *C. apella, Callimico goeldii* and *Saguinus labiatus/S. fuscicollis*, or *Procolobus rufomitratus* and *Cercopithecus ascanius*, home ranges are of different sizes. In this case, home ranges of groups of one of the participating species (*S. boliviensis, C. goeldii, P. rufomitratus*) overlap with the home ranges of several groups of the other species (*C. apella, S. labiatus/S. fuscicollis, C. ascanius*) (Podolsky 1990; Porter 2001; Teelen 2007). This pattern also has implications for the costs and benefits of mixed-species groups (see Sect. 15.3).

The time spent in mixed-species groups varies considerably, not only between different species combinations but also between populations and different groups (see Table 15.1). In mixed-species groups formed by members of the genus *Saguinus*, the time spent in association may vary between almost 100% and as little as 19% (Heymann and Buchanan-Smith 2000). Similarly, different species of *Cercopithecus* may spend almost all their active time (i.e. between leaving and entering a sleeping site) in mixed-species groups or associate less frequently

266 E.W. Heymann

Table 15.1 Examples for variation of time spent in mixed-species groups

Species combination	% of time spent in mixed-species groups	Source of variation	References
Saguinus mystax + Saguinus fuscicollis	89–93	Group	Smith et al. (2005)
Saguinus labiatus + Saguinus fuscicollis	43–57	Group	Pook and Pook (1982)
	50–70	Group	Porter (2001)
	0–63	Season	Rehg (2006)
Saguinus imperator + Saguinus fuscicollis	19		Windfelder (1997)
Callimico goeldii + Saguinus fuscicollis + Saguinus labiatus	13–89	Season	Porter (2001)
	24–100	Season	Rehg (2006)
Callimico goeldii + Saguinus fuscicollis	21–22	Group	Porter (2001)
	0–12	Season	Rehg (2006)
Callimico goeldii + Saguinus labiatus	0–7	Season	Rehg 2006
Cercopithecus ascanius + Cercopithecus mitis	18–74	Population	Cords (1990)
	0–30	Population	Chapman and Chapman (2000)
Cercopithecus mitis + Cercopithecus ascanius	11–49	Population	Cords 1990
	22–25	Population	Chapman and Chapman (2000)
Cercopithecus ascanius + Procolobus tephrosceles	3–50	Population	Chapman and Chapman (2000)
Procolobus tephrosceles + Cercopithecus ascanius	12–32	Population	Chapman and Chapman (2000)
Procolobus tephrosceles + Cercopithecus mitis	0–9	Population	Chapman and Chapman (2000)
Cercopithecus diana + Cercopithecus campbelli	56–87	Group	Wolters and Zuberbühler (2003)
Cercopithecus diana + Procolobus badius	31–72	Season	Wachter et al. (1997)

(Gautier-Hion et al. 1983). Differences in body size (as a proxy for ecological differences) and the degree of overlap in the plant portion of the diet seem to be determinants of the permanency of congeneric mixed-species groups. For example, *S. mystax* and *S. fuscicollis* differ strongly in body size and spend most of their time in association, whereas *Saguinus imperator* and *S. fuscicollis* differ much less in size and spend more time in single-species than in mixed-species groups (Heymann 1997). The time spent in mixed-species groups in different populations of *C. ascanius* and *C. mitis* and in different months of the year within the same population of these two species increases with the amount of overlap in the plant diet (Struhsaker 1981; Cords 1990). This is in contrast to heterogeneric mixed-species groups of *C. diana* and *P. badius*, where the time spent in association does not correlate with diet overlap (Wachter et al. 1997).

Mixed-species groups of birds (called "mixed flocks" by ornithologists) generally consist of many more species than primate mixed-species groups. In temperate zones, 10–15 (and in the tropics, as many as 100) different bird species can constitute these mixed flocks (see Greenberg 2000 for review). Similar to primates, it is almost always only one group per species that participates in mixed-species bird groups. However, in contrast to primates, the number of individuals per species is generally much lower; usually, only a pair or a family associate with other species, and sometimes only solitary individuals join mixed flocks (Terborgh 1990). Some species may participate more consistently in mixed flocks and may be more attractive to other species ("nuclear species" or "core species") while others join and follow less consistently ("attendant species"; Greenberg 2000). Home range or territory size seems to be the principal factor limiting participation in mixed flocks (Powell 1979; Pomara et al. 2007). Groups of species with smaller home ranges appear to attend mixed flocks only when groups of species with larger home ranges pass through the area. Noteworthy is that this pattern is similar to some mixed-species groups in primates.

15.3 Benefits and Costs for Primates in Mixed-Species Groups

15.3.1 Benefits

The benefits of mixed-species groups in primates can be grouped into two categories: benefits related to the reduction of the predation risk, and benefits related to an increase in foraging and feeding efficiency.

15.3.1.1 Reduction of Predation Risk

A reduction of the per capita predation risk mainly results from the increase in group size through associating with heterospecifics (see Table 15.2). Demonstrating the action of the "dilution effect" and the "confusion effect" (see Caro 2005 for examples of these effects in other animals) would require comparing predation rates between single-species and mixed-species groups. For several reasons, this is often difficult, if not impossible, however. First, successful predation events are rarely observed, hampering any meaningful statistical comparison. Second, species might tend to associate when the predation risk is high, and to live in single-species groups when the predation risk is low. This problem could only be overcome by estimating the hunting efforts of predators. In fact, Noë and Bshary (1997) have shown that *P. badius* associate more often with *C. diana* during seasons of the year when chimpanzees – a principal predator of *P. badius* but not of *C. diana* – are more likely to hunt.

A reduction in the predation risk can also be obtained through the "improved detection effect," as large groups are more likely to detect an approaching

Table 15.2 Potential benefits of primate mixed-species group

Reduction of predation risk	"Dilution effect": Risk of being attacked and preyed upon decreases with increasing group size "Confusion effect": Confusion of an attacking predator increases with increasing group size Vigilance-related effects: "Detection effect": probability of detecting a predator increases with groups size due to increased vigilance ("more eyes see more") Differential and complementary species-specific vigilance Eavesdropping on other species' alarm calls Joint defence against predators (mobbing, attacks)
Increased foraging and feeding efficiency	Access to habitats and resources that are not available to single-species groups Increased encounter rates with resources: Probability of detection of resources increases with group size Exploitation of other species' knowledge of resource distribution in habitat Scrounging resources detected by the other species Exploitation of feeding residuals of other species Exploitation of prey flushed by other species Increased rates of feeding and foraging (Avoiding visits to resources exhausted by other species (Increased resource defence)
Reduced risk of parasitism	Risk of being attacked by blood-sucking insects decreases with group size

predator than small groups. This effect is the result of more individuals being vigilant at any point in time in a larger group ("more eyes and ears") and a larger space being surveyed by a larger group. Benefits of improved detection can be asymmetric between species if individuals of one species show consistently higher levels of vigilance than individuals of the other species (Smith et al. 2004).

Vigilance-related effects can be more easily tested than the previously mentioned dilution and confusion effects by comparing rates of vigilance in and out of association, although this may become difficult in species that are almost permanently associated. Benefits of improved detection might also be confounded by species tending to associate in relation to the current predation risk (see above). Additionally, apart from protection against predators, vigilance may function against potential conspecific competitors. However, in this case individual rates of vigilance are expected to increase in mixed-species groups, as has in fact been observed in *C. ascanius* (Chapman and Chapman 1996).

The results of various studies have provided support for an antipredator function of associations. First, experiments that examined vigilance in and out of association in captive *S. labiatus* and *S. fuscicollis* revealed that when in association, more time was covered by at least one individual being vigilant compared to single-species

15 Coordination in Primate Mixed-Species Groups

groups, and at the same time the per capita costs of vigilance (time spent vigilant) were reduced (Hardie and Buchanan-Smith 1997). Similarly, wild *C. diana* and *S. mystax*, respectively, increased the time spent being vigilant when not associated with *Cercopithecus campbelli* and *S. fuscicollis*, respectively (Wolters and Zuberbühler 2003; Stojan-Dolar and Heymann 2010). Second, ranging at different forest strata can provide an additional advantage of mixed-species group formation. It has been shown in associations of tamarins and of guenons that species ranging lower in the forest are more likely to detect terrestrial predators, whereas species ranging in higher strata are more likely to detect aerial predators (Gautier-Hion et al. 1983; Peres 1993). Third, individuals living in mixed-species groups can also benefit from eavesdropping on the alarm-calling behaviour of the other species. For example, *C. diana* respond to the alarm calls of associated *C. campbelli* (Zuberbühler 2000), and *S. mystax* and *S. fuscicollis* mutually understand and respond to each others' alarm calls (Kirchhof and Hammerschmidt 2006). Finally, forces can be combined in association to attack and dissuade predators such as through joint attacks or mobbing of adult males from the different associated species (Eckardt and Zuberbühler 2004).

15.3.1.2 Increased Foraging and Feeding Efficiency

The formation of mixed-species groups may also render benefits in terms of increased foraging and feeding efficiency. These benefits can result from access to habitats and resources that are not accessible while ranging in single-species groups, or from increased encounter rates with food resources (Table 15.1).

Several observations support these predictions. For example, *Callimico goeldii* expand their habitat use when in association with *S. fuscicollis* and *S. labiatus*, and as a consequence have higher rates of fruit feeding while in association (Porter and Garber 2007). The arboreal *P. badius* and *C. diana* descend to lower forest strata and the forest floor more often while associated with the terrestrial *Cercocebus atys*, which gives them access to termite mounds (McGraw and Bshary 2002). It is conceivable that these foraging benefits are an indirect consequence of anti-predator benefits: The presence of heterospecific individuals provides increased safety in habitats that are usually avoided or used very infrequently.

Increased encounter rates with food resources may simply result from increased group size, as the higher number of individuals in an association increases the likelihood of detecting a food resource that can also be exploited by heterospecific group members. However, this has not yet been demonstrated.

It is also conceivable that members of one species may also have a superior knowledge of the location and availability of food resources, which can then be exploited by the other species in the association. This is particularly likely when the associated species differ in the size of their home ranges. In this case, the species with the smaller home range is expected to have better local knowledge and can therefore be exploited as a "guide" by the other species. *Cercopithecus ascanius* may use *C. mitis* as guides to food resources in cases where they have the larger home ranges, but the pattern is reversed in areas where *C. mitis* has the larger home ranges (Cords 1987). *Saimiri boliviensis* are led into large food patches by *C. apella* and *Cebus albifrons* (Podolsky 1990), these two species having much smaller home ranges than *S. boliviensis*. Similarly, *C. goeldii* – with large home ranges (150 ha) – probably exploit the knowledge about the location and abundance of food resources of *S. labiatus* and *S. fuscicollis*, who have much smaller home ranges (Porter 2001). In these cases, the net benefits of association are clearly asymmetrically distributed between species.

But even when home range size is similar or identical, as with *S. mystax* and *S. fuscicollis*, species may benefit from resource detection by others. *Saguinus mystax* initiate more feeding bouts than *S. fuscicollis*, which is obviously a benefit to the latter (Peres 1996). Conversely, *S. mystax* – travelling on average higher in the canopy than *S. fuscicollis* – scrounge on small resources in the lower forest strata detected by *S. fuscicollis*. In the extreme case, this scrounging may lead to the exclusion of the detecting species from the food resource (Peres 1996; Heymann, personal observations).

Associated species may also exploit dropped feeding residuals or prey flushed by the other species. *Saimiri boliviensis* gain access to the pulp of hard palm fruits that are opened, only partially eaten, and then dropped by *Cebus* (Terborgh 1983). *Saguinus fuscicollis* obtain a substantial proportion of their prey through capturing insects that escape from *S. mystax* (Peres 1992a; Heymann, personal observations).

Other potential benefits related to foraging and feeding have been seen with respect to avoided visits to exhausted food resources that might have otherwise occurred if travelling in single-species groups, as well as joint resource defence. The former is conceivable, but principally not testable and thus has no heuristic value. The latter is unlikely, since aggressive interactions during the defence of resources against other mixed-species groups usually take place within, not between, species.

15.3.1.3 Other Potential Benefits

Based on the correlation between temporal patterns of association and the activity of blood-sucking insects, Freeland (1977) suggested that mixed-species groups formed by mangabeys, *Cercocebus albigena*, with other primates in the Kibale Forest (Uganda) are a means of reducing the number of insect bites received by individual mangabeys. This suggestion has not received any further testing, however.

15.3.2 Costs of Mixed-Species Groups

The potential costs of mixed-species groups have received considerably less empirical attention than the benefits. This is not surprising, as it can be reasonably assumed that whenever species associate regularly with each other, the benefits

15 Coordination in Primate Mixed-Species Groups 271

must exceed the costs or there would be counter-selection against mixed-species group formation. Nevertheless, it is likely that – as is true in other forms of sociality – mixed-species groups do incur some costs, including increased direct (e.g. interference) and indirect (e.g. scramble) feeding competition, increased conspicuousness to potential predators, and higher risks of parasite transmission (Danchin et al. 2008) (Table 15.3).

In mixed-species of tamarins, the smaller species (*S. fuscicollis, C. goeldii*) are occasionally supplanted from food resources by the larger species (*S. mystax, S. labiatus* or *S. imperator*), but the rate of such interactions appears to be very low (Terborgh 1983; Heymann 1990; Peres 1996; Porter 2001). In an experimental study of *S. fuscicollis* and *S. imperator,* Bicca-Marques and Garber (2003) offered food on feeding platforms, measured the time spent on the platform and the number of individuals per species on the platform, and compared visits in single- and mixed-species groups. While the time spent on the platform was decreased in mixed-species groups for both *S. fuscicollis* and *S. imperator*, the number of individuals visiting the platform decreased in mixed-species groups for *S. fuscicollis*, but not for *S. imperator*. This indicates that the foraging costs of mixed-species group formation are less severe for the latter species. *Cercopithecus diana* and *C. campbelli* increase their daily travel path length when associated, although this has been interpreted as an anti-predator benefit of their association rather than a foraging cost (Wolters and Zuberbühler 2003). *Cercopithecus nictitans* have a narrower breadth of the feeding niche when associated with *C. diana*, which may reflect both interference and scramble competition (Eckardt and Zuberbühler 2004).

As previously mentioned, mixed-species groups might also be more conspicuous to predators than single-species groups, but this cost is certainly exceeded by the

Table 15.3 Potential costs of primate mixed-species group

Increased feeding competition	Direct or contest or interference competition:
	Some individuals can exclude others from resources due to superior strength and dominance; per capita food intake decreases with group size for individuals of the smaller/subordinate species
	Indirect or scramble or exploitation competition:
	Individuals have reduced access to resources because these have already been exploited by others; per capita food intake decreases with group size, but in the same way for all group members
Increased risk of predation	Larger groups produce more movement and noise and thus become more conspicuous to potential predators
	Calling to establish or to maintain association makes callers more conspicuous to potential predators
Increased risk of parasite transmission	Risk of transmission of directly transmitted parasites increases with increasing group size
Increased energy expenditure	Feeding competition results in longer daily path length, because groups have to travel further to obtain a sufficient amount of food
	Maintaining the association requires additional travelling when foraging goals differ between species
	Energetic costs of call production to establish or to maintain association

benefits resulting from mechanisms that reduce the predation risk in association (see Sect. 15.3.1.1).

Costs may also result from the behavioural efforts of establishing and maintaining association. For instance, mixed-species group establishment is often realised through the emission and exchange of loud calls (see Sect. 15.4.1), which could be energetically costly and make callers more conspicuous to acoustically orienting predators. Maintaining the association may require travelling to resources exploited by only one of the associated species. This is particularly likely in heterogeneric associations, where species differ more strongly in their food requirements (e.g. associations between the frugivorous-insectivorous *C. diana* and the folivorous *P. badius*) than in congeneric associations.

15.3.3   Consequences of Symmetry and Asymmetry of Net Benefits

Evidently, the balance between the benefits and costs of living in mixed-species groups must be tipped towards the benefits. Nevertheless, the magnitude of the net benefit can obviously vary between species; in other words, there can be an asymmetry in the benefits, as shown by examples provided in Sects. 15.3.1 and 15.3.2. This has obvious implications for questions of coordination in mixed-species groups. Members of species with a higher net benefit can be predicted to be more highly motivated to establish association and to take a more active role in inter-specific coordination.

15.4   Coordination in Mixed-Species Groups

One of the basic problems of group living – the need for coordination between individuals from different age/sex classes with different social and reproductive interests and strategies, physiological and metabolic needs, and foraging strategies – is acuminated in mixed-species groups. Here, not only the interests and needs of different age-sex classes of one species have to be reconciled, but also those of individuals from two or more species.

In studies of coordination in single-species groups, usually individual contributions to coordination are examined. In contrast, studies on coordination in mixed-species groups have generally focussed on the contribution of species instead of individuals (Cords 2000). This is surprising since it can be reasonably assumed that the balance of benefits and costs of mixed-species groups varies between different age-sex classes despite the expected net benefit for all individuals.

There are two principal contexts in which there is a need for coordination between species (see also Cords 2000):

1. Establishing/re-establishing the association. Associated species usually spend the night in different sleeping sites that can be some distance apart, making it

necessary to establish or re-establish association the next morning (e.g. Porter 2001; Smith et al. 2007). Also, associated species may become separated after travelling in different directions or through disruptive events such as inter-group encounters or predator attacks.

2. Maintaining association. Animals have to decide in which direction to travel, which food resources to visit, and how long to stay in a patch. Also, they have to decide when and where to rest and to sleep. For congeneric associations (e.g. *S. mystax – S. fuscicollis*, *C. diana – C. campbelli*), similar ecological and physiological requirements may make coordination less costly than for hetero-generic associations (e.g. *C. diana – P. badius*), where different dietary strategies (frugivory-insectivory vs. folivory) and different physiological processes (digestion of easily digestible fruit pulp vs. stodgy leaves) might actually dictate different optimal travel routes and activity budgets.

15.4.1 Coordination Through Inter-specific Vocal Communication

When associated species are ecologically similar and have a large overlap in their diet, coordinated travelling may simply be a by product of the convergence of optimal travel routes imposed by the local distribution of food resources. In this case, no communication would be expected to take place between species. However, most studies on primate mixed-species groups have noted that communication actually takes places between members of different species, further supporting the contention that many mixed-species groups do not simply represent the result of random encounters. Specifically, loud calls are used as a means of establishing/ re-establishing association (see Cords 2000 for review).

Loud calls are often given in the early morning before or shortly after leaving a sleeping site (e.g. Gautier and Gautier-Hion 1983; Heymann 1990; see Cords 2000 for additional references). These calls can be exchanged before the association is established. In some mixed-species groups of *Cercopithecus,* adult males seem to initiate the association through inter-specific loud calling (Gautier and Gautier-Hion 1983). This can be understood from the special role of adult males in these primates in the defence against predators. Through association with other species, additional males can be recruited for joint defence against predators without increasing reproductive competition. Nothing is known about individual or at least sex-specific contributions to loud calling in other associations.

In associations between *Cercopithecus pogonias* and *Cercopithecus neglectus*, loud calling is more often initiated by males of the former species (Gautier and Gautier-Hion 1983). The authors did not link this to any obvious asymmetries in the benefits obtained from mixed-species group formation. However, it could be predicted that the tendency to take an active role, such as by initiating loud calling or calling more frequently, in establishing an association should be more strongly developed in members of those species for which the net benefit of associating is higher. In line with this prediction, *C. goeldii* initiates association with *S. labiatus* and *S. fuscicollis* in the morning through loud calling in two thirds of records (Porter 2001). When a group of *C. goeldii* – which has much larger home ranges than *S. labiatus* and *S. fuscicollis* – abandons its association with one *Saguinus* group, it gives loud calls during travelling until it encounters another *Saguinus* group (Porter 2001). In associations between *S. fuscicollis* and *S. mystax*, the former species probably obtains a higher net benefit (Peres 1992a, b). Nevertheless, no clear support for a more active role of one species in establishing association through loud calling has been found (Heymann 1990), although Koch (2005) reported a trend towards *S. mystax* more often initiating calling, contradicting the net benefit prediction. Strong asymmetries in loud calling have been reported for mixed-species groups of *Cercopithecus* (Gautier-Hion 1988), but have not been linked to differential net benefits of association.

Through observational studies alone, it is difficult to tease apart whether apparently mutual loud calling is motivated by the interest in establishing association with another species or by an interest to communicate with neighbouring groups of the same species. Even if there is coincidental counter-calling between associates, this could potentially result from simultaneous but independent responses to loud calls from members of neighbouring groups of the respective species. However, support for a role of loud calling in establishing association has been provided through an experimental study of *S. fuscicollis* and *S. imperator*. Both species loud-called in response to playbacks of loud calls from the other species and approached the speaker following playbacks (Windfelder 2001). While *S. fuscicollis* responded slightly stronger to playbacks, *S. imperator* showed a higher tendency to approach the speaker (Windfelder 2001).

The distance by which two species are separated may influence the need for coordination and thus the tendency for loud calling. In mixed-species groups of *S. fuscicollis* and *S. mystax*, no loud calls are given on mornings when the single-species groups had used sleeping sites that were less than 20 m apart (Heymann 1990). When the species are close together, visual information or low-pitched vocalisations may suffice for rapidly establishing an association. When the species are further apart, loud calling will be necessary for establishing association. In line with this assumption, the two species took significantly longer to establish association than when no loud calling occurred (Heymann 1990). Obviously, the need for coordination is stronger when the two species are separated by greater distances.

Based on this finding, Koch (2005) examined the possibility that rather than different strengths of motivation for initiating calling, differential information on the whereabouts of the other species could be responsible for which species initiates loud calling. In mixed-species groups of *S. fuscicollis* and *S. mystax*, the single-species groups separate at variable times before they enter into their respective sleeping sites. Depending on how long before retiring the groups separate, or whether one species is present while the other species retires, information on the location of the sleeping site of the other species should vary. The species with less information should be more motivated to initiate loud calling. Three different conditions could be distinguished (1) *S. mystax* has information about the location of the sleeping site of *S. fuscicollis,* but not vice versa; (2) *S. fuscicollis* has information about the location of the sleeping site of *S. mystax,* but not vice versa; (3) neither species has information about the location of the sleeping site of the other species. In condition (1), *S. fuscicollis* should initiate loud calling more often; in condition (2), *S. mystax* should initiate loud calling more often; and in condition (3), both species should initiate calling equally often. However, these predictions were not supported by the data (Koch 2005). This may indicate either that the assumptions underlying this hypothesis were wrong or that the observer's rating of what the species may know about the whereabouts of the other species does not accurately reflect the actual situation.

After prolonged resting periods, both *S. fuscicollis* and *S. mystax* usually utter low-pitched vocalisations ("contact calls") before starting to move; this happens whether or not the species are associated (personal observations). While these vocalisations probably serve a function in intraspecific coordination, it is currently not known whether they also function in the coordination of movements of the mixed-species group.

During travelling, associated species may have to decide upon the direction of travel. Loud calling and countercalling during travel, even when in spatial proximity, may reflect coordination and decision making, but the possibility that loud calling is stimulated independently in the associated species through listening to loud calls from neighbouring groups not perceived by observers is difficult to exclude. Male *C. pogonias*, giving more loud calls than males from the associated species, may have a prominent role in the coordination of travel (Gautier and Gautier-Hion 1983). The difficulty of separating intra- and interspecific functions has hindered further analyses of the role of vocalisations in interspecific coordination during travel so far. Experimental approaches such as playback experiments are unlikely to render solutions. When travelling together, species do have information on the other species. Thus, playbacks create a situation in which existing information and information simulated by the playback can be contradictory, may create confusion, and therefore elicit inappropriate responses.

It is noteworthy that in some mixed-species groups no evidence has been found (or reported) for a role of loud calls in coordination, which might be related to the stability, permanency, and composition of the mixed-species groups concerned (see Cords 2000 for review). Loud calling should be used in more stable and permanent mixed-species groups, or where particular groups are always associated with each other, but Cords (2000) also pointed to the fact that this does not fit all mixed-species groups. Specifically, mixed-species groups of *C. goeldii* with *Saguinus* are not very permanent and stable, and a single group of *C. goeldii* may associate with several groups of *S. fuscicollis/S. labiatus* at different times, but nevertheless loud calling is employed for establishing association (see above). This seems to indicate that the relationship among association patterns (stability, permanency, and composition of the mixed-species groups), the net benefits, and the coordination/communication effort are quite complex. When the degree of association is low, this may mean for one species that the net benefit is low and thus little effort (loud calling) is spent establishing and maintaining association (this could be the case for *Cercopithecus cephus*; Gautier and Gautier-Hion 1983). However, it may also mean that the establishment and maintenance of mixed-species groups are constrained such as through the temporal use of very specific habitats (as, in the case of *C. goeldii,* bamboo forests that are not entered by potential association partners), but that whenever the opportunity for establishing a mixed-species group arises, corresponding efforts are made.

The diversity of findings with regard to whether or not loud calls are used for interspecific coordination may reflect different stages in the evolution of inter-specific communication, as suggested by Kostan (2002). These stages range from "unidirectional assessment" to "symmetric communication"; which stage is reached actually depends on the benefits and costs of the interspecific interaction (Kostan 2002). It is also feasible that our less than clear understanding of the findings can be due to communication motives that have yet to be identified.

15.4.2 Is There Really Interspecific Coordination?

Loud calling and countercalling as described in the previous section clearly indicate that there is a strong attraction between species or at least of one species to another. But if coordination is defined as "the goal-dependent management of interdependencies by means of hierarchically and sequentially regulated action in order to achieve a common goal" (Chap. 1), is there any evidence for coordination in mixed-species groups? At an abstract deductive level, common goals can be defined in mixed-species groups as the results of those activities that lead to benefits to individuals from participating species (e.g. predator avoidance, increased foraging efficiency). On a more concrete level, common goals can be resources to be visited or routes to be taken. It is, however, more difficult to identify whether there are actions that are hierarchically and sequentially regulated. Calling and countercalling may be seen as sequential actions, but as has been pointed out, they are not necessarily hierarchical.

It is conceivable that despite the benefits that can be achieved through the formation of mixed-species groups, the needs for coordination and the rules for coordination are much simpler than in single-species groups. Individuals of gregarious species usually depend on living with conspecifics for survival and reproduction, making sociality obligatory. The formation of mixed-species groups can bring substantial benefits that may directly enhance survival and indirectly also enhance reproduction. However, mixed-species groups are unlikely to be a condition for survival, and they are definitely not a condition for reproduction. Thus, mixed-species groups can be a facultative form of sociality, and selection pressures on effective interspecific coordination are likely to be much weaker than on intraspecific coordination.

Finally, the definition of coordination itself might be conceptually less appropriate for mixed-species groups because different levels of goals are present, which may more easily come into conflict compared to single-species groups. On an abstract level, members of species participating in mixed-species groups can be said to have a general common goal, namely, to obtain the benefits of mixed-species groups. This common goal might vary in response to environmental fluctuations (e.g. predator density or attack rates; seasonal variation in food abundance) but should always be present – otherwise, there would be no motivation for forming mixed-species groups. On a more concrete level, specific goals, such as visiting specific resources or taking specific travel routes, may coincide between species in a mixed-species group if, for instance, a limited number of optimal travel routes synchronise and synlocalise the species (so-called pseudo-coordination). If the participating species do not converge on concrete goals, conflict may arise – a situation where the need for coordination would be the strongest.

15.4.3 Inter-specific Coordination in Mixed-Species Groups of Birds

The formation of mixed-species groups is a widespread phenomenon, but detailed behavioural studies that may cast some light on the patterns and mechanisms of coordination are very rare (Stensland et al. 2003). As with primates, vocalisations seem to play a key role in bird mixed-species groups (for a review, see Greenberg 2000). This is suggested by the observation that mixed-species groups of birds begin to assemble during the dawn chorus (Munn and Terborgh 1979). Furthermore, particular bird species give loud calls in the morning that may attract other species (Munn 1985). High vocalisation rates of the "nuclear species" (see Sect. 15.2.3) could promote the cohesion of mixed-species groups during travel (Greenberg 2000). But as with primates, it might be difficult to disentangle intra-specific from inter-specific functions of loud calls and other vocalisations. Since bird mixed-species groups generally include many more species than primate mixed-species groups (sometimes 30 or more; see Sect. 15.2.3), any attempt to disentangle these functions is practically impossible.

It has also been suggested that conspicuous visual displays of some bird species may attract others into mixed species and may also facilitate maintenance of association (Moynihan 1962), but this hypothesis has received little support (Greenberg 2000).

An interesting case of "coordination" has been reported by Goodale and Kotagama (2006) for drongos, *Dicrurus paradiseus*. These birds mimic the songs and contact calls of other birds that are participating in mixed-species groups. The vocal mimicry attracts other birds into the association more strongly than drongo calls alone. Researchers consider this to be "behavioural management" of other species by drongos in an overall mutualistic relationship (Goodale and Kotagama 2006).

15.5 General Conclusions

Although mixed-species groups of primates are amongst the best-studied mixed-species groups (Stensland et al. 2003), it is still very difficult to paint a general picture of inter-specific coordination. It is established that coordination is principally through vocalisations, specifically loud calling. While there might be some observational bias (vocalisations are more easily observed and recorded than facial expressions or olfactory signals), it is plausible to assume that under the conditions of reduced visibility in tropical rainforests – the places where practically all mixed-species groups of primates exist – vocalisations are better suited for coordination than other modes of communication. In this respect, coordination between species does not obviously differ from communication within species. Furthermore, the same vocalisations (mainly loud calls) are employed for coordination within and between species. Interestingly, within-species loud calls are used in both within-group and between-group communication. Whereas this observation indicates that interspecific communication in the context of mixed-species group coordination can build upon available mechanisms, it makes the analysis of this coordination even more complicated. In many instances it can be difficult or impossible to determine whether the intended receivers of loud calling are members of the same group, members of another group of the same species, or members of another species. This added complication may be one of the reasons why detailed studies on coordination in primate mixed-species groups are still very rare.

We can be rather certain that coordination efforts in primate mixed-species groups are done in an effort to obtain the benefits of such groups, be it reduced predation risk, increased foraging and feeding efficiency, or reduced insect bites. These motivators fit the coordination definition of "goal-dependent management of interdependencies" (see Chap. 1), but, as has been discussed, it is more questionable whether they are "hierarchically and sequentially regulated actions in order to achieve a common goal." Observed imbalances or asymmetries in benefits may mean that the goals achieved can be less than common. It is exactly this point where the study of coordination in mixed-species groups might contribute to the understanding of coordination in humans.

Appendix: Index of Scientific and Common Names of Primates Mentioned in the Text

Scientific name	Common name
Callimico goeldii	Goeldi's monkey
Cebus albifrons	White-fronted capuchin
Cebus apella	Brown capuchin
Cercopithecus ascanius	Red-tailed guenon
Cercopithecus campbelli	Campbell's monkey
Cercopithecus diana	Diana monkey
Cercopithecus nictitans	Putty-nosed monkey
Cercopithecus pogonias	Crowned guenon
Procolobus badius	Red colobus
Saguinus fuscicollis	Saddleback tamarin

(*continued*)

Scientific name	Common name
Saguinus imperator	Emperor tamarin
Saguinus labiatus	Red-bellied tamarin
Saguinus mystax	Moustached tamarin
Saimiri boliviensis	Bolivian squirrel monkey

References

Bicca-Marques JC, Garber PA (2003) Experimental field study of the relative costs and benefits to wild tamarins (*Saguinus imperator* and *S. fuscicollis*) of exploiting contestable food patches as single- and mixed-species troops. Am J Primatol 60:139–153

Caro T (2005) Antipredator defenses in birds and mammals. University of Chicago Press, Chicago

Chapman CA, Chapman LJ (1996) Mixed-species primate groups in the Kibale forest: ecological constraints on association. Int J Primatol 17:31–50

Chapman CA, Chapman LJ (2000) Interdemic variation in mixed-species association patterns: common diurnal primates of Kibale National Park, Uganda. Behav Ecol Sociobiol 47:129–139

Cords M (1987) Mixed-species association *Cercopithecus* monkeys in the Kakamega Forest, Kenya. Univ Calif Publ Zool 117:1–109

Cords M (1990) Mixed-species association of East African guenons: general patterns or specific examples? Am J Primatol 21:101–114

Cords M (2000) Mixed species association and group movement. In: Boinski S, Garber PA (eds) On the move: how and why animals travel in groups. University of Chicago Press, Chicago, pp 73–99

Danchin E, Giraldeau L-A, Cézilly F (2008) Behavioural ecology. Oxford University Press, Oxford

DiFiore A, Suarez SA (2007) Route-based travel and shared routes in sympatric spider and woolly monkeys: cognitive and evolutionary implications. Anim Cogn 10:317–329

Eckardt W, Zuberbühler K (2004) Cooperation and competition in two forest monkeys. Behav Ecol 15:400–411

Freed BZ (2006) Polyspecific associations of crowned lemurs and Sanford's lemurs in Madagascar. In: Gould L, Sauther ML (eds) Lemurs: ecology and adaptation. Springer, New York, pp 111–131

Freeland WJ (1977) Blood-sucking flies and primate polyspecific associations. Nature 269:80–81

Garber PA (2000) Evidence for the use of spatial, temporal, and social information by some primate foragers. In: Boinski S, Garber PA (eds) On the move: how and why animals travel in groups. University of Chicago Press, Chicago, pp 261–298

Gautier J-P, Gautier-Hion A (1983) Comportement vocal des mâles adultes et organisation supraspécifique dans les troupes polyspécifiques de cercopithèques [in French]. Folia Primatol 40:161–174

Gautier-Hion A (1988) Polyspecific associations among forest guenons: ecological, behavioural and evolutionary aspects. In: Gautier-Hion A, Bourlière F, Gautier J-P, Kingdon J (eds) A primate radiation: evolutionary biology of the African guenons. Cambridge University Press, Cambridge, pp 452–476

Gautier-Hion A, Quris R, Gautier J-P (1983) Monospecific vs. polyspecific life: a comparative study of foraging and antipredatory tactics in a community of *Cercopithecus* monkeys. Behav Ecol Sociobiol 12:325–335

Goodale E, Kotagama SW (2006) Vocal mimicry by a passerine bird attracts other species involved in mixed-species flocks. Anim Behav 72:471–477

Greenberg R (2000) Birds of many feathers: the formation and structure of mixed-species flocks of forest birds. In: Boinski S, Garber PA (eds) On the move: how and why animals travel in groups. University of Chicago Press, Chicago, pp 521–558

Hardie SM, Buchanan-Smith HM (1997) Vigilance in single- and mixed-species groups of tamarins (*Saguinus labiatus* and *Saguinus fuscicollis*). Int J Primatol 18:217–234

Haugaasen T, Peres CA (2009) Interspecific primate associations in Amazonian flooded and unflooded forests. Primates 50:239–251

Heymann EW (1990) Interspecific relations in a mixed-species troop of moustached tamarins, *Saguinus mystax*, and saddle-back tamarins, *Saguinus fuscicollis* (Primates: Callitrichidae), at the Rio Blanco, Peruvian Amazonia. Am J Primatol 21:115–27

Heymann EW (1997) The relationship between body size and mixed-species troops of tamarins (*Saguinus* sp.). Folia Primatol 68:287–295

Heymann EW, Buchanan-Smith HM (2000) The behavioural ecology of mixed-species troops of callitrichine primates. Biol Rev 75:169–190

Holenweg A-K, Noë R, Schabel M (1996) Waser's gas model applied to associations between red colobus and Diana monkeys in the Taï National Park, Ivory Coast. Folia Primatol 67:125–136

Janson CH (2000) Spatial movement strategies: theory, evidence, and challenges. In: Boinski S, Garber PA (eds) On the move: how and why animals travel in groups. University of Chicago Press, Chicago, pp 165–203

Kirchhof J, Hammerschmidt K (2006) Functionally referential alarm calls in tamarins (*Saguinus fuscicollis* and *Saguinus mystax*) – evidence from playback experiments. Ethology 112:346–354

Koch C (2005) Re-establishment of interspecific associations after separation – interspecific communication in mixed species troops of *Saguinus fuscicollis* and *S. mystax*. Diploma thesis, University of Göttingen, Göttingen

Kostan KM (2002) The evolution of mutualistic interspecific communication: assessment and management across species. J Comp Psychol 116:206–209

Krause J, Butlin RK, Peuhkuri N, Pritchard VL (2000) The social organization of fish shoals: a test of the predictive power of laboratory experiments for the field. Biol Rev 75:477–501

McGraw WS, Bshary R (2002) Association of terrestrial mangabeys (*Cercocebus atys*) with arboreal monkeys: experimental evidence for the effects of reduced ground predator pressure on habitat use. Int J Primatol 23:311–325

Moynihan M (1962) The organization and probable evolution of some mixed species flocks of neotropical birds. Smithson Misc Coll 143:1–140

Munn CA (1985) Permanent canopy and understory flocks in Amazonia: species composition and population density. Ornithol Monogr 36:683–710

Munn CA, Terborgh JW (1979) Multi-species territoriality in neotropical foraging flocks. Condor 81:338–347

Noë R, Bshary R (1997) The formation of red colobus–diana monkey associations under predation pressure from chimpanzees. Proc Roy Soc Lond B 264:253–259

Parrish JK, Viscido SV, Grünbaum D (2002) Self-organized fish schools: an examination of emergent properties. Biol Bull 202:296–305

Peres CA (1992a) Prey–capture benefits in a mixed-species group of Amazonian tamarins, *Saguinus fuscicollis*, and *S. mystax*. Behav Ecol Sociobiol 31:339–347

Peres CA (1992b) Consequences of joint-territoriality in a mixed-species group of tamarin monkeys. Behaviour 123:220–246

Peres CA (1993) Anti-predation benefits in a mixed-species group of Amazonian tamarins. Folia Primatol 61:61–76

Peres CA (1996) Food patch structure and plant resource partitioning in interspecific associations of Amazonian tamarins. Int J Primatol 17:695–723

Podolsky RD (1990) Effects of mixed-species association on resource use by *Saimiri sciureus* and *Cebus apella*. Am J Primatol 21:147–158

Pomara L, Cooper RJ, Petit LJ (2007) Modeling the flocking propensity of passerine birds in two neotropical habitats. Oecologia 153:121–133

Pook AG, Pook G (1982) Polyspecific association between *Saguinus fuscicollis*, *Saguinus labiatus*, *Callimico goeldii* and other primates in north-western Bolivia. Folia Primatol 38:196–216

Porter LM (2001) Benefits of polyspecific associations for the Goeldi's monkey (*Callimico goeldii*). Am J Primatol 54:143–158

Porter LM, Garber PA (2007) Niche expansion of a cryptic primate, *Callimico goeldii*, while in mixed species troops. Am J Primatol 69:1340–1353

Powell GVN (1979) Structure and dynamics of interspecific flocks in a mid-elevational neotropical forest. Auk 96:375–390

Powell GVN (1985) Sociobiology and adaptive significance of interspecific foraging flocks in the neotropics. Ornithol Monogr 36:713–732

Quérouil S, Silva MA, Cascão I, Magalhães S, Seabra MI, Machete MA, Santos RS (2008) Why do dolphins form mixed-species associations in the Azores? Ethology 114:1183–1194

Rehg JA (2006) Seasonal variation in polyspecific associations among *Callimico goeldii*, *Saguinus labiatus*, and *S. fuscicollis* in Acre, Brazil. Int J Primatol 27:1399–1428

Smith AC, Kelez S, Buchanan-Smith HM (2004) Factors affecting vigilance within wild mixed-species troops of saddleback (*Saguinus fuscicollis*) and moustached tamarins (*S. mystax*). Behav Ecol Sociobiol 56:18–25

Smith AC, Buchanan-Smith HM, Surridge AK, Mundy NI (2005) Factors affecting group spread within wild mixed-species troops of saddleback and mustached tamarins. Int J Primatol 26:337–355

Smith AC, Knogge C, Huck M, Löttker P, Buchanan-Smith HM, Heymann EW (2007) Long term patterns of sleeping site use in wild saddleback (*Saguinus fuscicollis*) and mustached tamarins (*S. mystax*). Am J Phys Anthropol 134:340–353

Stensland E, Angerbjörn A, Berggren P (2003) Mixed species groups in mammals. Mamm Rev 33:205–223

Stojan-Dolar M, Heymann EW (2010) Vigilance of mustached tamarins in single-species and mixed-species groups – the influence of group composition. Behav Ecol Sociobiol 62:325–335

Struhsaker TT (1981) Polyspecific associations among tropical rain-forest primates. Z Tierpsychol 57:268–304

Teelen S (2007) Influence of chimpanzee predation on associations between red colobus and red-tailed monkeys at Ngogo, Kibale National Park, Uganda. Int J Primatol 28:593–606

Terborgh J (1983) Five new world primates. Princeton University Press, Princeton, NJ

Terborgh J (1990) Mixed flocks and polyspecific associations: costs and benefits of mixed groups to birds and monkeys. Am J Primatol 21:87–100

Wachter B, Schabel M, Noë R (1997) Diet overlap and polyspecific associations of red colobus and Diana monkeys in the Taï National Park, Ivory Coast. Ethology 103:514–526

Waser PM (1982) Polyspecific associations: do they occur by chance? Anim Behav 30:1–8

Waser PM (1987) Interactions among primate species. In: Smuts BB, Cheney DL, Seyfarth RM, Wrangham RW, Struhsaker TT (eds) Primate societies. University of Chicago Press, Chicago, pp 210–226

Whitesides GH (1989) Interspecific associations of Diana monkeys, *Cercopithecus diana*, in Sierra Leone, West Africa: biological significance or chance? Anim Behav 37:760–776

Windfelder TL (1997) Polyspecific association and interspecific communication between two neotropical primates: saddle-back tamarins (*Saguinus fuscicollis*) and emperor tamarins (*Saguinus imperator*). PhD thesis, Duke University, Durham, NC

Windfelder TL (2001) Interspecific communication in mixed-species groups of tamarins: evidence from playback experiments. Anim Behav 61:1193–1201

Wolters S, Zuberbühler K (2003) Mixed-species associations of Diana and Campbell's monkeys: the costs and benefits of a forest phenomenon. Behaviour 140:371–385

Zuberbühler K (2000) Interspecies semantic communication in two forest primates. Proc Roy Soc Lond B 267:713–718

Index

A
Action regulation theory, 139, 141, 151
Actions, 12, 16, 19–23, 25, 26, 29, 76, 79–83, 85–88, 202, 204, 205, 211
Action team, 75–88
Adaptation, 76, 80, 83–87
Anaesthesia, 75–88
Animal, 229–231, 234–242
Anticipation, 156, 157, 159, 174
Average deviation score
  absolute knowledge, 169
  agreement, 169

B
Baboons, 226
Behavioural management, 277
Behavioural patterns, 26
Benefits, 267–276, 278
Birds, 263, 266–267
  dawn chorus, 277
  Dicrurus paradiseus, 277
  loud calls, 277
  maintenance, 277
  nuclear species, 267, 277

C
Callimico goeldii, 265, 266, 269–271, 273–276, 278
Cebus, 265
  albifrons, 270, 278
  benefits, 269–270
  Saimiri boliviensis, 265, 270, 279
  Saguinus fuscicollis, 265–266, 268–271, 273–275, 278
  Callimico goeldii, 265, 266, 269–271, 273–276, 278
  Saguinus mystax, 265, 266, 269–271, 273–275, 279

Cercocebus
  albigena, 270
  atys, 269
  costs, 269, 270
Cercopithecus, 269–271, 273–275
  ascanius, 265, 266, 268, 269, 278
  benefits, 269–271
  campbelli, 266, 269, 271, 273, 278
  cephus, 275
  countercalling, 274–276
  diana, 265–267, 269, 271–273, 278
  establishment, 264, 265, 272–276
  Saguinus fuscicollis, 265–266, 268–271, 273–275, 278
  Callimico goeldii, 265, 266, 269–271, 273–276, 278
  Saguinus imperator, 266, 271, 274, 279
  interference, 271
  Saguinus labiatus, 265, 266, 268–271, 273–275, 279
  maintenance, 265, 275, 277
  mitis, 265, 266, 270
  neglectus, 273
  nictitans, 271, 278
  pogonias, 273, 275, 278
  scramble, 271
  Procolobus tephroscel, 266
Children, 246, 249–251, 254, 256–257
Chimpanzees, 246–248, 250–255, 257, 267
CMCM. *See* Coordination mechanism circumplex model
Coding units, 211, 212
Cognition, 240–242
Cohesion, executive manager, 62
Collective action regulation, 182
Collective decision making, 224, 226, 227
Common ground, 249, 252–254, 256

Communication, 76, 77, 79–83, 200, 203–204, 211, 235, 241–242
  processes, 182, 185, 187, 190, 191
Communicative, 245–246, 248, 249, 251, 252, 254–257
Comparative studies, 223–224, 226, 227
Competition, 255, 271, 273
Concept analysis
  matrices, 165
  multidimensional scaling (MDS), 161, 164, 165, 172
  pathfinder, 161, 165, 172
  proximity, 165
  quadratic assignment procedures (QAP), 161
Confusion, 267, 268, 275
  effect, 267–268
Cooperative, 251–257
Coordination, 37–52, 264, 272–278
  of behaviours, 19–20
  concepts, 119–131
  entities, 12, 15–17, 27–28
  explicit, 156, 170, 178, 180, 181, 183–186, 188, 190, 191, 193, 194
  of goals, 17–18
  implicit, 178, 179, 181–183, 188, 190, 193–195
  levels, 12, 16, 28–30
  of meanings, 18–19
  mechanisms, 12, 20–23, 26
    coding agreement, 65–66
    implicit and in-process coordination, 59, 71
    two-category systems, 58
  outcomes, 120–123, 130
    adaptation, 121–123
    coordination success, 120–123
    effectiveness, 121–123
    team learning, 121–123
  patterns, 12, 23–27
  processes, 23, 26, 120, 121, 123–131, 177–181, 186, 188, 194
  requirements, 14, 15, 19, 26, 30, 93–111
  synchronizing, 120–121, 124, 126
Coordination mechanism circumplex model (CMCM)
  agreement, 58–59, 61, 64–67, 69
  coder agreement, 61, 62, 64, 66, 68, 69
  explicit in-process coordination, 58–59, 68
  explicitness/implicitness, 58–65, 67, 71
  explicit pre-process coordination, 59, 68
  pre-process/in-process, 59, 61, 62, 69, 71

temporal phase (pre-process/in-process), 59
  timing, 68, 71, 72
  two-category taxonomies, 60–61
Costs, 265, 267–272, 276
Criteria
  conventional, 145, 146
  expansive, 146
  impoverished, 145, 146
  for team coordination potential, 138–143, 145
Cycles, 95–97, 101, 103, 104, 108

D
Decision, 230–238, 240–242
Decision making, 199–215
Definition, group movements, 48–50
Democracy
  aggregate, 234
  interactive, 234
Design problems, 184, 193–194
Despotism, 232, 234
Detection effect benefits, 267–269
Dilution, 267–268
  effect, 267–268
Distributed leadership, 24
Dogs, 255–257

E
Eavesdropping
  alarm calls, 268, 269
  Cercopithecus campbelli, 269
  Cercopithecus diana, 269
  Saguinus fuscicollis, 268–269
  Saguinus mystax, 269
Elicitation methods
  card sorting, 161–163
  interviews, 160–162
  observation, 160–162
  process tracing, 161, 162
Enculturated apes, 255
Entities, 76, 77, 79, 81, 83–86
  action, 79, 80, 87, 88
  help, 16
Episodes, 95–97, 103, 108
Evaluation, 200–202
Evolution, social behaviour, 4, 8
Exchanges, 18, 20
Explicit, 76, 80–88, 120–130, 199–215
Explicit coordination, 19, 21, 22

F
Feedback, 122–126, 128–130
Fish, 263

Index 285

Followers, 46, 49–51
Followership, 41, 46, 47, 49–51
Foraging, 265, 267–272, 276, 278
Functions, 75–79, 82, 83, 85–87

G
Gaze-shift following, 245–246, 256
Gestures, 248–257
Great apes, 224, 226, 246–247, 250, 256
Group agreement, 161, 163–164, 166–167, 171
Group cohesion, 4–7, 225
Group coordination, 11–32
Group decision making, 5–7, 14, 15, 22, 24, 25, 27, 29
Group decisions
  despotic decision-making, 47
  self-organized, 47
  shared, unshared, 47
Group living, 223–225, 227, 230–231
Group movements, 37–52, 223, 225–227
Group performance, 4, 7, 94–96, 98, 102, 107, 109–110
Group process
  analysis, 94, 95, 97, 110, 111
  cycles, 95, 97, 108
  episodes, 95, 97, 108
  phases, 95, 97, 108
  stages, 97, 108
Group size, 267–269, 271
Group task, 12, 14, 15, 17, 18, 20–22, 26, 28
Guenons, 269, 278

H
Habitats, 268, 269, 276
  Callimico goeldii, 269, 276
  use
    Cercopithecus diana, 269
    Cercopithecus mitis, 270
    Procolobus badius, 269, 272
    Saguinus fuscicollis, 269–270
    Saguinus labiatus, 269–270
Health care, 75–88
Heterospecific associations, 227
Hidden leadership, 49
Hierarchical task analysis (HTA), 95, 97, 98, 100, 101, 108, 110, 111
High-risk, 76, 77, 83, 86–88
Human group performance, 51
Humans, 245–257
Hunting, decision making, 231

I
Iatrogenic, 87
  errors, 79

Iatrogenic injuries, 76, 83
Imperatives, 250, 251, 256
Impersonal, 125
  coordination, 121, 122, 125
Implicit, 76, 80–83, 85–88, 120–128, 130, 202–204, 215
  coordination, 19, 21, 22, 26
Incentives, 46
Inclusive model of group coordination, 76, 85, 87
"Indices of sharedness", 171
Information, 12, 14–18, 20–22, 24, 25, 27, 76, 79–88, 200–205, 209, 213–214
  exchange, 122–125, 128
Initiator, 41, 48–50
Innovation
  autonomy, 140, 147, 148, 150–152
  feedback, 141, 142, 148–152
  involvement in problem setting, 140, 142, 146–148, 150–152
  organizational support for innovation, 142, 146, 148–151
  potential for team self-regulation, 137, 141
Input-process-outcome model
  implicit pre-process coordination, 59
Input-process-output (IPO) model, 95–97, 102, 107
Instruments, 121, 122, 125
Integrated model of coordination for action teams in health care, 75–88
Intention, 58–63, 69–71
Intentionality, 59, 70–71, 231, 241
  substitutes, 70
Interaction, 200–205, 215
Interdisciplinary, 39, 51–52
(Inter-)personal coordination, 121, 122, 125
Interrater reliability, 212
Interspecific, 263, 264, 272–278
Interspecific communication
  benefits, 273–276
  costs, 276
  countercalling, 274–275
IPO model. *See* Input-process-output (IPO) model

J
Job analysis, 137, 143, 144

K
Knowledge, 155–174

L
Lag sequential analysis, 211, 214, 215

286   Index

Leadership, 19, 20, 24, 26, 27, 31, 38, 40–46, 48, 49, 80–85, 87, 88, 107–109, 122, 125, 126, 129, 130
Lemurs, 263
Levels
  of coordination mechanisms, 61, 65–66, 68, 69, 71
  implicitness *vs.* explicitness, 69
  pre-process *vs.* in-process, 69
Life history, 224–225
Likert-type questionnaires
  shared mental model (SMM) index, 158, 173
  within-group agreement, 166

M

Macro, meso and micro levels, 61, 62, 66–72
Mechanisms, 38, 46–48, 51, 76, 77, 80, 81, 83–85, 87, 88, 199–215
Mental models, 158–160, 162–170, 172, 173
  shared, 81, 178–186, 188, 190–195
  team, 177–195
Mental states, 245, 250, 257
Meso-level agreement, 61, 62, 65–67, 69
Methodological approach
  conventional conditions, 145
  expansive conditions, 145
  expert ratings, 143
  impoverished conditions, 145
  interviews, 143, 144
  observations, 143, 144
  video recordings, 143–145
Methods, 156–158, 160–166, 170–173
  observation, 121, 128–131
  questionnaires, 121, 129–131
Micro-analytically, 199–215
MICRO-CO, 199–215
  substitutes, 62–67
Micro-level process analysis, 62
Mixed flocks, 266–267
  attendant species, 267
  core species, 267
  nuclear species, 267
Mixed-species groups, 227, 263–278
  benefits, 267–276, 278
  Callimico goeldii, 265, 266, 269–271, 273–276, 278
  Cercopithecus campbelli, 266, 269, 271, 273, 278
  Cercopithecus diana, 265–267, 269, 271–273, 278
  Cercopithecus pogonias, 273, 275, 278
  congeneric, 265–266, 272, 273

  contact calls, 275, 277
  coordination, 264, 272–278
  costs, 265, 267–272, 276
  countercalling, 274–276
  criterion distance, 264
  establishment, 264, 265, 272–276
  heterogeneric, 265–266, 272
  interference, 271
  loud calls, 272–278
  maintenance, 264, 265, 271–273, 275, 277
  non-randomness, 264–267
  Procolobus badius, 265–267, 269, 272, 273, 278
  randomness, 264
  Saguinus fuscicollis, 265–266, 268–271, 273–275, 278
  Saguinus mystax, 265, 266, 269–271, 273–275, 279
  scramble, 271
  Waser's gas model, 264–265
Multisystem team, 78
Multiteam system, 79

N

Nonroutine events, 82, 84, 86, 87
Nontechnical skills, 79, 107

O

Object-choice paradigm, 248, 251, 253
Observation, 211, 215
Operationalization, group movements, 48–51
Ostensive cues, 245, 254–257
Outcome performance, 95, 102, 106

P

Parasite, 271
Patterns, group movements, 39–40
Performance, 75–79, 81, 83–88, 203, 204
Perspective taking implicitness, 60
Planning, 178, 181, 186, 188, 190–191, 194
Playbacks
  coordination, 274, 275
  Saguinus imperator, 274
Pointing, 245, 248–252, 254–256
Polyspecific associations, 263–264
Predation, 231
  risk, 267–269, 272, 278
Pre-, in-and post-processes, 122, 124–126
Primates, 245–248, 254–257
Problem definition, 180, 184, 186, 188–190, 194
Problem solving
  process, 177–178, 181, 193–194

Index 287

Process, group movements, 40
Process losses, 13–14, 200, 202
Process performance measures, 95, 101–110
Procolobus, 265–267, 269, 272, 273, 278
    costs, 265, 272
    pattern, 265
    rufomitratus, 265
Product development, 138, 139, 143, 145,
    146, 148
Pseudo-coordination, 276–277

Q
Quorum
    combined, 234
    response, 233, 236
    sensing, 233–235

R
Random group resampling (RGR), 171
Recording rule, 211
Redfronted lemurs, 39, 42, 49–51
Referent, 248–257
Reflection, 181, 186, 188, 190–191
Representation
    cognitive, 177, 179
    internal, 179
    methods
        qualitative, 161
        quantitative, 161, 164
    shared, 178, 180, 182
RGR. *See* Random group resampling
Road traffic
    implicit in-process mechanism, 69
    substitutes, 61, 62, 64, 67

S
Safety, 75–77, 79, 84–88, 203
Saguinus, 265–268, 270, 274, 275, 278–279
    benefits, 268–270, 273–274
    fuscicollis, 265–266, 268–271,
        273–275, 278
    heterogeneric, 265, 266
    imperator, 266, 271, 274, 279
    labiatus, 265, 266, 268–271, 273–275, 279
    mystax, 265, 266, 269–271, 273–275, 279
Saimiri
    Cebus paella, 265, 270
    boliviensis, 265, 270, 279
Sampling rule, 211
Scrounging, 268, 270
Self-organization, 232–233
Semiotics
    explicitness, 60, 71
    semantic specialisation, 60, 70

Shared, 158–160, 162–173
    cognition, 159
    collective action, 242
    intentionality, 224, 226
    knowledge, implicit, 156, 158
Shared mental models, 18, 21, 23, 24, 31, 59,
    201, 203
Sharedness, 177–195
Shared understanding, 178–181, 183, 185, 188,
    190–193
Signals, 38, 40, 42, 44, 46, 47, 50, 235,
    238–241
Social cognitive, 245, 246, 255
Social groups, 37
Social influence, 59
Social learning
    barks, 239–240
    grunts, 239–240
    hamadryas, 231, 236–238
    non-vocal, 238
    notifying, 238
    olive, 237, 240
Social systems, 224–225
Solution ideas, 184, 186, 188, 190
Specificity, 252–255, 257
Steering, 24, 27, 29–31
Structuration process, 29
Synchronisation of actions, 19, 20, 30
SYNSEG, 211

T
Tacit behaviours, 121, 122, 126, 127
Tacit coordination
    Dickinson and McIntyre, 126
    implicit, 121, 126, 127
    mutual adaptation, 126
    performance monitoring, 126
Takeover, 49
Tamarins, 269, 271, 278–279
    Callimico goeldii, 271
    Saguinus fuscicollis, 265–266, 268–271,
        273–275, 278
    Saguinus imperator, 266, 271, 274, 279
    Saguinus labiatus, 265, 266, 268–271,
        273–275, 279
    Saguinus mystax, 265, 266, 269–271,
        273–275, 279
Task-related performance, 51
Tasks, 76, 78–86, 88, 200–205, 207, 211,
    212, 215
    analysis, 94, 95, 97–107, 109–111
    requirements, 93–111
Taskwork, 95, 98, 107, 109, 110
Taxonomy, 199–215

Team attitudes, 121, 122, 127
  cohesion, 122, 126, 127
  collective, 122, 127
  group efficacy, 127
  orientation, 122, 127, 128
  trust, 122, 127
Team climate, 180–188, 191, 192, 194
Team coordination potential, 137–152
Team development, 178–185, 187, 190, 193, 194
Team knowledge (TK), 119, 121–123, 127, 128, 130
Team mental models, 159, 160, 163–166, 168, 169, 172
Team performance, 178, 179, 183, 184
Team situation models, 159–160, 171, 172
Team specific agreement, 167, 170–171
Teamwork, 94–99, 107, 108, 110
Temporal dimension, 23, 28
Temporal perspective, 120, 123–126
Termination, 38, 40, 41, 49, 50

Theory of Mind, 241
TK. *See* Team knowledge
Traffic road, 61–64, 66, 70
Transactive memory, 159, 160, 163, 167, 169, 170, 172, 183
Travel calls, 46, 48, 51
Traveling types, 48–49
Triadic nature, 256

U
Understanding, 156–157, 159, 160, 162, 165, 171, 172, 174

V
Validity, 212, 215
Verreaux's sifakas, 42, 46, 50, 51
Vigilance, 268–269
Visual or acoustical displays, 42, 44, 46
Vocalizations, 274, 275, 277, 278
Vocal mimicry, 277